Accelerating the Socio-Ecological Transition

Mohamed Cheriet
Editor-in-Chief

Jean-François Boucher
Luciana Gondim de Almeida Guimarães
Jean-Marc Frayret
Guest Editors

Accelerating the Socio-Ecological Transition

Strategies and Innovations for Sustainable Development

Springer

Editor-in-Chief
Mohamed Cheriet
Systems Engineering Department
École de Technologie Supérieure
Montréal, QC, Canada

Guest Editors
Jean-François Boucher
Department of Fundamental Sciences
Université du Québec à Chicoutimi
Chicoutimi, QC, Canada

Luciana Gondim de Almeida Guimarães
Management Department
Universidade Potiguar
Natal, Brazil

Jean-Marc Frayret
Department of Mathematics and Industrial Engineering
Polytechnique Montréal
Montréal, QC, Canada

ISBN 978-3-031-82898-0 ISBN 978-3-031-82896-6 (eBook)
https://doi.org/10.1007/978-3-031-82896-6

This work was supported by École de technologie supérieure.

This Springer imprint is published by the registered company Springer Nature Switzerland AG
The registered company address is: Gewerbestrasse 11, 6330 Cham, Switzerland

Foreword

Across our interconnected globe, we face very real and urgent human challenges, such as eradicating hunger and poverty, and providing access to essential services such as high quality education and healthcare. The paramount challenge of our time is balancing efforts to address these challenges with the equally urgent need to protect the climate and natural ecosystems, and to do so in a way that is fair for everyone. This is why in 2015, at the United Nations, 193 countries adopted the Sustainable Development Goals (SDGs) as a shared blueprint for peace and prosperity, both for people and the planet, now and into the future.

The transformations that are needed to make the SDG vision a reality are enormous in scale, and will require deep structural changes across all sectors in society. The SDGs are profoundly interconnected, requiring understanding of the complex links between natural, human, and technological systems. In order to avoid negative spillover effects across different systems, the necessary transformations cannot be improvised as we go. We need to develop sophisticated, scientifically rigorous and long-term pathways for action that take into account relationships between whole systems in order to identify and address potential trade-offs. These pathways for action will also need to be co-designed among different stakeholders. The participation of a broad range of communities and sectors can generate the necessary public support to ensure effective implementation.

Unfortunately, a growing number of people today are unable to lead a decent life. The 2024 *Sustainable Development Report*[1] offers sobering insights. None of the 17 SDGs are on track for being achieved by 2030. Globally, only 16% of SDG targets will be achieved on time, while efforts to achieve many more have stagnated since 2020. SDG 2 (Zero Hunger), SDG 11 (Sustainable Cities and Communities), SDG 14 (Life Below Water), SDG 15 (Life on Land), and SDG 16 (Peace, Justice, and Strong Institutions) are particularly off-track. This reversal of previously positive trends has been driven by multiple factors, most notably the COVID-19 pandemic, which has exacerbated existing inequalities while upending global health and the

[1] Sustainable Development Report 2024 https://dashboards.sdgindex.org/

economy, and by the escalating climate crisis. The climate emergency has shown its severity in the last 3 years. The year 2023 was the hottest on record, and all continents had drastic episodes of droughts, floods, or catastrophic forest fires. At this writing, 2024 is on track to be hotter, and the coming years hotter still.

Compounding these challenges are a range of growing geopolitical tensions and armed conflicts, some of which involve countries with nuclear capabilities, which increase the risk of a nuclear escalation. These tensions and conflicts have driven inflation, undermined food security, and complicated efforts to work together on common goals. In short, the world faces an extraordinarily difficult scenario, and this is contributing to hopelessness and apathy in many places where the population deserves instead to look forward with expectations of a better future.

Fortunately, an increasing number of people and governments are being alerted to these challenges and are developing pathways to help effect the transformations needed to overcome them. Often, they are doing so by placing sustainable development at the center of focus. To a large extent, this seems to be the case of the province of Quebec, where different actors are piloting ambitious initiatives that can drive change, and provide examples for action in other regions of the world. This is why this book is a very important read. It presents sophisticated analyses of key socio-ecological transitions needed to avoid the worst scenarios. The diverse contributions within highlight the importance of accelerating action and moving from theory to implementation through comprehensive strategies that take all dimensions of sustainable development into account. The different chapters recognize the interconnectedness of these challenges and propose holistic perspectives that address existing barriers to action.

This important book reminds us that in the face of deep challenges, researchers and universities are well-positioned to support the necessary transitions. They develop new technologies, business models, and governance frameworks; train future leaders to be globally conscious and innovative; and have a proven track record of working with diverse stakeholders, including governments, the private sector, civil society, and international organizations. In summary, this book serves as a testament to the critical role that informed research and strategic planning play in navigating the complexities of our global sustainability challenges.

New York, NY, USA María Cortés Puch

Preface

Climate change, environmental degradation, and the overconsumption of natural resources pose real and growing threats to societies around the globe. Events related to climate change are becoming more frequent and severe, such as forest fires that ravage entire regions, extreme weather events like hurricanes and heatwaves, and devastating floods that displace communities. In response to these escalating dangers, the call for a socio-ecological transition has become urgent, recognized over recent decades as essential for ensuring the well-being of both current and future generations. This transition requires not only a shift in thinking but also the production of new knowledge and tools to transform economic, social, cultural, and environmental paradigms.

Academia plays a critical role in this transformation. Universities, research centers, and interdisciplinary research groups are uniquely positioned to lead the way in this transition by generating innovative solutions and guiding public policy.

In the province of Quebec, one prominent player in this movement is CIRODD (the Interdisciplinary Research Centre on Operationalization of Sustainable Development). As the first strategic organization in sustainable development in Quebec, CIRODD has been at the forefront of fostering collaboration between researchers from diverse fields since its inception in 2013. By bridging gaps between disciplines such as environmental science, economics, engineering, and social sciences, CIRODD has set ambitious objectives that push the boundaries of interdisciplinary work. For example, CIRODD projects have brought together engineers, urban planners, and sociologists to tackle complex issues like sustainable urbanization and renewable energy adoption.

After more than a decade of accumulating knowledge and experience, this book emerges as a culmination of CIRODD's efforts and reflects the progress Quebec society has made in adopting sustainable practices.

The book delves into the fundamental principles and key issues involved in implementing a socio-ecological transition, with a particular focus on the province of Quebec. It tackles this vast subject from the perspectives of both research and education, offering strategies to accelerate this transition across various economic sectors. For example, it discusses how renewable energy projects in Quebec's rural communities are revolutionizing local economies while reducing environmental impact.

Each chapter addresses critical research questions that prompt readers to reflect on the socio-ecological transition from different angles and sectors. For instance, one chapter focuses on the role of academia, asking: What role can universities and research institutes play in advancing social and ecological transitions? and How can research initiatives drive progress while maintaining rigorous academic standards?

Additionally, readers will find discussions on operationalizing the socio-ecological transition through concrete examples. For instance, the book explores successful policies and practices implemented in Quebec, such as its zero-emission vehicle policy and the integration of sustainable construction methods in urban planning.

Key topics also include: What methodologies can effectively accelerate the socio-ecological transformation of Quebec's economic sectors, from manufacturing to transportation? and What innovative approaches have been tested and proven to work in Quebec's quest for sustainability?

The primary goal of this book is to equip college and university educators with the concepts, methodologies, and references they need to integrate sustainable development into their curricula and research. Furthermore, it aims to provide professionals and decision-makers with real-world examples of practices designed to fast-track socio-ecological transformation. Finally, the book offers the general public accessible insights into how they can contribute to the socio-ecological transition, empowering them to be active participants in the fight against climate change.

As co-editors, we are proud to present a unique collection of contributions from some of CIRODD's most prominent members. This volume offers deep insights into newly developed approaches for operationalizing socio-ecological transitions at both the local and global levels, across sectors like industry, mobility, and education.

Chicoutimi, QC, Canada — Jean-François Boucher
Natal, Brazil — Luciana Gondim de Almeida Guimarães
Montréal, QC, Canada — Jean-Marc Frayret

Acknowledgments

The editors extend their heartfelt thanks to everyone involved in this project. We are particularly grateful to the authors and reviewers whose contributions were crucial during the revision process. Their support has been instrumental in bringing this book to fruition.

We would like to express our warmest thanks to CIRODD for its financial support and to all the members of its implementation team for their technical support, which made this work possible.

We acknowledge the invaluable efforts of the reviewers, whose insights significantly enhanced the quality, coherence, and presentation of the content.

Below is a list of the reviewers whose contributions we deeply value:

Amin Chabane, École de Technologie Supérieure
Anne Marchand, Université de Montréal
Armin Jabbarzadeh, École de Technologie Supérieure
Cristine Hermann Nodari, Universidade Potiguar
Etienne Berthold, Université Laval
Francisco Roberto Farias Guimaraes Junior, HEC Montréal
Ivanka Iordanova, École de Technologie Supérieure
Jean-François Boucher, Université du Québec à Chicoutimi
Jean-Marc Frayret, Polytechnique Montréal
Lais Karla da Silva Barreto, Universidade Potiguar
Luciana Gondim de Almeida Guimarães, Universidade Potiguar and École de Technologie Supérieure
Matthieu Gruson, Université du Québec à Montréal
Nathalia de Paula, École de Technologie Supérieure
Rim Larbi, École de Technologie Supérieure
Sérgio Junio da Silva, Université Laval
Thierry Lefèvre, Université Laval

Contents

Chapter 1
Introduction

Mohamed Cheriet

CIRODD is a strategic cluster funded by the Quebec Research Funds (FRQNT-SC: Nature and Technology, and Human Sciences). It brings together around 100 professors from 15 universities and 3 technical high schools across Quebec. CIRODD's core mission is to accelerate societal transformation toward a sustainable future by fostering innovation and transdisciplinarity.

With ambitious interdisciplinary goals, CIRODD aims to create unprecedented synergies among researchers and stakeholders from diverse fields. This collaboration is essential for driving the socio-ecological transition that our society urgently needs. By leveraging technological advances, CIRODD ensures that the solutions developed align with the needs and aspirations of key stakeholders. Our role is to catalyze, inform, advise, mobilize, and support decision-makers, ensuring that their actions are rooted in the best scientific knowledge and practices.

CIRODD's research program is structured around three interconnected axes: **Evaluation, Design, Integration** (Axe 1), **Systemic Approaches and Accelerators** (Axe 2), and **Collaboration with Field Actors** (Axe 3). Through these axes, along with various financial support programs such as Levier, Exploration, and Synergie, CIRODD supports and values the interdisciplinary work of its members. In 2023, we launched a new program aimed at fostering transformational projects—those that embrace a transdisciplinary approach, align with sustainable innovation principles, and contribute to achieving at least one of the United Nations' Sustainable Development Goals (SDGs). These projects, like others supported by CIRODD, share a common focus: **accelerating** the operationalization of sustainable development.

Recognizing the importance of sharing our experiences and insights, we have compiled this book to serve as a reference for other societies. Although we have not yet reached our ultimate goal, we have made significant progress and learned

M. Cheriet
CIRODD, École de Technologie Supérieure, Montreal, QC, Canada
e-mail: mohamed.cheriet@etsmtl.ca

M. Cheriet et al. (eds.), *Accelerating the Socio-Ecological Transition*,
https://doi.org/10.1007/978-3-031-82896-6_1

valuable lessons. This book is the product of a collective effort to disseminate the knowledge and results we have acquired over time.

The book is structured into two main parts:

1. **Engaging Society: Co-Creation and Mobilization Strategies for Socio-Ecological:** This section explores practical interventions and citizen mobilization strategies. It delves into transition projects, urban planning, and models for knowledge co-construction, drawing on theoretical, practical, and experiential knowledge. The role of research and the arts in facilitating this transition is also examined.
   Chapters:
2. *Operationalizing Urban Ecology and Circular Economy in Montreal's South-West Neighborhoods Using an Interdisciplinary Approach*
3. *Co-construction as Implementation: The Experience of Circular Economy in Quebec*
4. *Intervention-Research as a Social-Ecological Transition Belt, Steering Wheel, and Engine*
5. *Chemins de Transition: An Innovative Method of Knowledge Mobilization to Accelerate the Socio-Ecological Transition in Quebec*
6. *Methodological Proposal for Graphic Design Education Aimed at Fostering Social and Environmental Responsibility*
7. *Exploring Emerging NPL and Machine Learning Methods in Climate Change Discourse Analysis on Social Media: A Systematic Literature Review*
2. **Transformative Tools: Methods for Navigating the Socio-Ecological Transition:** This section focuses on the tools necessary for facilitating the socio-ecological transition. It presents models and methods for characterizing complex systems and showcases digital practices for designing circular and sustainable business models.
   Chapters:
8. *Performance Indicators for Sustainable Remanufacturing Closed-Loop Supply Chains*
9. *Advancing Sustainability Through Digital Maturity: An Open Approach for Evaluating Quebec Organizations' Environmental Responsibility*
10. *Navigating the Shift: Strategies Beyond "Build It and They Will Come" for Sustainable Mobility in Quebec*
11. *Fostering Sustainability Through Digital Evolution: Evaluating Industry 5.0 Preparedness in Quebec's Regional SMEs*

Through these contributions, this book offers valuable insights and practical tools for advancing the socio-ecological transition, aiming to contribute to a more sustainable future. We hope that readers will find both inspiration and actionable knowledge within these pages, empowering them to contribute meaningfully to the ongoing efforts toward sustainability. Whether you are a researcher, practitioner, or decision-maker, may this book serve as a guide and a catalyst in your journey toward fostering a more resilient and equitable world.

Mohamed Cheriet, in the Department of Systems Engineering at ÉTS (Université du Québec, Montreal). He is the former Tier 1 Canada Research Chair in Sustainable and Smart Echo-Cloud Computing, Founder and Director of the Synchromedia Laboratory, and General Director of CIRODD, a strategic research cluster focused on socio-ecological transition through innovation, funded by the provincial funding organism, Fonds de Recherche de Québec (FRQ).

Dr. Cheriet is an expert in artificial intelligence, pattern recognition, and machine learning, with applications in intelligent and sustainable networks, cloud computing, and image processing. His work bridges advanced research and practical impact in digital sustainability. He has authored over 550 peer-reviewed publications and delivered more than 70 invited talks. He is Fellow of the International Association of Pattern Recognition -IAPR (2016), the Canadian Academy of Engineering - CAE (2017), the Engineering Institute of Canada - EIC (2018), and Engineers Canada - EC (2019)

Chapter 2
Operationalizing Urban Ecology and Circular Economy in Montreal's South-West Neighborhoods Using an Interdisciplinary Approach

Daniel Pearl, Amy Oliver, Cécile Bulle, Titouan Greffe, and Claudiane Ouellet-Plamondon

Abstract As a result of deindustrialization and insufficient proactive investments by governments, industrial wastelands on the edges of towns and cities are multiplying. These sites illustrate the instability of mono-functional zoning and inadequate city regulations to ensure long-term resilience, particularly evident along once-industrialized waterfronts like Montreal's Lachine Canal. Since its closure to commercial navigation in 1974, the Lachine Canal's programming has focused on recreational activities, largely neglecting infrastructure and adjacent industrial areas. However, these post-industrial areas present significant opportunities to rethink the city and create sustainable, resilient neighborhoods. In the context of climate change and resource depletion, rehabilitating both the under-used building stock and the largely vacant sites with mixed programming, appropriate density, and green infrastructure, can regenerate urban fabrics and the earth's productive capacity. The post-industrial landscapes along the Lachine Canal possess several assets that the City of Montreal could leverage in its quest for resilience in the face of climate change, which it can achieve by leveraging industrial heritage, promoting urban biodiversity, and incorporating principles of circular economy, urban ecology and urban metabolism. This chapter documents an Action Research process focused on the sustainable redevelopment of neighborhoods along the Lachine Canal. It draws on interdisciplinary and inter-university research involving architecture, life cycle assessment, and construction engineering from three Montreal universities

D. Pearl (✉) · A. Oliver
Université de Montréal, Montréal, QC, Canada
e-mail: daniel.pearl@umontreal.ca; amy.oliver@umontreal.ca

C. Bulle · T. Greffe
Université du Québec à Montréal, Montréal, QC, Canada
e-mail: bulle.cecile@uqam.ca; greffe.titouan@courrier.uqam.ca

C. Ouellet-Plamondon
École technologie supérieure, member of the Université du Québec network, Montréal, QC, Canada
e-mail: claudiane.ouellet-plamondon@etsmtl.ca

M. Cheriet et al. (eds.), *Accelerating the Socio-Ecological Transition*,
https://doi.org/10.1007/978-3-031-82896-6_2

and builds upon previous inter-university endeavors going back to 2019. The approach highlights the importance of interdisciplinary collaboration and the role of universities in promoting urban ecology and circular economy through collaborative research and demonstration projects. This interdisciplinary approach aims to bridge gaps in architectural and engineering education and fosters co-learning, contributing to the creation of resilient and sustainable urban environments.

Keywords Circular economy · Urban ecology · Interdisciplinary · Ecosystemic urbanism · Life cycle analysis · Territorial development · Action research · Urban diagnosis · Adaptive reuse · Magic threshold · Carbon footprint

1 The Problem: Current Issues Preventing Proactive and Inclusive Post-industrial Revitalization of Montreal's Southwest Neighborhoods

As a result of deindustrialization and the lack of proactive investments on the part of governments, industrial wastelands on the edges of towns and cities are multiplying. These large tracts of land and neglected buildings are no longer home to their original activities and have been slowly weakened by a build-up of soil contamination and airborne pollution caused by their past functions (Merle & Perrin, 2018). They bear witness to the instability of monofunctional zoning and the lack of city regulations ensuring sufficient environmental standards to equip municipalities with longer-term resilience. Along once industrialized waterfronts, the lack of resilience and biodiversity protection is more glaring than ever. In Montreal, the Lachine Canal and its adjacent neighborhoods are no exception. Since its closure to commercial navigation in 1974, the Lachine Canal's programming has focused on recreational activities, while the riverbanks, beyond repairs to its concrete retaining walls, have been neglected, and the adjoining industrial areas have mostly been ignored. The lack of forward-looking investment in renewing infrastructure can be easily seen along the length of the canal.

Yet, paradoxically, these urban voids and dilapidated infrastructure represent real opportunities, known as "magic thresholds," to rethink the city and create sustainable, resilient neighborhoods. In a context of climate change and natural resource depletion, the rehabilitation of these brownfield sites with green infrastructure represents an opportunity to create complex living environments in response to our planetary limits (Sarni, 2010). Through their renewal and transformation, they can become real catalysts for regenerating the urban fabric and the earth's productive capacity (Vigier et al., 2023).

With its vast post-industrial expanses, Montreal's South-West borough possesses several assets that the metropolitan region could leverage in its quest for resilience in the face of climate change. To do so, we need to harness the full potential of the area's industrial heritage, starting, of course, with the Lachine Canal. The canal could become a focal point for urban biodiversity, while also facilitating the transport of goods, this time by promoting eco-friendly mobility. In areas that are virtually being rebuilt, we need to think about circularity at the scale of major infrastructure and implement urban metabolism principles. Why not harness the free heat waste from industry to power new developments? Why not rethink our water management, for example, finding a more local source to supply industries, rather than using drinking water from the city's overused network? Why not immediately add long-term resilience to major infrastructure, which can yield \$5–6\$ in value for every dollar invested, as opposed to waiting for very costly disasters (C2ES, 2019)? Finally, we need to rethink what to do with all post-industrial buildings and rehabilitate or transform them based on a circular economy model and preserve embodied energy, while encouraging the creation of post-industrial jobs in this sector.

While it is essential to rethink the future for the South-West of Montreal and see brownfield sites as opportunities for creating affordable and mixed neighborhoods supporting more sustainable lifestyles, it is also vital to do so from a lifecycle perspective. Indeed, if we confine ourselves to a municipality's emissions at the city scale, we run the risk of missing out on the greatest opportunities to reduce our environmental footprint, and of identifying strategies that would have only displaced the problem elsewhere. For example, it would theoretically be possible to drastically reduce a city's carbon footprint by moving all emitting industries out of the city, but in reality, the resulting impact would be unchanged. It is essential for land-use planning scenarios to consider the way in which such planning will promote lifestyles that are genuinely preferable for the environment (Vigier et al., 2023), which is what life-cycle analysis makes possible, as displacing impacts is not real progress.

A recent CIRAIG study presented an inventory of greenhouse gas (GHG) emissions resulting from Greater Montreal's consumption (Agez & Patouillard, 2022). This study highlights that while some major contributors to emissions do indeed take place in the Greater Montreal area (daily car journeys, domestic combustion), a significant proportion of emissions occur elsewhere (food, etc.). Total GHG emissions were estimated at an average of 12.8 t CO_2-eq per capita, dominated by household consumption and more specifically the use of personal vehicles, food (mainly meat, restaurants, and dairy products) and the use of fossil fuels at home for heating and cooking. According to IPCC figures, our individual 1.5 °C-compatible carbon footprint budget is 2 t CO_2-eq per capita. Since industrial decarbonization is insufficient to reach 1.5 °C-compatible level of individual carbon footprint (Cap et al., 2024), a drastic change in lifestyles and consumption patterns in Montreal is therefore necessary. We can keep these orders of magnitude in mind when regenerating post-industrial neighborhoods and plans for more sustainable and local lifestyles.

An unavoidable question is: who will catalyze a sustainable transformation of the post-industrial neighborhoods along the Lachine Canal while promoting local employment, a circular economy, and local environmental, economic, and cultural resilience? A second unavoidable question: what programs and what financial aids are needed to transform these post-industrial zones, which, before the decontamination and infrastructure revitalization, are sometimes the only affordable land in our cities? It is important to have a strategy at all levels of government to ensure long-term affordability, as these areas offer rare opportunities to develop post-industrial mixed-use neighborhoods with affordability, circular economy, and the cohabitation of light industrial, commercial, and residential uses.

This chapter describes an Action Research process conducted between 2019 and the present (and towards 2025) focusing on the sustainable and resilient redevelopment of neighborhoods along the Lachine Canal. Based on several years of teaching and research in the master's program in architecture at Université de Montréal, in life cycle assessment at UQAM, and in environmental and construction engineering at École de technologie supérieure (ÉTS), this chapter explains an approach based on the creation of a network of actors and scientists around interdisciplinary and inter-university research projects. The projects in question illustrate the importance of integrated design thinking when creating more sustainable projects, such as the redevelopment of urban brownfield sites or the revitalization of monofunctional, suburban-type neighborhoods. There are gaps in the training of architects in this area, and the projects described in this chapter aimed to offer an interdisciplinary approach to Master of Architecture students at the Université de Montréal, enabling co-learning with students and researchers from other universities and disciplines. This chapter demonstrates that universities can play an important role in operationalizing urban ecology and the circular economy through Action Research and demonstration projects.

1.1 Towards Solutions: Urban Ecology, Life Cycle Analysis, Circular Economy, and "Magic Thresholds"

While several approaches exist for creating sustainable, regenerative, or resilient neighborhoods, this chapter focuses on a few key concepts that are at the heart of the Action Research projects described in this chapter. These concepts represent core research areas for the professors involved and are posited to be some of the most profound tools for transformative change at a neighborhood scale.

1.1.1 Ecological Urbanism

There are many lessons to be learned from the ecosystemic experiments in Barcelona, Spain, carried out over the past couple of decades thanks to the Urban Ecology Agency of Barcelona (UEAB) and its former director, Salvador Rueda. This multidisciplinary agency has gained notoriety for its "superblock" proposal. Their idea, simply put, is to take space away from cars and give it to pedestrians by concentrating vehicular traffic on main arteries in a 3 × 3 block grid. The agency's strength lies in its urban sustainability diagnosis methodology, which addresses the physical, social, economic, cultural, and natural dimensions of the city in equal measure, using some 50 quantitative indicators, the majority of which can be mapped using Geographic Information System (GIS) software. The agency seeks to reveal the inherent synergies that link each of these dimensions together in order to develop urban density, social mix and cohesion and to create resilient, synergistic urban plans. The agency's mission is to profoundly understand the context in question: the environment, but also its neighborhoods, its networks and the region of which it is a part, and all of this in an interdisciplinary, transcalar approach.

1.1.2 Life Cycle Assessment

Life cycle assessment (LCA) is a systemic approach that enables the standardized evaluation (ISO 14044) of the environmental balance of the life cycle of a product, service, company, or process, i.e., from the acquisition of raw materials to production, use, and end-of-life (cradle-to-grave) (Jolliet et al., 2010). It is endorsed by the United Nations Environment Programme and the European Union. LCA can therefore help to understand the environmental performance of development strategies in the sector under study, to assess the environmental relevance of the circular economy strategies envisaged, and to avoid certain unexpected shifts in impacts when implementing certain innovations (Chabas et al., 2023).

Life cycle assessment is traditionally applied at the scale of a product or service, but more and more approaches exist for using it at a larger scale, or for analyzing lifestyles. For example, Lesage et al. (2007) compared different options for rehabilitating urban brownfield sites in Montreal (Shop Angus), taking into account the contamination of the site, the impact of decontaminating it, and the environmental benefits of redeveloping a central Montreal district rather than continuing with urban sprawl—demonstrating that the latter benefits far outweighed the impacts associated with rehabilitating the site. Loiseau et al. (2013) adapted the LCA methodological framework to make it suitable for territorial development analysis and applied it to the case of the Thau Basin in southern France (Loiseau et al., 2014). To do so, they adopted an eco-efficiency approach based on quantifying the multiple functions rendered by the region (job creation, food production, etc.) and the environmental impacts associated with each regional development scenario, the idea being to maximize the services while minimizing its environmental impact. More recently, a study looked at the carbon footprint of consumption in Canada and its provinces and territories from a life-cycle perspective (Patouillard et al., n.d., submitted for publication), highlighting, for example, that the annual carbon footprint of an average Quebecer is of the order of 14.5 tons of CO_2-eq according to this "consumption" perspective. However, according to Quebec's carbon balance sheet, which only takes into account emissions that take place on a provincial scale, without taking into account imports linked to our consumption, it would only be 9.5 tons of CO_2-eq. These differences highlight the importance of better understanding life-cycle boundaries, so as to promote genuinely sustainable lifestyles without simply shifting the problem elsewhere.

1.1.3 The Circular Economy

The circular economy is among the approaches that have the most promising potential to reduce the pressure our society puts on resources, which is often accompanied by a reduction in overall environmental impacts, but not always (André, 2024; Geissdoerfer et al., 2017; Harris et al., 2021). The environmental consequences associated with different circular economy strategies are often highly contextual and depend on numerous parameters (energy required, recycled content, distance from the recovery site, etc.) (Chabas et al., 2023). For example, in Quebec, although operational energy remains important (especially when it is related to minimizing peak loads), our carbon footprint is sometimes largely related to the embodied energy of materials, since around 98% of energy in Quebec is renewable, thanks to hydroelectricity (Tirado-Seco et al., 2014). We cannot understand energy in Quebec without also looking at ecosystem-based urban planning and life cycle assessment (LCA), and the importance of the long-term resilience of our systems (Chabas et al., 2023).

The operationalization of circular economy strategies depends on several factors: zoning changes, health improvements (in connection with noise, pollution, air quality, water quality, etc.), the availability of significant subsidies to assist private and third-sector revitalization projects (pilot projects), high functional transportation linked to public transport and highly efficient (yet unobtrusive) local delivery strategies, encouraging the local green economy, and comprehensive carbon calculations that go beyond simple operational carbon assessment.

While the reuse of residual materials can be beneficial for the environment, it sometimes has unintended undesirable consequences (impact shifts). The life-cycle perspective is therefore inseparable from circular economy, and life-cycle impact quantification tools must be put to good use with the aim of reducing environmental impacts in an optimized way (Lonca et al., 2018; Wiprächtiger & Hellweg, 2024). As such, LCA is a relevant approach for sustainable urban planning (Chabas et al., 2023).

1.1.4 Magic Thresholds

Underused, contaminated, and landlocked sites, especially if they are located somewhat close to residential areas or high-quality green spaces, for our pedagogical purposes, are defined as "magic thresholds," as they have the latent potential to profoundly elicit "*transformative change.*" To draw a parallel with biological environments, ecotones represent thresholds that are rich and conducive to increased biodiversity. In other words, these ecotones between built-up areas are also "magic thresholds" where conditions can emerge whose possibilities are multiplied in relation to the individual context of each zone. These thresholds can then offer greater resilience and withstand more arduous conditions. They are places where dynamic public spaces can be created, whose richness comes from the appearance of a certain incoherence. These areas offer surprising possibilities for encounters/junctions, on a human scale, between established contexts and those in the making, as well as being part of a logic of ecological connectivity, and on a city-wide scale through their landscaped and living character (fertile ground of the possible). By way of example, railroad corridors, which are often also true green corridors, enable the development of nectar-rich (melliferous) plants, attracting and encouraging insect populations and a whole range of fauna adapted to urban conditions, which can offer the development of significant biodiversity. Thanks to these "magic/urban thresholds", where the right intervention can catalyze the profound changes needed to build urban resilience (Pearl & Beauvais-Sauro, 2020), it is possible to witness the rebirth of a neighborhood by increasing its social diversity and enhancing its resilience, without destroying its existing fabric. The aim is to create places where it is possible to live, whether for a young aspiring urban farmer, an artist, or a young family, each within walking distance of a business incubator or local services, as well as to understand the different ecological systems in place in order to enhance and integrate them into urban structures as genuine vectors of social and ecological cohesion.

1.2 From **Design Research** *to* **Action Research** *in Montreal's South-West Neighborhoods*

Urban ecosystemic synergies between existing and future industries, green infrastructure, industrial heritage, and community needs for a series of neighborhoods bordering the Lachine Canal and the Aqueduct Canal, in Montreal, were explored collaboratively by various teams of Master of Architecture students under the supervision of different professors from several disciplines. Some of these Montreal sites are the subject of a design studio theme offered to Master of Architecture students at the Université de Montréal, entitled "Co-constructing our collective future". This *Design Research Studio* focuses on how the revitalization of post-industrial neighborhoods can one day lead to promising sustainable living environments. This theme consists of three courses taken over a calendar year: a research studio and a theoretical seminar in the winter term, followed by the Final Thesis Studio Project in the fall term. The courses cover a wide range of topics, at several scales, so that students can develop a broad ecosystemic perspective: urban ecology, heritage, landscape architecture and the natural environment, green infrastructure, social mix and cohesion, urban metabolism, etc. The two studios, taught by an interdisciplinary team of architects, landscape architects and structural engineers, lead students to explore different urban strategies in a holistic *Urban Metabolic Approach* and at multiple simultaneous scales. This enables them to open their horizons to the real benefits of living in stimulating, sustainable and more equitable urban environments. Over the years, the approach has covered several neighborhoods along the Lachine Canal in succession.

The research studio is of particular importance, as it teaches students how to diagnose neighborhoods using the ecosystemic and holistic methodology of the Urban Ecology Agency of Barcelona (UEAB). Students work in teams of 5 or 6 students over a period of 5–7 weeks to map the current state of the site/neighborhood in question, at varying scales, using a series of quantitative and intersubjective indicators. Lectures by experts from various disciplines, ranging from water management to urban agriculture and the reuse of building materials, are invited to feed the students' reflections during this diagnostic exercise. After this preliminary diagnostic analysis, students work on a short exercise in order to identify which "magic threshold" sites are most critical to transforming the neighborhood. The magic thresholds, which are more at the scale of the building or a group of buildings with their associated public spaces and infrastructure, must necessarily connect to the major urban-scale strategies identified in the diagnosis. While many students are interested in exploring circular economy and the reuse of building carcasses and their building materials, and transcalar synergies, the master's program does not currently offer in-depth opportunities to hone in on any specific theme (as in programs offered at UQAM and ÉTS). Finally, students propose sustainable and resilient master plans by applying the sustainable strategies that emerge from the urban diagnosis.

This pedagogical approach attempts to fill several gaps in architectural education. The link between academic and practical fields has historically been a point of frustration for those moving from one to the other, with the result that graduates too often feel that they have not been adequately prepared for professional practice (Cole & Pearl, 2007). With specific regard to sustainable design education, it is recognized that high-performance buildings depend on an integrated design process (IDP) (Azari & Kim, 2016; Busby, Perkins & Will, 2007; Ibrahim, 2015; Leoto & Lizarralde, 2019; Van der Ryn & Cowan, 2010), which is difficult to recreate in the classroom. In a pan-Canadian initiative called "Greening the Curriculum" between 2002 and 2010, a committee at the Université de Montréal developed four strategies to transform architecture education to be more transdisciplinary. With regard to the classic design studio format, this committee proposed hiring professors from different disciplines, grouping students from different disciplines in the same studio, and regularly inviting expert speakers from different disciplines to contribute to the studio (Cucuzzella et al., 2010). The committee also stressed the need to incorporate activities outside the program and to promote transdisciplinary research activities (ibid).

While incorporating Action Research[1] strategies within the school year curriculum proved to be difficult, they were more successfully implemented during the summer sessions in the form of workshops with various stakeholders (local community groups, NPOs, private companies, professionals and specialist consultants, citizens, issue

[1] Action Research is a collaborative research method used for informing or improving practice (Reason & Bradbury, 2008) that involves action, evaluation, and critical reflection. One of its strengths lies in its focus on developing solutions to practical problems and its ability to empower practitioners by getting them involved in research and implementation activities (Meyer, 2000).

tables, municipal officials, activists, artisans) and collaborative action research projects with UQAM and ÉTS.[2] In particular, the 2019 cohorts (UdeM, McGill, and EPM) and 2022 cohorts (UdeM, ETS, and UQAM), were able to benefit from interdisciplinary and inter-university Action Research projects during the summer to work with local stakeholders, post-professional masters' students and PhD candidates, in exploring circular economy and life-cycle analysis issues.

The next section describes the approach used in the architectural research studio, moving from research (and diagnosis), to design, then to action and operationalization. The students focused on post-industrial areas in the southwest of Montreal, along the Lachine Canal and the Aqueduct Canal. Initially, they carried out a graphic diagnosis in the form of maps and statistics, inspired by the Urban Ecology Agency of Barcelona's methodology and zooming in on smaller areas with more uniform characteristics. The focus was on a few emblematic buildings, leftover ambient spaces, and broken urban infrastructure along the borders of the historic Lachine Canal. The sites were also analyzed in terms of potential to become true, green infrastructure with increased biodiversity and incorporating phytotechnology. In the second phase, students reflected on the redevelopment of these post-industrial areas using a multi-scale approach. Then, during the summer sessions, some students used their studio explorations to feed into Action Research projects with local stakeholders and LCA engineering students from UQAM and ÉTS to validate and further develop their working hypotheses in terms of ecological footprint.

2 How Can we Radically Transform the South-West's Post-industrial Neighborhoods? Urban Diagnosis and Preliminary Proposals

Over the past 6 years, the courses from the "*Co-constructing Our Collective Future*" Master's in Architecture stream have used different areas in the South-West borough of Montreal to test the applicability opportunities of urban greening, urban ecology, magic thresholds, and circular economy. Figure 2.1 shows these different zones, which include three parallel waterways in close proximity to each other, each with a significant history and rich industrial heritage. Two of these zones are highlighted as being part of interdisciplinary and inter-university research projects in

[2]This collective teaching has been funded by three inter-university and interdisciplinary applied research grants since 2019: (1) funding by the Trottier Energy Institute in 2019–2021 led by École Polytechnique de Montréal, Université de Montréal and McGill University, entitled "4th generation heat networks for sustainable neighbourhoods"; (2) a grant in 2022–2023 with CERIEC led by ETS, Université de Montréal and UQAM entitled "Analysis of circular economy strategies at the scale of the Lasalle/Ville Saint-Pierre/Lachine Est neighbourhoods"; (3) and a LEVIER grant with CIRODD (2024–2025), which has just begun, entitled "Un levier pour opérationnaliser l'écologie urbaine et l'économie circulaire à l'échelle du territoire dans le Sud-Ouest de Montréal selon une approche inter et transdisciplinaire".

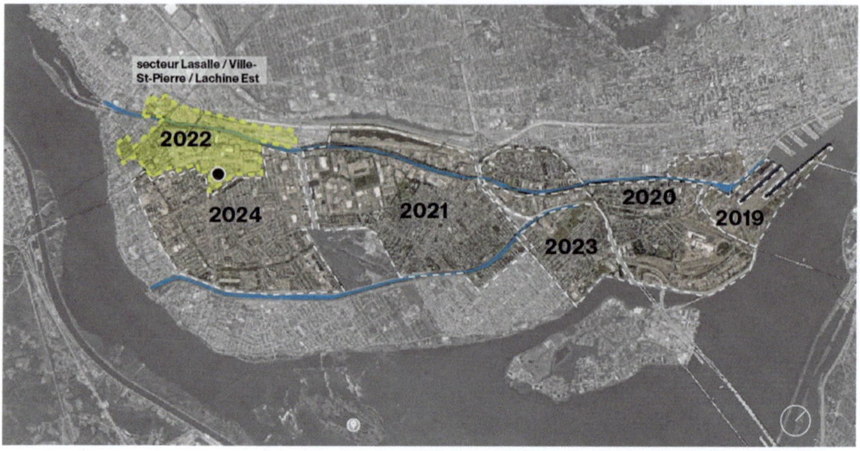

Fig. 2.1 Map of study areas over the past 6 years. One of the two areas that have been the subject of research projects outside the classroom, described in Sect. 2.1, is highlighted in yellow. The CIL industrial building rehabilitation project analyzed in Sect. 3 is shown with the black dot inside the 2022 study area. (Source: Oliver, 2024)

2019 and 2022: the first is the Bridge-Bonaventure sector (2019) and the second, top left and in yellow, the Lasalle-Ouest/Ville Saint-Pierre/Lachine Est sector (2022). The latter sector posed several urban challenges but offered a fertile ground for thinking about how to recycle industrial buildings while heavily encouraging circular economy and building rehabilitation and transformation.

2.1 Lasalle/Ville Saint-Pierre/Lachine Est (2022)

In 2022, in its fourth edition, the "*Co-constructing Our Collective Future*" studio explored the regenerative potential for a site along the Lachine Canal and the St-Pierre interchange, between the boroughs of Lachine and LaSalle, on the verge of becoming an industrial wasteland. The area is currently a commercial and industrial zone and is a prime example of the failures related to monofunctional zoning. Still, it represents a real opportunity to build a greener, more resilient neighborhood that can incorporate natural resources. What's more, this zone represents an opportunity to demonstrate the benefits of the circular economy in every respect (Figs. 2.2 and 2.3).

The students went through a comprehensive diagnostic exercise of the area (and four sub-areas illustrated in Fig. 2.2), again inspired by the UAEB's *Ecological Urbanism* methodology. As shown in Fig. 2.5, in LaSalle and Ville-Saint-Pierre, the exclusively industrial zones bordering the canal have created a physical barrier that disconnects the two neighborhoods' residents to both the Lachine Canal and each other. Students were able to point to several weaknesses of the area:

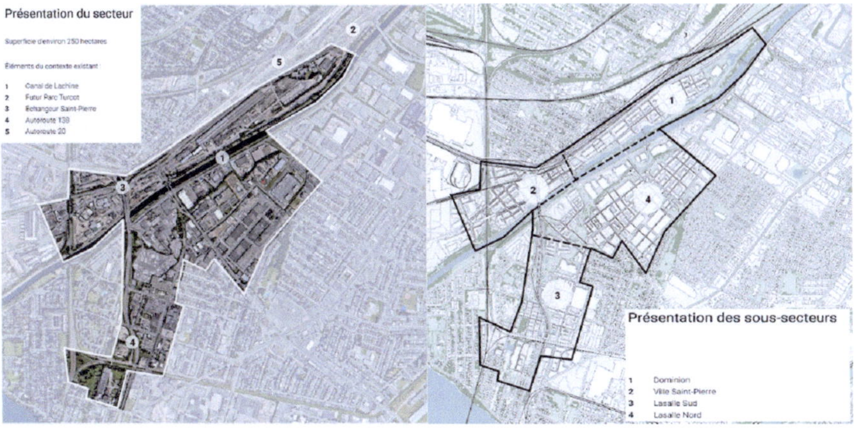

Fig. 2.2 Presentation of the study area (whole class) and sub-areas or zones (teams of 3–4 students) in the Architecture 2022 research studio. The site runs along the Lachine Canal and the St Pierre interchange, between the boroughs of Lachine and LaSalle. (Source: Masters' class ARC6802-F, School of Architecture, University of Montreal, 2022)

Fig. 2.3 Mapping of buildings in the selected industrial zone (left without GPS, right with GPS) (Source: Masters' class ARC6802-F, School of Architecture, University of Montreal, 2022)

- Physical isolation: Marked by the intensive urbanization of the second half of the twentieth century and the construction of the St-Pierre interchange, the area is now landlocked by transport infrastructure, dominated by cars, large delivery trucks, and cargo trains, and devoid of any significant plant coverage.
- Monofunctional zoning: The surrounding suburban-type neighborhoods are characterized by a plethora of shopping malls and big-box stores, recalling the original car dependency planning of a half-century ago.
- Linear economy: The industries on the site support a linear economy that neglects the impacts of their GHG emissions as well as their waste. In other words, the study area is in need of a major transformation (Chabas et al., 2023).
- Poor air quality caused by the overwhelming presence of industry (including trucks and cars), and the lack of greenery to help reduce airborne particulates.

- Significant heat islands and low biodiversity: the low percentage of the tree canopy (see Fig. 2.2): 8.15% on the site studied (Ville de Montréal, 2020), compared with around 20% in 2015 for the city of Montreal as a whole (Lapierre & Pellerin, 2018) speaks to the dire need for more green infrastructure such as tree plantations and water access points.
- Inadequate rainwater management: The entire site has a minor to moderate vulnerability to heavy rainfall, which will become even more vulnerable over time. At present, during snowmelt or heavy rainfall events, the network is regularly inundated. This leads to overflows and discharges of wastewater directly into the St. Lawrence River, the Lachine Canal and/or other waterways, without any treatment (Fondation Rivières, 2020). Thus, due to the high impermeability of the site, paved with concrete, asphalt, and large industrial buildings, as well as the proximity of the Lachine Canal, the risk of flooding is constantly present.

On the other hand, a number of opportunities for transforming the area into a more vibrant, human-scale living environment were also identified by the Master's students:

- The Lachine Canal, and the landscapes that border it, occupy an important place in Montreal's collective memory. It is home to a rich industrial and landscape heritage that is an important part of Montreal's industrial history, such as the LaSalle-Coke crane, the Dominion Car & Foundry, and the Gauron-Lafleur bridge (ibid). Revitalizing the area's industrial and landscape heritage, which have now been abandoned, is essential to reviving the site, which currently leaves little room for human activity. The re-use of derelict land for a variety of transformative projects would enhance the quality of life in the surrounding area, while recalling the canal's industrial past. Moreover, these industrial sites each have their own dormant inherent potential that can be taken advantage of when adaptively reused, whether related to structural modifications to accommodate the addition of mezzanines or vertical extensions or, structural grid flexibility, which could allow for easier changes of use in the future.
- Thermal mass and/or cross-sectional design openings in these industrial sites can be modified to enable cross-ventilation (including chimney stack effect), or more natural daylight in.
- Beyond the more obvious historical value, there is also the energy value, particularly embodied energy, which refers to the amount of energy consumed by a building itself, but also by its components, throughout its life cycle over time (Office québécois de la langue française, 2022). In the face of depleting natural resources, what is already present on a site must become the first resource to be analyzed and mobilized at every scale of the project.
- There is an opportunity to create new, high-quality public spaces and improved access to a wide range of services in underdeveloped sectors.

Following this ecosystemic diagnostic work (see an example in Fig. 2.4), the architecture students proposed sustainable master plans (at the scale of the overall studio sector, about 200–250 ha), while simultaneously developing a smaller (but

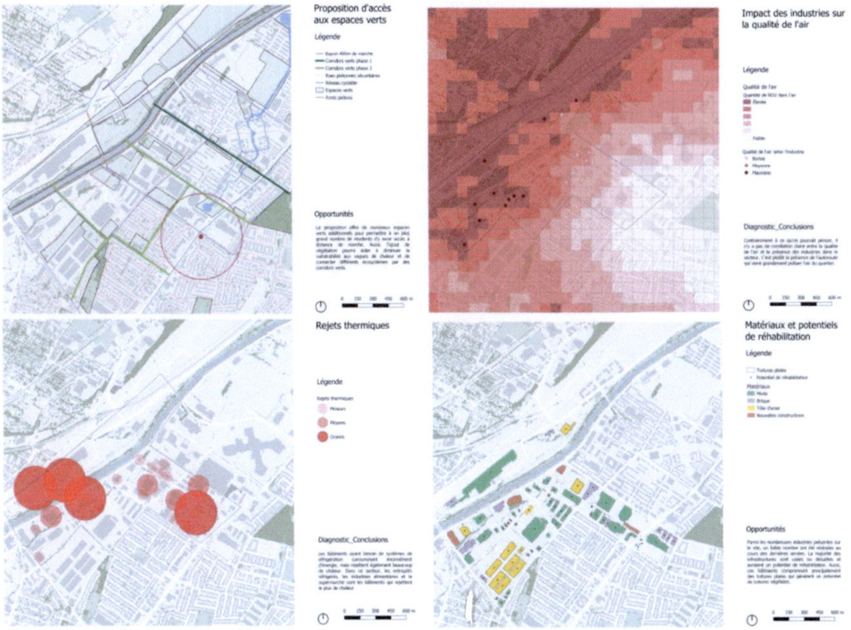

Fig. 2.4 Urban Diagnosis based on the Urban Ecology Agency of Barcelona's Ecological Urbanism framework for the Dominion/North Lasalle North zone, with indicators for access to green spaces, impact of industries on air quality, thermal waste, and rehabilitation potential. (Source: Janie Ouellette, Olivier Pesant and Jordane Castonguay, 2022)

more detailed) site plan, by group (ranging from 30 to 60 ha) and then, individual thesis project masterplans (from about 2 to 6 ha) were proposed. Some of these projects went beyond the studio boundaries and were the subject of an additional collaborative research project with UQAM and ÉTS (see Sect. 3). The further developed proposals were based on a collective reflection, connecting different projects and time scales. The idea was to plan a gradual transformation, over several decades, using a phased approach, starting with rethinking the place of heavy industry and truck transport through the city. Several fundamental themes fed into this vision for a renewed district, such as the integration of industrial and landscape heritage, the role of a circular economy and green infrastructure (including the introduction of resilient measures to combat future flooding and prolonged droughts), the provision of affordable spaces for artists and artisans, the strengthening of the social fabric, and the creation of sustainable and flexible spaces (Fig. 2.5).

The revitalization of the area under investigation was accompanied not only by densification, but also by significantly increasing complexity, with a variety of housing and daily services within walking distance. By offering a variety of housing typologies and tenure types, the needs of various groups would be addressed. The aim was to strengthen social cohesion by bringing together a mixed population, then adding inclusive and safe public spaces that encourage citizen engagement and

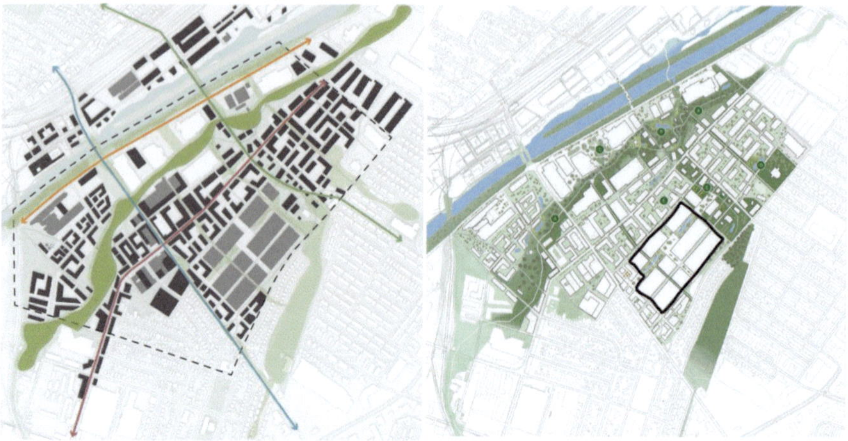

Fig. 2.5 Master plan. The revitalization project for the North Lasalle area, south of the Lachine Canal, proposed by a team of 5 students. On the left, the proposed new densification (black buildings) and building renovation strategy (dark gray buildings). On the right, the proposed biodiversity and greening strategy. Note the green corridor strategy that runs along an existing embankment parallel to the Lachine Canal. The Montreal CIL complex is identified in black on the right-side map (the site analyzed in Sect. 3). (Source: Marylou Filiatrault, Janie Ouellette, Charlie Proteau, Alexandra Thibodeau-Gagnon, Godefroy Vallette)

interaction, including the development of a green corridor along an existing natural slope. These urban proposals were made in tandem with drastic improvements in public transport (e.g., adding shuttles, streetcars, bus lines, and/or increasing the frequency of existing lines). This is crucial to reducing the number of cars in the area and encouraging daily life activity. In fact, these pivotal projects, strategically located in "magic thresholds," are discussed immediately below.

3 An Example of Life Cycle Assessment Applied to a "Magic Threshold": The CIL Industrial Complex Rehabilitation Project in Montreal

As explained in this chapter, it is important for built environment professionals, including architects, to have a basic understanding of LCA and carbon footprint issues at multiple scales, from the city scale (which includes citizens' daily activities in terms of good consumption, transportation choices, energy consumption, etc.) to the building scale, and all the way down to specific material components and assemblies. At this smaller scale, built environment professionals should be using both quantitative (ecological footprint) tools and qualitative measures to best understand the myriad of issues while evaluating the usefulness of rehabilitating nondescript[3] industrial buildings. This is where questions involving embodied energy

[3] Buildings with no exceptional architectural or heritage quality.

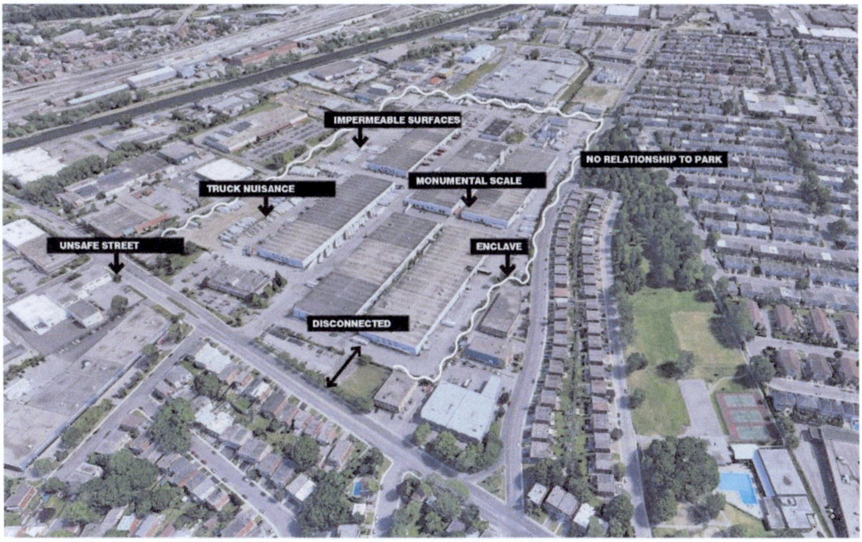

Fig. 2.6 Aerial image of the Cité industrielle Lasalle (CIL) complex. Note the heavy presence of impermeable surfaces (asphalt and concrete). (Source: MaryLou Filiatrault, 2022. Published with permission)

become substantive. The exercise of evaluating building materials and their various assembly options, with respect to LCA was explored in partnership, between students from three different universities.

As a test case, a group of buildings on the Cité Industrielle de Montréal (CIL) site was selected for LCA analyses with respect to various rehabilitation scenarios within the context of a 2022 UdeM Masters' studio Action Research project. The CIL complex in Montreal acts as a divide between the residential area to the south and the Lachine Canal to the north. This building complex was selected for a number of reasons: there was sufficient access to technical drawings, which enabled a more precise analysis; there was knowledge of its past use as a factory that stored ammunition, and thus the building carcass already contained a large amount of poured-in-place concrete (and embodied energy), including an extensive underground infrastructure network; the buildings, in general, have a very wide and flexible structural grid; and the building's lack of natural light presented an interesting challenge with respect to future use and how to most effectively alter this condition (Figs. 2.6 and 2.7).

The student's project centered around transforming the site into one based on the circular economy, and more specifically, in relation to food and urban agriculture. She also strove to better connect the site to the neighborhood, with a rainwater-focused axis to the south-west of the site on Dollard Street (in blue in the diagram below), an educational and green axis to the north-east of the site (in green) and a central connection that also deals with water and public space (in orange), giving access to the renovated hangar buildings, which would house the programming related to the circular economy (Fig. 2.8).

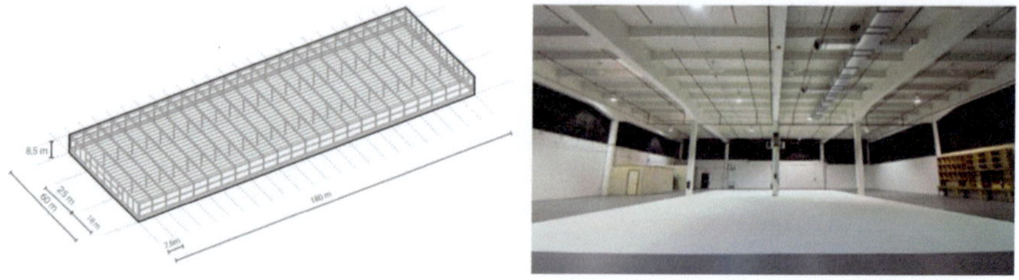

Fig. 2.7 Structure of industrial hangars (left). Photo of inside of buildings (right). (Source: MaryLou Filiatrault, 2022. Published with permission)

The LCA analyses included a professor and a doctoral student at UQAM evaluating how to best make use of the existing concrete already present in the project, while a second professor and her student at ÉTS assessed the carbon footprint of different building material choice scenarios with respect to renovating the buildings' cross-sections.

3.1 Conceptual Analysis of the Reuse of Material in the Industrial Building Rehabilitation of the CIL Industrial Site

As the students' projects focused on circular economy and proposed incorporating reused materials, a conceptual study of residual materials management processes in the construction sector was carried out by a project group comprising architecture students, engineering students from ÉTS, and professors from both universities. The study explored the various techniques for renovating and rehabilitating old industrial buildings and integrating them into the current climate and specifically looked at the potential for reusing construction, renovation, and demolition (CRD) materials. This section presents a preliminary study of resource management for the CIL industrial site (Auffan, 2022).

The CIL Lasalle industrial site is covered with large horizontal industrial buildings (Figs. 2.9 and 2.10). The main idea explored for rehabilitation and renovation would be to retain and redevelop the industrial space into a space welcoming ecological activities, and to build a brand-new market and community center, with the aim of integrating the neighborhood into this now highly industrial space. The aim of the community hub is to transform one of the main heat islands into a green space to reduce the ecological impact of the heat generated by these areas. Indeed, the entire area of the Lasalle industrial estate is a gigantic parking lot.

To develop structures for agricultural production, the Master of architecture student proposed roof-mounted or ground-mounted hoop houses (a portable greenhouse-type structure with an arched frame). The project also aimed to reuse materials along with another Master of architecture student's project located on the Fleichmann

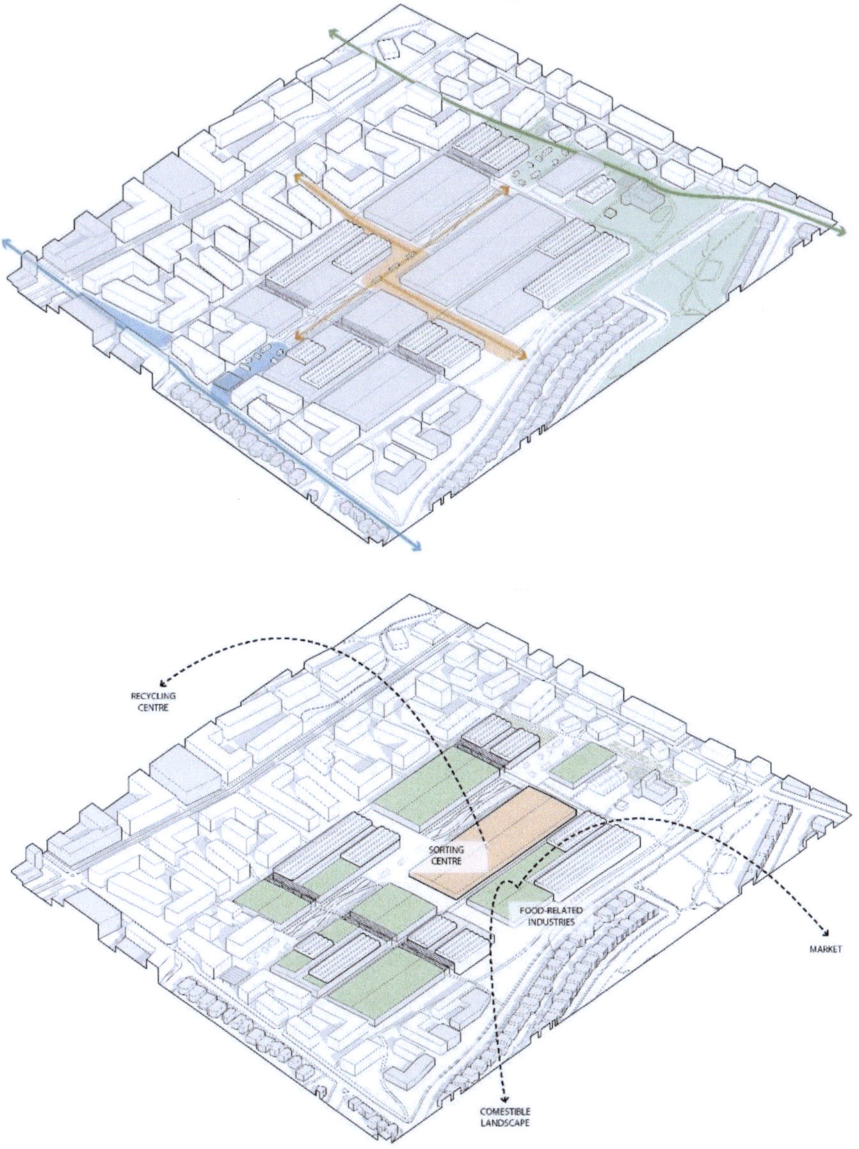

Fig. 2.8 Diagram of different connective axes with the surrounding neighborhoods (top) and circular economy synergies (bottom). (Source: MaryLou Filiatrault, 2022. Published with permission)

Yeast factory site. Among other things, sourcing the same materials facilitates the process of recovering and transporting them from the sorting centers to the work-sites. This reduces the carbon impact of transporting both materials and residual materials. One proposal would be to create growing plots directly on the ground,

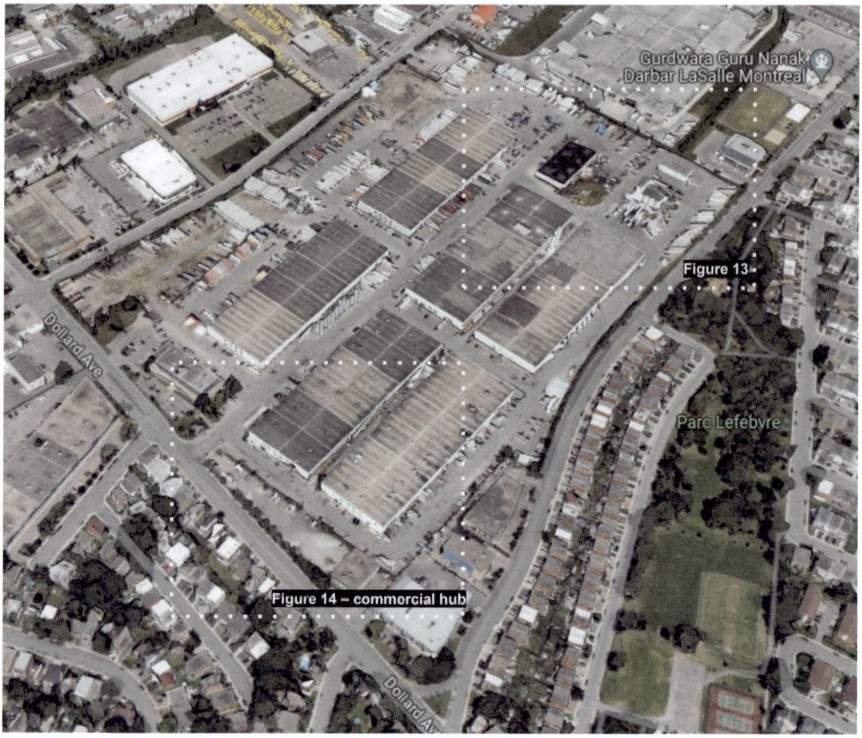

Fig. 2.9 Aerial image of existing site (top). (Source: Google Maps, 2024)

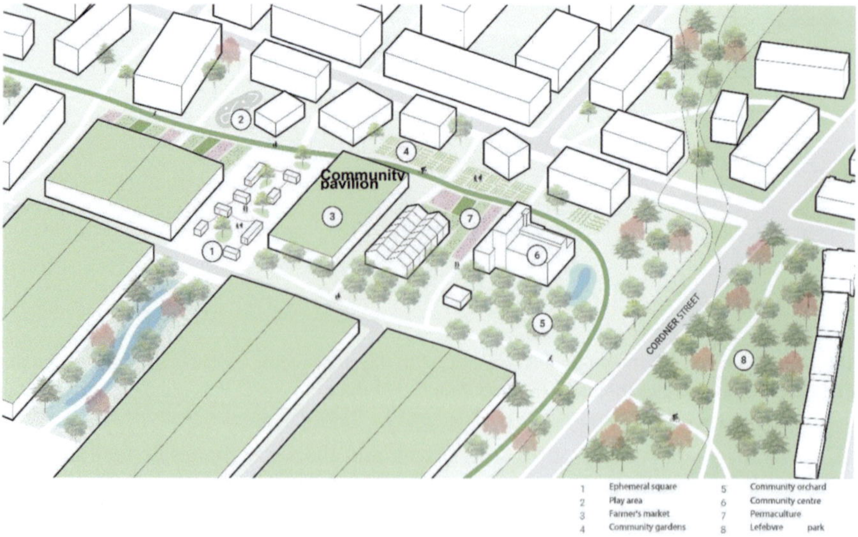

Fig. 2.10 Project to convert an existing parking lot into an urban park with community gardens, community orchard, permaculture, and greenhouses. (Source: Marylou Filiatrault. Published with permission)

Table 2.1 Identification of materials required for three urban agricultural scenarios at the LaSalle industrial estate scenarios

Scenario	Title	Materials	Functions
A	Greenhouse: Hoop houses	Metal (steel) or PVC	Roof
		Wood	Framing
		PVC—UV-stabilized	Pipes
B	Planters and ground plantation alley	Wood	Planters
C	Planters and ground plantation alley	Ribbed sheet metal	Planters

and, on the other hand, to install wooden or ribbed and folded sheet metal trays. Table 2.1 summarizes the various possible scenarios.

In the process of transforming the site, the research team from ÉTS and Université de Montréal planned to remove all bitumen and recycle the bituminous aggregates, in order to reuse them to renovate road axes in other areas of this project, such as the public square in the commercial zone, or to make paths and cycle tracks in the park (Fig. 2.10). Asphalt aggregates (bituminous aggregate in Québec), according to SN EN 13108-8, "come from the milling of asphalt layers, the crushing of slabs extracted from asphalt pavements, pieces of asphalt slabs, asphalt waste or surplus from asphalt production". In this way, the materials cycle within the worksite could be optimized from both an economic and environmental point of view (KIES, 2021).

3.1.1 Community Pavilion

In the case of the community pavilion (building number 3 in Fig. 2.10), the proposal is to use an existing building on the site and redevelop it. The only modification to be considered for this building would be a reinforcement of the load-bearing structures. Indeed, to install a greenhouse on the roof, structural calculations must be made to determine whether the existing structure and foundations in concrete can accept the additional loading of the greenhouses, or whether an independent mezzanine strategy should be employed, avoiding extra loading on the foundations. If we assume that reinforcement work is necessary, the most likely scenario would be the latter (a new independent mezzanine structure). Adding steel reinforcement to the concrete beams is one of the cheapest and most environmentally friendly solutions, even in the long term (assuming the steel reinforcement does not need to be fireproofed).

3.1.2 Public Square

The space chosen for the public square is already open and ready to be developed. The main aim would be to create more space and fit it out with benches and other elements of a public square. It will also involve the removal of tar from certain areas to eliminate other heat islands present on the site, and the development of green, welcoming spaces for local residents.

3.1.3 Public Market

As far as the public market is concerned (see Figs. 2.11 and 2.12), there were two options. The first idea was to demolish the floors of the existing building chosen to house the market and create a space with substantial ceiling height. This would make it possible to accommodate and build a green space sheltered from the harsh weather conditions of the Quebec winter, while retaining the idea of a local market. A second idea was to keep the two storeys and create a space like New York's Chelsea Market, with its many local shops and restaurants. This idea would work well with having local shops.

3.1.4 Commercial Hub

The aim of this area is to create a link between the community, the project, and existing local businesses. The aim is to involve local residents in the use of these specially created spaces to create a sense of community. The architecture student

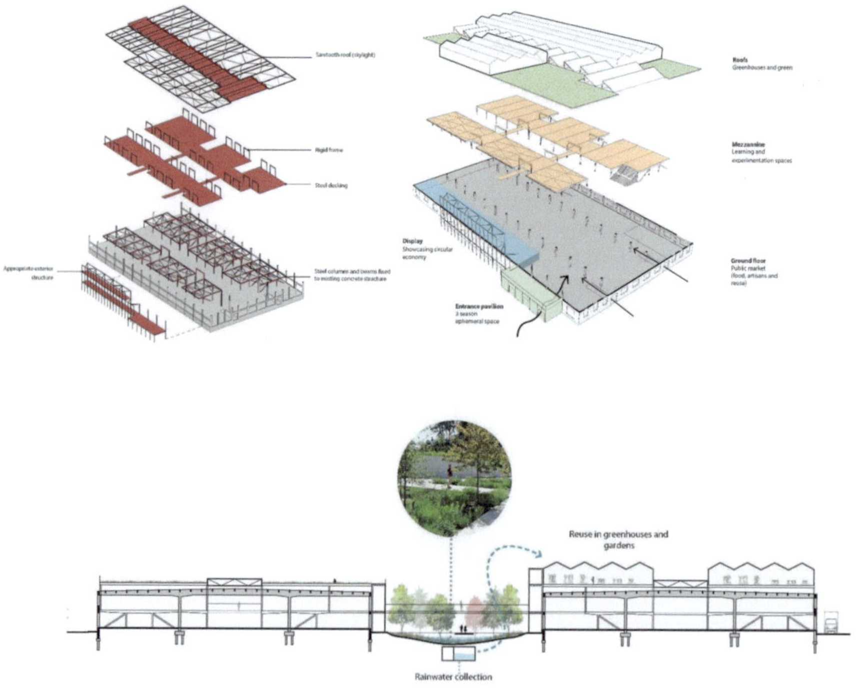

Fig. 2.11 Exploded axonometric highlighting the new structural additions in red, versus the existing, in gray (top left). Exploded axonometric showing programmatic elements, including the public market on the ground floor (top right). Section drawing showing the rooftop greenhouses supported by the new secondary structure, so that the existing foundations do not take on additional loading (bottom). Source: Marylou Filiatrault. Published with permission

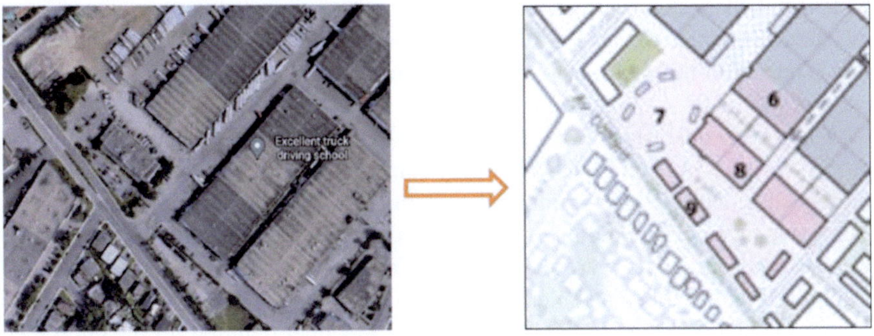

Fig. 2.12 Commercial hub. Project to transform the commercial hub on Dollard Street on the south-west part of the site (right figure: 7-Public square; 8-Public market; 9-Local shops) (Source: Marylou Filiatrault. Published with permission.)

proposed creating a public square, a public market for locally grown and produced products, and local shops.

This study enabled the ÉTS students and professors to contribute to research into processes for managing residual materials from CRD worksites at two specifically chosen sites within the Montreal South-West district rehabilitation project. In the case of the CIL industrial complex, the repetitive nature of the buildings made deconstruction more interesting and viable than demolition.

3.2 *Comparative Study of Fiber-Reinforced Polymer Concrete Structure as Part of an Industrial Rehabilitation Project at Montreal's CIL Complex*

The CIL industrial complex's large industrial buildings offer untapped large-scale programming flexibility potential for effective adaptive reuse while also providing an opportunity to integrate new green infrastructure and some basic urban design strategies. Instead of first demolishing the complex and then rebuilding something new to meet future programmatic requirements, rehabilitation was fully favored, and explored, for its multiple benefits (Lentier, 2012). The rehabilitation included substantial changes to the building section and roof plane in order to introduce hybrid natural ventilation opportunities, direct sunlight, and indirect daylighting (see Fig. 2.14). The student project proposed to take advantage of the vast network of roofscapes to design a vegetated roof, including greenhouses and planters, and potential changes to the existing structure were analyzed by a master's student at ÉTS (Legru, 2024). The existing reinforced concrete structure would require significant reinforcement to support the increased structural loading. Numerous reinforcement techniques for reinforced concrete exist, but this study focuses solely on the use of fiber-reinforced polymers (FRP). This technique is becoming increasingly widespread thanks to its effectiveness in reinforcing existing structures. The various mechanical characteristics of FRP reinforcements are widely discussed in the literature. The aim of this study is to make an environmental comparison of three different types of FRP.[4]

A life-cycle analysis of 3 reinforcement systems was carried out by students and professors at UQAM and ÉTS. The study compares a carbon-fiber-reinforced polymer system, a glass-fiber-reinforced system, and a flax-fiber-reinforced system. Life-cycle inventory modeling of the 3 systems was carried out using SimaPro software. Background data were mainly taken from the ecoinvent version 3.8 cut-off database. For flax fiber technical textiles, life cycle inventory data from Gomez-Campos et al. (2021) was used. The amount of flax fiber required for the reinforced polymer was calculated using the equivalence factor from Deng and Tian (2015) (Fig. 2.13).

Once the 3 systems were modeled, it was possible to carry out the life cycle inventory calculation, which takes stock of the substances extracted from and emitted to the environment over the life cycle of the 3 systems. However, the result is a list of thousands of substances that do not have the same environmental impact. For example, carbon dioxide (CO_2-eq) has an impact on climate change, but not on water toxicity. On the other hand, glyphosate, a well-known pesticide, has an impact on water toxicity, but not on climate change. To determine the contribution of CO_2-eq to climate change and the contribution of glyphosate to water toxicity, we need what are known as characterization factors. We used the characterization factors of Impact World+ (IW+) version 2.0 (Bulle et al., 2019) in this study. This method provides several levels of analysis with its "footprint" and "expert" versions. The "footprint" version provides 5 indicators: carbon footprint, water footprint, fossil resource footprint, other impacts on human health (including impact pathways on human health apart from climate change and water footprint), and other impacts on ecosystem quality (including impact pathways on ecosystem quality apart from climate change and water footprint). The "expert" version presents 2 indicators: human health and ecosystem quality (including climate change and water footprint impacts) (Figs. 2.14 and 2.15).

In light of the results obtained using the IW+ expert method, the best solution for limiting impacts on human health and ecosystem quality is between glass-fiber-reinforced and flax-fiber-reinforced polymers. However, given the uncertainties in both inventory and impact modeling, it is not possible to recommend one system over another.

In light of these results, the LCA and engineering students were able to make the following recommendations to the architecture students: carbon-fiber-reinforced polymer still seems to be the worst choice and should therefore be avoided. The other two alternatives (glass fiber or flax fiber-reinforced polymer) are very similar in terms of environmental impact if the flax fiber is manufactured in China (which

[4] The structural analysis is covered in a forthcoming Canadian Society for Civil Engineering conference paper.

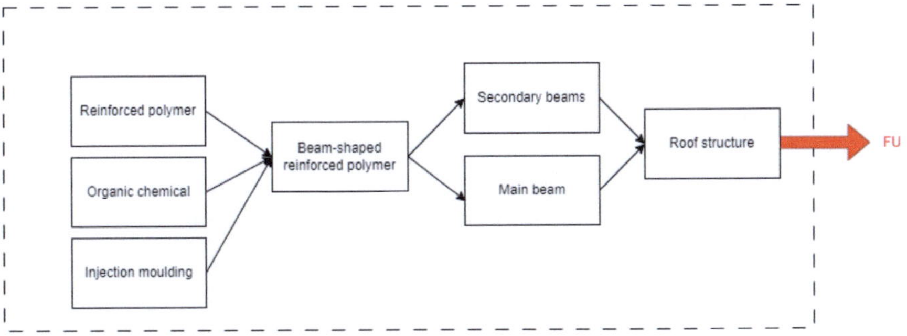

Fig. 2.13 Generic boundary of the 3 systems under study. *FU* functional unit

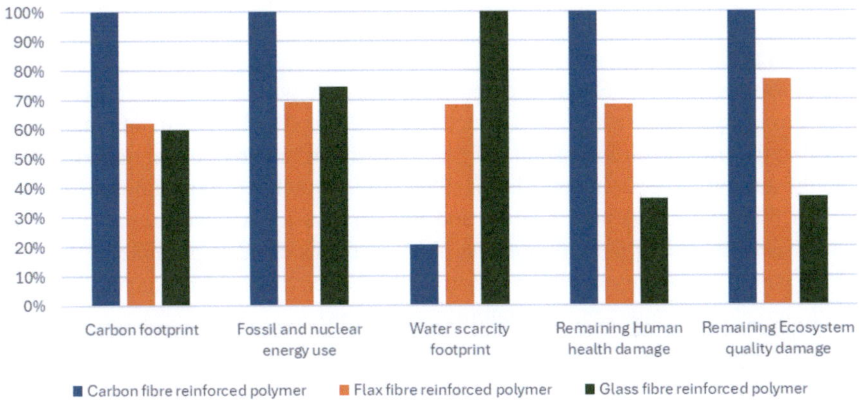

Fig. 2.14 Comparison of total potential impacts for the 3 systems for the 5 final indicators of the Impact World+ footprint method

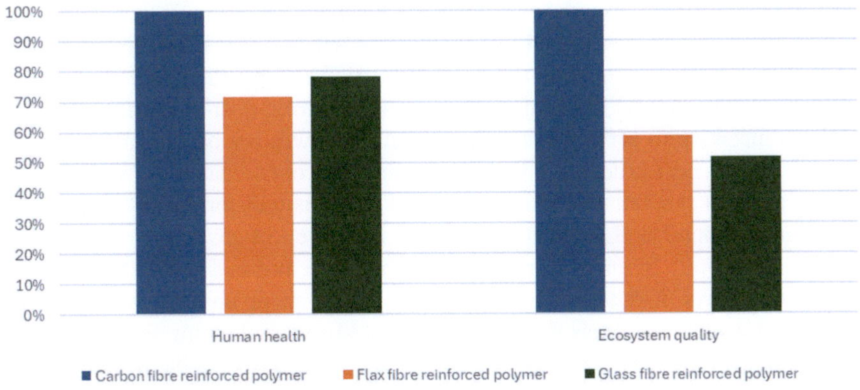

Fig. 2.15 Comparison of total potential impacts for the 3 systems for the 2 indicators of the Impact World+ expert method

is currently the case). If a supplier of flax fiber-reinforced polymer beams is able to source flax in Quebec, this would become the best-performing option in terms of environmental impact. The Master of Architecture student was therefore able to incorporate these findings into her final design.

3.2.1 Brownfield Redevelopment Project with Bioretention Basins

In addition to the LCA study described above on concrete reinforcement, a second LCA study was conducted by an ÉTS doctoral student on the proposed bio-retention basins on the CIL complex site. The analysis work involved studying breaking up the asphalt mix and transporting it off-site. The basins are formed by first removing the asphalt, and then excavating sufficient soil. The soil would then be transported off-site (normally about 60 km away). The next step would include installing storage tanks and the restoration of the site with planting, and hard surface landscaping (bitumen asphalt and gravel).

For this analysis, the SimaPro calculation software was used (with the ecoinvent v3.8 inventory database and the Impact World+2.0 method). The main conclusion was that the *transportation* off-site of materials like bitumen or concrete to ameliorate the area was the main contributor (95%) to the total eco-footprinting (for both carbon footprint, ecosystem quality, and human health). A lower-impact solution would involve finding parallel projects in Montreal as close as possible to the site, which would reuse the excavated materials, to reduce transport distances, and as a result, reduce the impact of the rehabilitation project.

4 Conclusion: The Community at the Heart of Developing Solutions

Post-industrial areas, like those around the Lachine Canal, represent incredible opportunities to experiment and test sustainable and resilient redevelopment concepts that can draw upon the concepts of industrial ecology, the circular economy, and "magic thresholds." Current owners of underutilized post-industrial buildings have no incentive to act now, since increased value will not occur until zoning is upgraded, new infrastructure is built, and land is decontaminated. As we asked at the start of this chapter, who will finance and lead this transformation? In the case of the areas around the Lachine Canal, the various levels of government have yet to invest significantly in major infrastructure. Therefore, this is a pivotal moment for reinventing ways of doing things before any siloed major investments are made. Opportunities and solutions do exist, however. Mixed-use zoning that allows for more urban complexity could be one avenue to explore. The implementation of a city-led ecological distribution/delivery system would be another. Third-party affordable developments led by NPOs or innovative funding support from the Federation of Canadian Municipalities could be a third. Other avenues would need to be developed with stakeholders (businesses, employees, citizens, artists, activists, etc.).

This chapter has described an Action Research process to operationalize the concepts of urban ecology and circular economy in an inter-university and interdisciplinary context, but including, where possible, the participation of several stakeholders involved in the areas under study. It also described a pedagogical approach designed to fill certain gaps in the teaching of green/ecological architecture in Canadian universities, notably: (1) the difficulty of recreating the integrated design process (IDP) that is recognized as a necessary process for the creation of high-performance buildings and built environments; (2) the lack of inter- or transdisciplinarity in architecture programs (both professors and the mix of students in courses and design studios). As a result, the pedagogical approach favors the creation of collaborative Action Research projects, such as those with Université de Montréal, UQAM and ÉTS, to enable the exchange of knowledge in sustainable architecture and planning, circular economy, and life cycle assessment. The next step will be to operationalize this work in the field (and explore the question of phasing) in close collaboration with a wide range of stakeholders, notably through an interactive exhibition in the study areas and participatory co-creation workshops with the communities concerned. Secondly, the aim will be to co-create innovative and realistic pilot and demonstration projects to support sustainable post-industrial development in the study areas, which guarantee sufficient ecological resilience, drawing on deliverables from previous research. This will be documented in a booklet.

The next phase of the Action Research project will therefore highlight community involvement and citizen participation, which are central to the co-construction of tomorrow's cities. Urban redevelopment projects can be seen as opportunities to include local populations in the creative and decision-making processes, by involving them in the co-creation of solutions adapted to their needs. The bottom-up/top-down approach combines the perspectives of communities and experts to build projects that meet the expectations of all stakeholders in a co-learning process. Experimentation makes it possible to test proposed solutions and identify the potential of certain green infrastructure during the development of a project. This can increase the impact of urban projects tenfold, as it allows communities to become more involved and stakeholders in the future of their territory, which has a far greater impact than the projects themselves.

Acknowledgments The authors would like to acknowledge the assistance from the Trottier Energy Institute, CERIEC, and CIRODD in the Action Research projects described in this chapter. These projects could not have been carried out without the help from these three organizations. The authors would also like to thank Edith Beauvais-Sauro, Marylou Filiatrault, Benjamin Gutzeit and Christophe Aubry, who contributed to certain sections as research assistants since 2020. The authors would like to thank the rest of the teaching team from the "Co-construction of our collective future" Master's program in Architecture: Michel Langevin (NIPPaysage), Jean-Marc Weill (C&E Ingénierie), Marianne Lafontaine-Chicha (NIPPasyage) Élène Levasseur (ASFQ), Édith Beauvais-Sauro (L'OEUF) and Peter Soland in 2019 (Civiliti). Finally, they would like to thank all of their students for their commitment to designing more sustainable and resilient neighborhoods.

References

Agez, M., & Patouillard, L. (2022). *Inventaire des émissions de gaz à effet de serre découlant de la consommation de la collectivité montréalaise pour l'année 2017*. Rapport technique CIRAIG pour le Bureau de transition écologique et de la résilience.

André, H. (2024). Opening the black box of the use phase in circular economy life cycle assessments: Environmental performance of shell jacket reuse. *Journal of Industrial Ecology, 28*(3), 542–555. https://doi.org/10.1111/jiec.13475

Auffan, B. (2022). *Analyse conceptuelle des divers processus de recyclage de matières résiduelles de construction, de rénovation et de démolition à travers deux études de cas de réhabilitation et rénovation de bâtiments industriels dans le secteur sud-ouest de Montréal* (p. 79). École de technologie supérieure.

Azari, R., & Kim, Y. W. (2016). Integration evaluation framework for integrated design teams of green buildings: Development and validation. *Journal of Management in Engineering, 32*(3), 04015053.

Bulle, C., Margni, M., Patouillard, L., Boulay, A. M., Bourgault, G., De Bruille, V., & Jolliet, O. (2019). IMPACT World+: A globally regionalized life cycle impact assessment method. *The International Journal of Life Cycle Assessment, 24*, 1653–1674.

Cap, S., de Koning, A., Tukker, A., & Scherer, L. (2024). (In) Sufficiency of industrial decarbonization to reduce household carbon footprints to 1.5°C-compatible levels. *Sustain Prod Consum, 45*, 216–227. https://doi.org/10.1016/j.spc.2023.12.031

Center for Climate and Energy Solutions. (2019). *Investing in resilience*. https://www.c2es.org/document/investing-in-resilience/

Chabas, C., Tollemer, M., Gonon, J., Filiatrault, M., Gutzeit, B., Castonguay, J., et al. (2023). À la croisée des savoirs: Mettre en oeuvre l'interdisciplinarité dans le cadre d'un projet d'aménagement urbain du secteur LaSalle/Ville Saint-Pierre/Lachine-Est. *Revue Organisations & Territoires, 32*(3), 55–70.

Cole, R. J., & Pearl, D. (2007). Blurring boundaries in the theory and practice of sustainable building design. In M. Horner, C. Hardcastle, A. Price, & J. Bebbington (Eds.), *Proceedings: SUE-MOT Conference 2007, International Conference on Whole Life, Urban Sustainability and its Assessment, 27–29 June 2007, Glasgow, UK* (pp. 408–418).

Cucuzzella, C., Pearl, D., & Mertenat, C. C. (2010). Greening the curriculum: A Canadian Academic National Forum. In *Proceedings from Engineering Education in Sustainable Development (EESD) Conference* (pp. 19–22).

Deng, Y., & Tian, Y. (2015). Assessing the environmental impact of flax fibre reinforced polymer composite from a consequential life cycle assessment perspective. *Sustainability, 7*(9), 11462–11483.

Geissdoerfer, M., Savaget, P., Bocken, N. M., & Hultink, E. J. (2017). The circular economy—A new sustainability paradigm? *Journal of Cleaner Production, 143*, 757–768.

Gomez-Campos, A., Vialle, C., Rouilly, A., Sablayrolles, C., & Hamelin, L. (2021). Flax fiber for technical textile: A life cycle inventory. *Journal of Cleaner Production, 281*, 125177.

Harris, S., Martin, M., & Diener, D. (2021). Circularity for circularity's sake? Scoping review of assessment methods for environmental performance in the circular economy. *Sustainable Production and Consumption, 26*, 172–186.

Ibrahim, H. (2015). Integrated design approach: A mode towards sustainability in building project. *Advances in Environmental Biology, 9*(4), 70–72.

Jolliet, O., Saadé, M., & Crettaz, P. (2010). *Analyse du cycle de vie: comprendre et réaliser un écobilan* (Vol. 23). PPUR Presses polytechniques.

KIES. (2021). *Recyclage des agrégats d'enrobés et recours aux enrobés tièdes*. Consulté 26 juillet 2022 https://www.kiesfuergenerationen.ch/resources/20210429-Best-Practice-Guideline_f_def.pdf

Lapierre, E. et Pellerin, S. (2018). Portrait des infrastructures vertes et des ouvrages phytotechnologiques dans l'agglomération de Montréal [Rapport]. Fondation Espace pour la vie. https://www.phytotechno.com/wpcontent/uploads/2019/02/Infrastructures-vertes-de-l%C3%AEle-de-Montr%C3%A9al.pdf

Legru, S. (2024). *Étude comparative de renforcement PRF sur une structure de béton armé dans le cadre d'un projet de réhabilitation industrielle de la C.I.L. de Montréal* (227). École de technologie supérieure.

Lentier, F. (2012). *Comparaison environnementale entre la démolition et la rénovation du bâtiment.* [Travail de fin d'études, Université de Liège Faculté des sciences appliquées]. https://matheo.uliege.be/bitstream/2268.2/2416/1/2011_2012_LENTIER_Fanny.pdf

Leoto, R., & Lizarralde, G. (2019). Challenges for integrated design (ID) in sustainable buildings. *Construction Management and Economics, 37*(11), 625–642.

Lesage, P., Ekvall, T., Deschênes, L., & Samson, R. (2007). Environmental assessment of brownfield rehabilitation using two different life cycle inventory models: Part 2: Case study. *The International Journal of Life Cycle Assessment, 12*, 497–513.

Loiseau, E., Roux, P., Junqua, G., Maurel, P., & Bellon-Maurel, V. (2013). Adapting the LCA framework to environmental assessment in land planning. *The International Journal of Life Cycle Assessment, 18*, 1533–1548.

Loiseau, E., Roux, P., Junqua, G., Maurel, P., & Bellon-Maurel, V. (2014). Implementation of an adapted LCA framework to environmental assessment of a territory: Important learning points from a French Mediterranean case study. *Journal of Cleaner Production, 80*, 17–29.

Lonca, G., Muggéo, R., Imbeault-Tétreault, H., Bernard, S., & Margni, M. (2018). Does material circularity rhyme with environmental efficiency? Case studies on used tires. *Journal of Cleaner Production, 183*, 424–435.

Merle, P., & Perrin, J.-L. (2018). Les friches industrielles: Une nouvelle ressource secondaire ? *Annales Des Mines—Responsabilité et Environnement, 91*(3), 34. https://doi.org/10.3917/re1.091.0034

Meyer, J. (2000). Using qualitative methods in health related action research. *British Medical Journal, 320*, 178–181.

Patouillard, L., Agez, M., de Bortoli, A., & Bulle, C. (n.d.). Consumption-based carbon footprint of Canadians: Exploring provinces' variability based on OpenIO-Canada, a regionalized input-output table with capital endogenization. *Submitted to the Journal of Industrial Ecology.*

Pearl, D., & Beauvais-Sauro, E. (2020). La construction de notre avenir collectif. *ARQ: Architecture & Design Québec: La construction reformatée, 193*, 16–19.

Busby, Perkins, and Will. (2007). Roadmap for the Integrated Design Process. BC Green Building Roundtable.

Reason, P., & Bradbury, H. (2008). *The Sage handbook of action research participative inquiry and practice* (2nd ed.). SAGE Publications.

Sarni, W. (2010). *Greening brownfields: Remediation through sustainable development.* McGraw-Hill Education.

Tirado-Seco, P., Martineau, G., Fallaha, S., Saunier, F., & Samson, R. (2014). *Comparaison des filières de production d'électricité et des bouquets d'énergie électrique.* Chaire internationale sur le cycle de vie, Centre interuniversitaire de recherche sur le cycle de vie des produits, procédés et services, Rapport technique.

Van der Ryn, S., & Cowan, S. (2010). *Ecological design* (10th ed.). Island Press. https://www.hoopladigital.com/title/11450910

Vigier, M., Ouellet-Plamondon, C. M., Spiliotopoulou, M., Moore, J., & Rees, W. E. (2023). To what extent is sustainability addressed at urban scale and how aligned is it with Earth's productive capacity? *Sustainable Cities and Society, 96*, 104655.

Wiprächtiger, M., & Hellweg, S. (2024). Circularity assessment in a chemical company. Evaluation of mass-based vs. impact-based circularity. *Resources, Conservation and Recycling, 204*, 107458.

Daniel Pearl is a half-time Full Professor at the School of Architecture (University of Montréal) since 2001, and founding board member of the Canada Green Building Council in 2003. Daniel's academic and professional research has involved action-research and the documentation of L'OEUF Architects' experimental practice, from institutional projects to large-scale community housing, such as "BENNY FARM" (rehabilitation and new construction) and "COOP COTEAU VERT" (2010) and "PLACE GRIFFINTOWN", (2025), in Montreal, Quebec. Daniel Pearl co-founded l'OEUF architects *(L'Office de l'Éclectisme Urbain et Fonctionnel)* in 1992, with Mark Poddubiuk, in order to concentrate his expertise in sustainable and community design. Daniel's current partners at L'OEUF are Sudhir Suri and Jennifer Benis. Driven by a profound sense of ethics and urgency, Daniel and L'OEUF have been focusing on how can "we" revive the craft of the "re-Making of Cities" via re-investing in Architecture, Landscape Design, the Public Realm and the dormant layers of our post-industrialized cultural history, often through the lens of Circular Economy.

Amy Oliver holds a PhD in planning from the Université de Montréal and divides her time between practice and teaching. She is a visiting professor at Université de Montréal's School of Architecture, where she co-teaches three master's courses on the theme of sustainable communities. In 2023, her team, led by Daniel Pearl, received an award for teaching excellence in sustainable development and socio-ecological transition. Her doctoral research focused on eco-neighborhoods, and more specifically on the challenges of implementing sustainability frameworks and tools in pilot projects. She is a member of the ARIAction research team, headed by Professor Isabelle Thomas, at the Université de Montréal. Outside academia, Amy also practices in the field of sustainable and resilient design.

Cécile Bulle is a professor in the department of strategy and corporate social and environmental responsibility of the management school at the University of Quebec in Montreal (ESG-UQÀM). She obtained her Ph.D. in chemical engineering at Polytechnique Montréal in 2007. Between 2017 and 2022, she has been the co-holder of the International Life Cycle Chair, and since 2022 she is the codirector of the International research consortium on Life Cycle Assessment and Sustainable Transition, the main research unit of CIRAIG, the International Reference Center on Life Cycle of products, processes and services, which is one of the most recognized research centers in the world in life cycle assessment. She is also a very active member of CIRODD 2.0 (member of the executive committee, member of the scientific committee and leading the CIRODD-ISE-RIISQ pole at UQÀM). Her research interests are life cycle impact assessment, in particular the impacts on biodiversity, the depletion of resources and the water footprint and the operationalization of the Life Cycle Assessment to community life (sustainable neighborhoods, etc.) and individuals (sustainable consumption, etc). She teaches life cycle assessment, sustainable building management, sustainable real estate and corporate social responsibility. She has been leading the IMPACT World + project, a major project involving several research teams around the world to develop the very first regionalized life cycle impact assessment method.

Titouan Greffe is a PhD candidate at the Institute of Environmental Sciences at UQAM and member of the CIRAIG, which is the International Reference Center for Life Cycle Assessment and Sustainable Transition. He is a member of the task force "Natural Resources" within the GLAM (Global Guidance for Life Cycle Impact Assessment Indicators and Methods) program hosted by UNEPSETAC Life Cycle Initiative which aims at developing a consensus-based harmonized life cycle impact assessment framework at the international level.

Claudiane Ouellet-Plamondon is a full professor in the Department of Construction Engineering at the École de technologie supérieure (ÉTS) in Montreal, member of the Université du Québec network. She holds the Canada Research Chair in Sustainable Multifunctional Construction Materials. She received a bachelor of biological engineering from Dalhousie University, a master's degree in biological sciences from Université de Montreal, and a PhD in engineering from the

University of Cambridge in the United Kingdom. She was a postdoctoral fellow at ETH Zurich, Switzerland. She was a visiting scholar at the Center for the Built Environment at the University of California, Berkeley during her sabbatical leave. Her research focuses on the multiscale study of construction materials from the characterization of the microstructure, circularity, design, and integration in buildings, infrastructure, and cities. She participates in international committees.

Chapter 3
Co-construction as Implementation: The Circular Economy Experience in Quebec—Consolidation Stage (2021–2024)

Stéphanie Jagou and Emmanuel Raufflet

Abstract The chapter explores Quebec's circular economy experience, documenting the roles of academia, government, organizations, and businesses between 2021 and 2024. Its four main messages are the following. First, Quebec has established several key research infrastructures which encompass numerous sectors and interdisciplinary collaborations, positioning Québec as a leader in North America's circular economy research landscape. Those research networks are interconnected and Quebec's focus on circularity also intersects with climate change and biodiversity conservation efforts. Second, Quebec's research endeavors have impacted policies related to the circular economy by influencing legislation, fostering industry partnerships, and supporting regional strategies. Third, Québec has successfully developed a common language around circular economy concepts, aided by a structured definition, visual aids, and an official lexicon. Fourth, Québec views the circular economy not merely as an operational model but as a societal transition project, aligning with broader sustainability goals beyond production and consumption efficiencies. Despite these achievements, Québec's circularity rate remains at 3.5%, indicating a substantial gap between theory and practice. While the circular economy offers pathways to sustainability, it may not sufficiently challenge broader economic growth and consumption-based models. Quebec's research community should critically explore alternative perspectives like degrowth and sobriety to address these systemic challenges more comprehensively.

Keywords Circularity · Transitioning · Research · Collaboration · Systemic · Bottom-up · Sustainability · Knowledge mobilization

S. Jagou (✉)
CERIEC, Montréal, QC, Canada
e-mail: stephanie.jagou@outlook.com

E. Raufflet
RRECQ, CIRODD and HEC Montréal, Montréal, QC, Canada
e-mail: emmanuel.raufflet@hec.ca

M. Cheriet et al. (eds.), *Accelerating the Socio-Ecological Transition*,
https://doi.org/10.1007/978-3-031-82896-6_3

As of 2024, the Québec research ecosystem in the circular economy is arguably one of the most advanced in North America in terms of research capacity and the mobilization of sectors (government, industry, and civil society). Québec research on circularity is overarchingly interdisciplinary and transdisciplinary. Meanwhile, the 2021 circularity index is 3.5%, meaning that 96.5% of the materials in the Québec economy and society are used only once. How can this apparent paradox be explained?

The purpose of this chapter is to discuss this contradiction. It does so by mapping and discussing the experience of the circular economy in the Province of Quebec (Canada), over the 2021–2024 period. This period can be termed *a stage of consolidation*, building on the stage of emergence (2014–2020), which consisted of five interrelated processes. Those were, namely, the building of an inter- and multidisciplinary community of researchers, the establishment of knowledge networks, working on the benefits related to a circular transition, and educating and communicating to a variety of audiences. The detailed evolution of the role played by academia, governments, organizations, and businesses in sparking the transition to a circular economy is documented in *Transitioning to a circular economy—learning from Quebec experience* (2021).[1]

This chapter is divided into five sections. The first section presents the research methods, scope, and limitations of this chapter. The second section maps the roles and missions of circularity-related research infrastructures in engaging with society and other sustainability-related research networks. The third section highlights the processes of engaging with government policies. The fourth section illustrates that the concept of circularity is now reaching a greater variety of publics, with many examples from several key sectors and with the growing offerings in education and communication channels. The last section provides a discussion of this paradox and identifies the risks related to the advancement of circularity in Québec's economy and society.

1 Research Methods, Scope, and Limitations

Data for this chapter were collected from both primary and secondary sources. As for primary sources, the main author organized a focus group[2] in April 2024 with 9 research and government stakeholders who discussed the accomplishments and limitations of the transition to a circular economy in Québec. In addition, some individual interviews were conducted to gather various perspectives and feedback on the diversity of experiences and perceptions of the circular movement in Québec (see Appendix 1: List of persons consulted and interviewed). These interviews were

[1] https://www.quebeccirculaire.org/library/h/transitioning-to-a-circular-economy-learning-from-the-quebec-experience-2014-2020.html

[2] Details are provided at the end of the chapter.

complemented by several communications with a number of stakeholders. Secondary sources included over 20 reports, documents, and web-based sources (see Appendix 2: References).

This chapter presents three main limitations. First, it focuses on the 2021–2024 period. For more specific reference to the previous stage, see *Transitioning to a circular economy—learning from Québec experience* (2021). Second, this chapter focuses strictly on the research–society interface, on how research organizations may have contributed to "operationalizing sustainable development" through the mobilization of diverse sectors of society (government, civil society, and the economy) around an umbrella concept (the circular economy). Third, it aims to illustrate the diversity and abundance of circularity-related initiatives to provide an account of the research–society interface in Québec. As such, it does not intend to provide an exhaustive account of all the initiatives, projects, and policies related to the circular economy; a more systematic report could play this role.

2 Circular Research Infrastructures: Strategies of Engagement

Mobilization around the umbrella concept of the "circular economy" builds on the co-construction of diverse scientific, managerial, technical, and social forms of knowledge convened and facilitated by a group of promoters between 2014 and 2020. The year 2021 has seen the acceleration of research initiatives in and around circularity, largely related to the establishment of a research infrastructure in the form of key research networks and centers.

This section introduces the deployment of this infrastructure and illustrates some important achievements, as well as the collaboration with other sustainability-related research networks.

2.1 Establishing a Research Infrastructure

2.1.1 The CTTEI: A Pioneer in the Ecosystem of CEGEP-Based Technology Transfer Centres

Based at CEGEP[3] Sorel–Tracy, the *Centre de transfert technologique en écologie industrielle* (CTTÉI) has aimed, for over 25 years, to increase the performance of companies and communities through research and development in industrial ecology. It has actively partnered with other academic entities for several initiatives.

[3] CEGEP is a French acronym that stands for *Collège d'enseignement général et professionnel,* that is, a general and vocational college. They are public institutions and represent the first level of higher education, dubbed *post-secondary education*, in Québec.

In addition, the CTTEI also convenes the community of practice Synergie Québec, which represents, as of 2024, over 3300 organizations in Québec for 23 symbioses and 850 completed synergies.

With a consortium of six locally based organizations, in 2021, the CTTEI also spearheaded the EFC Québec (Economy of Functionality and Cooperation Québec) initiative with the objective of experimenting with these two circular strategies. Pilot projects were led over 3 years with 20 organizations to address climate change and increase businesses' competitiveness. Celebrating its 25 years of existence in 2024, Claude Maheux Picard, CTTEI's general manager, stated that the organization's "*efforts to raise awareness among businesses, organizations, and government bodies have paid off, and today we can say without hesitation that we have helped position Québec as a leader in the circular economy in Canada.*"

As of 2023, other CEGEPs and CEGEP-based Technology Transfer Centres (*Centre de transfert technologique*—CTT) have added circularity as a research theme. Several collaborate in the province with the Québec Circular Economy Research Network (*Réseau de recherche en économie circulaire du Québec—RRECQ)*, which will be presented later in the chapter. For example, the CEGEP of Victoriaville, with its new Technology Access Center (TAC), promotes research in organic agriculture as well as the woodworking and furniture industries. Gaspésie-based CIRADD[4] has also developed projects with RRECQ. The 2022 creation of the *Cité de l'innovation circulaire*[5] in Victoriaville (*Centre-du Québec*) illustrates yet again how the concepts and strategies have found ground in the region and how scientific research can also be pursued in local CEGEPs and cities, away from the City of Montréal, which concentrates both its population and a significant part of its research capacity on a circular economy.

2.1.2 CERIEC: Taking on the Circularity Transition Challenge

The Centre for Intersectoral Studies and Research on the Circular Economy (*Centre d'études et de recherches intersectorielles en économie circulaire*—CERIEC) was established in 2020 at ÉTS Montréal (*École de Technologie Supérieure*) and relies on an approach of action-research partnerships and interdisciplinarity. It built the ELEC—the Circular Economy Labs Ecosystem (*Ecosystème de laboratoires en économie circulaire*)—in 2021 with the financial support of Groupe Desjardins and following the idea of a sector-specific lab proposed by the EDDEC Institute[6] in 2019.

[4] The *Centre d'initiation à la recherche et d'aide au développement durable* (CIRADD) is a specialized research center for sustainable development affiliated with the *Cégep de la Gaspésie et des Îles*. It is part of the Synchronex network—https://www.ciradd.ca/?lang=en

[5] https://www.regionvictoriaville.com/page/1538/cite-de-linnovation-circulaire.aspx

[6] The IEDDEC was created on *Campus Montréal* in 2014, and its mandate ended in 2019. The CERIEC then accepted the challenge to pursue the transition to a circular economy in Québec.

The aim of the living lab approach is to increase the circularity of key industries by providing an open-ended experimentation area. This network has the ambition to ultimately comprise eight or nine interconnected laboratories focusing on key circularity sectors, such as construction, agri-food, mines, plastics, or specific strategies, such as the product–service system or the social/solidarity-based economy. The ensuing projects aim to develop and test solutions so they can ultimately become mainstream. With the support of the ELEC steering committee, the creation of laboratories is prioritized according to the importance of the issues at stake in the Province of Québec, the circularity potential, as well as government priorities and potential financial support. Because industries may be mature in their respective transitions, each lab is designed as a catalyzer with a limited lifespan of 2–3 years.

Participants recognized that university leadership and convening provide reassuring and neutral guidance for the sharing of benefits and spin-offs to all. The lab's managers' constant support of the project teams is an essential element of success in keeping the community motivated and active. The aim of the labs is to help reduce and optimize the use of resources by demonstrating that certain circular economy strategies are relevant, workable, and profitable. Their detailed documentation facilitates their replication and scaling up.

The ELEC holistic approach is guided by co-creation and field experimentation, and it unfolds in three stages. First, a variety of sector players and researchers develop a vision of a desirable circular future, as well as identify obstacles to achieving this vision of circularity. Second, they co-define solutions to be tested with experimentation projects. Third, these tested solutions are discussed together either in the lab per se (between solutions) or with an *interlab* approach (across sectors).

The ELEC intends to mesh the scientific approaches and field results of all its labs to cross-pollinate knowledge and emulate the necessary systemic approach that characterizes the transition to circularity. A first *interlab* event was held in 2023. It underlined that, to face complex societal challenges—in which we do not know which sectors, disciplines, or professions will bring effective, efficient, and appropriate solutions—an intersectoral approach is necessary.

As of April 2024, three laboratories have been deployed: the Construction Lab (*labCo)* in the construction sector, launched in 2021, and the Food Systems Lab (*Systèmes Alimentaires*), launched in the summer of 2023. The textile lab will be launched in the spring of 2024.

The *Construction Lab* aims to increase circularity in the construction industry, which accounts for 25% to 40% of global greenhouse gas emissions. In its first phase (2021–2024), it has convened 40 co-creation workshops that have contributed to building a shared vision, a set of priorities, and identifying key obstacles for a more circular construction industry. In addition, 19 pilot projects were devised to overcome these obstacles (see Fig. 3.1). A second phase, furthering knowledge transfer and education, is deemed to start in 2024.

> The solutions come from the stakeholders, are driven by them, and respond to their challenges—this is the relevance of a co-creation, industry-based approach. When we started the lab, we didn't know in advance what we were going to work on. It was defined by the collective. (Hortense Montoux, Project Coordinator, CERIEC)

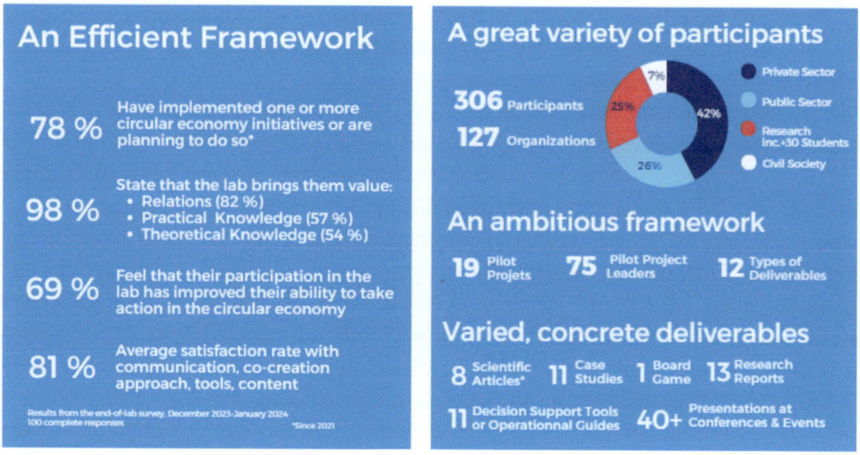

Fig. 3.1 Achievements of the Construction Lab (2021—2024)

Fig. 3.2 RRECQ in a nutshell (2023)

2.1.3 RRECQ: More Integrated Research

The RRECQ was established in 2021 as a research network between HEC Montréal, *École de technologie supérieure*, *Université Laval*, and *Polytechnique Montréal*. Supported by the ETS-based CERIEC, the RRECQ is a network of researchers that aims to strengthen the capacity to deploy circularity strategies at the industrial and territorial levels, supported by coherent government policies, to further a sustainable societal transition. Unique to Canada and arguably the first of its kind in the world,[7] the RRECQ convenes over 270 researchers from 26 Québec universities and colleges, as well as 23 Canadian and international universities,[8] from over 50 research areas (see Fig. 3.2).

[7]According to the Ellen McArthur Foundation.

[8]As of spring 2024.

Funded by FRQSC (*Fonds de recherche du Québec—Société et Culture*; five million CAD$/2021–2026), RRECQ convenes and supports interdisciplinary projects on circularity-related areas. It is organized around four axes: (1) Change and Transition Management, (2) Planning Optimization, (3) Resource and Product Maximization, and (4) Policy Levers. Each research axis is chaired and co-led by diverse Québec universities.

In 2023, an internal consultation led to the identification of 13 flagship transversal projects and themes (TPS). Organized into five poles, those TPSs focus on implementation approaches in industrial sectors, territories, and organizations; tools and methods to develop circularity, monitor, and measure it; resources and support, which addresses impacts on work, employment, and knowledge transfer; and the observation and evaluation of circularity, itself, and its potential rebound effects. These TPSs aim to remove the boundaries between the four research axes of the RRECQ by approaching projects in a coordinated way from different disciplinary angles, ultimately offering a holistic vision of the issues at stake.

Interdisciplinarity and transdisciplinarity are central to the RRECQ. The network crystalizes the rising need for more integrated research and better-meshed initiatives and networks. Of the 19 Construction Lab field experimentation projects, 15 were led or supported by RRECQ members (Fig. 3.3).

Who: Researchers, *Université Laval, Ministère de l'Environnement et de la Lutte contre les Changements Climatiques, de la Faune et des Parcs (MELCCFP),* and *the Ministère de la Santé et des Services sociaux* (MSSS).

Challenge: How can the flow of materials consumed and discarded in a hospital be mapped? This research project aims to identify issues related to the supply, logistics, consumption, and management of hospital residual materials and their causes. The results of this project will form the knowledge base for possible solutions to help make healthcare more sustainable and circular.

What makes this project singular? First, a systemic and comprehensive assessment of the consumption and management of materials has led to (i) a complete characterization of inputs, specifically identifying the composition and weight of purchased items and their various packaging layers, (ii) observation and interviews with health professionals (nurses, doctors, orderlies, logistics, hygiene, sanitation, procurement, managers, etc.), and (iii) a rigorous and detailed characterization of outputs. Second, the multi- and interdisciplinary approach mobilized different disciplines in a collaborative manner (e.g., accounting, healthcare management, clinical staff, healthcare facility managers, ministry employees, sustainable development, circular economy consultants, etc.).

Next stages (as of 2024): These new data will enable the team and hospital management to identify promising and structure-related actions that prioritize the reduction and reuse of materials and, ultimately, their recycling.

Fig. 3.3 A holistic approach to research: A material flow analysis of a hospital context

2.2 Interlinking Networks

Other academic research initiatives are blooming, aiming to bridge the gap between science and its application in the field and targeting themes such as the socioecological transition, sustainable development, or a green economy, all closely linked to the implementation of circularity. This will contribute over time to the development of a more holistic approach to the socioecological transition in Québec.

Moreover, circularity is a narrative related to the economic transition to sustainability. Connecting with other research networks and centers focused on sustainability contributes to generating knowledge relevant to achieving a just transition toward life and respecting planetary boundaries. In line with this, partnering locally with other Québec-established scientific networks such as the CIRODD, IEDS (*Institut en environnement, développement et société [Institut EDS], Université Laval*), the GERAD,[9] and the CSBQ,[10] as well as internationally with *Circulades*–AIFREC,[11] the RRECQ aims to instigate a systematic scientific approach that fosters actionable knowledge to accelerate the transition to a more circular economy.

The mission of the Interdisciplinary Research Centre for the Operationalization of Sustainable Development (CIRODD)[12] is to accelerate the transformation of society to support the socioecological transition through sustainable innovation and transdisciplinarity. This academic network, which represents 18 post-secondary institutions in Québec, has funded eight applied circularity-related research projects in the 2021–2024 period.

The links between the green economy and circularity are highlighted with the recently created *Mouvement Entreprises Vertes Québec* (MEVQ), coordinated by the *Institut EDS*. This mobilization and action project is based on the realities and needs of business companies based in Québec City's industrial parks, using innovation to integrate and accelerate the transition in business in models and practices in the face of the energy transition and the shift to a green economy.[13]

Chemins de Transition, a knowledge mobilization and transfer project launched in 2020 by the *Université de Montréal* and *Espace pour la Vie*, has developed a participatory foresight method to meet the major challenges of the socioecological transition in Québec. The circular economy emerged as an early key element common to all the trajectories drawn up by *Chemins de transition* in their three challenges. Digital, food, and territorial challenges have benefited from a four-stage

[9] GERAD, *Groupe d'études et de recherche en analyse des décisions*, https://www.gerad.ca/fr

[10] CSBQ, *Centre de la science de la biodiversité du Québec*, https://qcbs.ca/fr/

[11] AIFREC, *Association Interdisciplinaire Française pour la Recherche en Économie Circulaire*, https://aifrec.fr/

[12] *Centre interdisciplinaire de recherche en opérationnalisation du développement durable* (CIRODD).

[13] As per its *Cadre conceptuel et indicateurs pour la mesure de l'économie verte* (2020), the *Institut de la statistique du Québec* clearly identified the implementation of circular strategies to develop a green economy.

methodology, including (1) a desk review and expert panels who document likely futures (including trends and weak signals), (2) stakeholder discussions of one-page simplified narratives of these likely futures, leading to a collective vision, (3) panels of experts who plot a course to achieve this vision, and (4) activities to share the content of the trajectories and the foresight method used to develop them.

While the focus of the first three challenges was not circularity, circular dimensions popped up in the analysis and discussion. For instance, if digital and ecological transitions converge, it is critical to extend the useful life of equipment and encourage its sharing. The trajectory of the food transition challenge emphasizes the fight against food waste through structuring circular economy projects. Circularity is also one of the major economic transformations identified in the vision and path developed to meet the challenge of living in the Province of Québec's territories in a more sober and resilient way. Since 2023, *Chemins de Transition* has also supported the crafting of a roadmap for a circular society in Québec in 2050 (see Fig. 3.4).

2.2.1 Connecting Circularity and Climate Change

International instances and organizations, such as UN Climate Change, WRI (World Resources Institute), or the Ellen MacArthur Foundation, argue for the potential of circularity to address climate change. According to Circle Economy's *Circularity Gap Report 2021*, "*Circular economy strategies have the potential to slash global greenhouse gas emissions by 39%.*" Increasingly, decarbonation has become one of the foci of the Province of Québec.

As early as 2019, Québec ZéN[14] was releasing the first Québec roadmap to carbon neutrality that reflected the work of over 190 individuals from 85 organizations,

Chemins de transition has been collaborating since 2023 with the RRÉCQ to assume the challenge of developing the RRECQ-supported roadmap to circularity. Close to 90 experts from a wide range of backgrounds and 250 diverse stakeholders were first mobilized to lay the foundations for a collective vision of Québec society's circular future and contribute to the drawing up of a prospective diagnosis. Starting in the fall of 2024, the actors involved in the first stages of this project (governments, businesses, investors, etc.) will reconvene with the objective of guiding and stimulating actions in terms of circularity to establish links between the current situation and the desired future by proposing trajectories with milestones to achieve this transition. To facilitate this transfer phase, the team has taken care to involve key networks from the outset of the process and to ensure a good fit with local or ministerial roadmap initiatives linked to the circular economy. The results should be unveiled in the fall of 2024. The three-year process was accelerated thanks to the lessons from other Chemin de transition challenges, and the project is planned to last only 18 months. The fact that, thanks to an accelerated process, the diagnosis was available early may have informed the government's initiative, which started years ago, to elaborate its own Québec roadmap, which was launched in April 2024.

Fig. 3.4 Roadmap for the transition to a circular economy in Québec by 2050

[14] "Québec ZéN" (*zéro net émissions*) is a civil society initiative aimed at transitioning Québec to carbon neutrality, initiated by the *Front commun pour la transition énergétique*.

including a dozen universities. According to their vision, circularity had become the industry norm, and the report advocated the implementation of circularity indicators to advance this transition.

In addition, science is increasingly establishing links between climate change mitigation and adaptation as well as the implementation of the strategies of the circular economy. This slowly permeates provincial policies, as some governmental institutions are contributing research of their own to inform decision-making. This is particularly illustrated, for example, in RECYC-QUÉBEC's *Study on materials for the energy transition: Current situation and possible solutions*,[15] whereby the analysis of best practices and proposed recommendations have been detailed according to the 12 circular economy strategies.

Another example is the report released in 2023 by Québec' Climate Change Advisory Committee (*Comité consultatif sur les changements climatiques*). "Decarbonizing Heavy Goods Transport: Building a Sustainable Path[16]" takes a strong position on circularity: "*The Committee proposes to adopt a roadmap for the development of the circular economy and to promote sobriety in consumption and production choices.*"

2.2.2 Linking the Circular Economy and Biodiversity

The 2022 UN Biodiversity Conference COP15 in Montréal highlighted the links between biodiversity conservation and the expected progress of the circular economy. For instance, a session entitled "The Circular Economy as a Crucial Tool for Biodiversity & Climate"[17] organized by European Circular Economy Stakeholder Platform (ECESP) members showcased how circularity should contribute to preventing biodiversity loss and fostering its restoration, as more than 90% of biodiversity loss is due to resource extraction and processing. In Québec, initiatives are only just emerging. As of early 2024, the RRECQ is elaborating on a partnership with the McGill-based Québec Centre for Biodiversity Science to develop both a conceptual and operational framework to strengthen the understanding and integration of the links between biodiversity conservation and CE to support decision-making. Additional research initiatives and financial means will probably be geared toward such research in the coming years.

---

[15] *Étude sur les matériaux de la transition énergétique: État de la situation et pistes de solutions, 2022.*

[16] "*Décarbonation du transport lourd de marchandises: construire une voie durable*", *Commission sur les changements climatiques.*

[17] Organized by the IUCN European Regional Office, Ellen MacArthur Foundation, and partners.

2.2.3 New Connections: Circularity, Life Cycle Assessments, and the Built Infrastructure

Life cycle assessments and measuring tools are at the core of the transition to a more circular economy. The International Research Consortium on Life Cycle Assessment and Sustainable Transition, established in 2023, aims to guide the transition to a carbon-neutral economy. It was initiated by CIRAIG[18] and four universities (*Polytechnique Montréal*, UQAM,[19] *École polytechnique fédérale de Lausanne*, and HES–SO Valais–Wallis of Switzerland). In all, it convenes 10 industrial partners to build operational and strategic research to accelerate the sustainable transition.

Finally, ÉTS Montréal launched the AdapT Institute in 2022 with the mission to develop innovative solutions for the design, construction, and operation of circular, climate-resilient infrastructures. It draws on a multidisciplinary team of hundreds of researchers from engineering, management, and health, humanities, and social sciences fields from universities across Canada as well as collaborations with several private, public, and non-profit organizations. It also collaborates with the CIRODD and will partner with the CERIEC to establish the second stage of the *LabCo*, namely, the *Laboratoire Construction+*.

In all, 2021–2024 has been a stage to consolidate the research infrastructure in and around circularity in Québec on two levels. While, for example, the historically industrial ecology-centered CTTEI has thematically expanded toward the functional economy, a circularity strategy, new research infrastructures have been established with a sector-wide or circularity tools focus. Furthermore, these new infrastructures have collaborated with existing sustainability-centered ones on common themes.

3 Engaging in Circular Local and National Policies

Throughout this phase of consolidation, several research-based briefs have been presented in parliamentary commissions, and some seem to have positively impacted provincial legislation and regulations. Among them, the brief submitted in early 2023 by CERIEC to the Transport and Environment Commission (*Commission des transports et de l'environnement*) has in part informed the new 2023–2028 version of the Government Sustainable Development Strategy (GSDS).[20] The first orientation of the GSDS, "Make Québec a center of innovation and excellence in the green and responsible economy," strongly relies on the implementation of a circular economy. The Circular Economy Roadmap (FREC) was

---

[18] International Reference Centre for the Life Cycle of Products, Services, and Systems (*Centre international de référence sur le cycle de vie des produits, procédés et services*).

[19] UQAM | *Université du Québec à Montréal.*

[20] https://www.quebec.ca/en/government/policies-orientations/sustainable-development/government-strategy

drawn-up by the *Ministère de l'Environnement, de la Lutte contre les changements climatiques, de la Faune et des Parcs* (MELCCFP) in collaboration with RECYC-QUÉBEC. This long-awaited roadmap was the subject of consultations with representatives of 40 ministries and agencies, as well as 63 external organizations (Nonprofit organizations, educational institutions, the research community, the private sector, and citizens). This Québec provincial government's roadmap, launched in the spring of 2024, responds to Objective 1.1.2 of the GSDS: "Accelerate the development of the circular economy." The government sees its implementation as "*a way of preserving the ecosystems that generate resources and preserve biodiversity, and a means of combating and adapting to climate change*."[21] This reflects the concept of circularity as a meta-concept, bridging borders with key environmental and societal issues. Three of the economic sectors prioritized—the construction and textile industries and biofood—are already subjects of ELEC acceleration labs. Heightened governmental collaboration and funding may derive from the implementation of the FREC, as the principle of collaboration underscores this roadmap.

Another illustration concerns the concerted effort between academics and the civil society organization Équiterre, which has contributed to the inclusion of many key principles in Québec's Bill 29, which aims "to protect consumers against programmed obsolescence and to promote the durability, repairability, and maintenance of goods."[22]

The brief submitted in September 2023 by Équiterre was based on the findings of a study published earlier in 2022. This pan-Canadian study was the fruit of collaboration with several partners and was coauthored with the Responsible Consumption Observatory[23] from ESG UQÀM. *Université de Montréal* was also a member of the monitoring committee. A joint CERIEC–RRECQ brief was also submitted on this occasion. As of the spring of 2024, following the signature of a partnership with the Ministry of Justice, the RRECQ is sitting on an expert committee to assist the Consumer Protection Office (*Office de la protection du consommateur)* in the development of regulations following the adoption of Bill 29.[24] Marc Journeault, professor at *Université Laval* and co-lead of the RRECQ, states: "*This collaborative, transdisciplinary approach aims to make Québec legislation an up-to-date reference in terms of scientific and technical data and to strengthen*

[21] *Mot du ministre—"…C'est une façon de préserver les écosystèmes qui (les) génèrent et la biodiversité, et un moyen de lutter contre les changements climatiques et de s'y adapter"—Traduction libre*—https://cdn-contenu.quebec.ca/cdn-contenu/adm/min/environnement/publications-adm/developpement-durable/strategie-gouvernementale/feuille-route-economie-circulaire.pdf

[22] At the occasion of consultations before the Economic and Labor Committee, in the fall of 2023, in an effort by the Ministry of Justice to update the Consumer Protection Act.

[23] *Observatoire de la consommation responsable*—https://ocresponsable.com/

[24] https://www.quebec.ca/nouvelles/actualites/details/projets-de-reglements-faisant-suite-a-ladoption-du-projet-de-loi-no-29-un-comite-dexperts-fera-des-recommandations-a-loffice-de-la-protection-du-consommateur-55049

Québec's position as a North American leader in the circular economy and legislation associated with extending the life of goods."

At the industry level, the material flow analysis conducted by the MUTREC team in 2019 is still cited as a reference in 2024. Of the various recommendations, several have been acted upon, for example, the Act respecting upholstered and stuffed articles was repealed in 2021. Now, there are several defibration facilities available. In addition, several applied research projects have been financed by the private sector and by research organizations, such as the industrial symbiosis project between Niedner and Alkegen–Texel, coordinated by the CTTÉI.

Academic influence has also been applied at the local government level. The Regional Circular Economy Roadmap Toolkit prepared by the CTTÉI and its collaborators for RECYC–QUÉBEC was released in 2022. It supported the realization of several regional and local circular roadmaps.[25] Finally, researchers and professionals serve on several government-created committees. Among those, a CERIEC professional sits on the Expert Committee on the Management of Construction, Renovation, and Demolition Residues[26] in the continuation of the work of the Construction Lab of ELEC (*labCo*).

4 Engaging with Civil Society

For researchers and practitioners, "speaking the same language" was identified very early on as a necessary stepping stone to co-construct a path to circularity. Over time, this has been translated several ways, moving away from the traditional publishing of scientific work, and now ranging from webinars to conferences and the development of sector- or subject-specific education.

Not only did the COVID-19 pandemic contribute to an evolution in public perceptions of the need for a transition to a circular economy, but it also helped to develop ways to reach wider publics more easily and in a more cost-effective fashion. This has translated into the blossoming of offerings in webinars and online conferences about circularity—too many to list. The 2021–2024 period witnessed the intensification of communication on and around circularity to wider audiences in several forms and formats.

[25] Many roadmaps have been or are being created (Chapais, Sherbrooke, *Montérégie*, *Outaouais*, *Ville de Montréal, Ville de Québec*). https://www.recyc-quebec.gouv.qc.ca/sites/default/files/documents/guide-methodologique-fdr-ec-english.pdf

[26] https://www.quebechabitation.ca/actualites/le-comite-dexperts-sur-la-gestion-des-residus-de-construction-renovation-et-demolition-reprend-ses-travaux-en-vue-de-deposer-une-feuille-de-route-au-gouvernement-du-quebec/

4.1 Webinars and Events Galore

Academia has played a key role in harnessing the power of online seminars and webinars for knowledge transfer to more varied audiences. In 2023 alone, the RRECQ hosted 24 circular economy webinars, reaching over 900 participants. A series of webinars[27] was presented, up to twice per month. The RRECQ also actively collaborates with ACFAS[28] to reach out to a wider community of researchers. Along the same lines, the *labCo* is organizing a series of webinars in the spring of 2024 to transfer the results and knowledge from its 19 projects.

Several outreach activities have aimed to reach general audiences in more creative ways. In a resolutely innovative outreach approach to educate a wider public, the results of the Équiterre 2022 pan-Canadian study on zero waste, conducted with Polytechnique Montréal and partially financed by the RRECQ, were presented with the exhibition *Zéro, de l'art à l'action*, aimed to raise public awareness about the need for eco-responsible behavior. In the iconic location of the Montréal Biosphere, 10 artists and collectives from a variety of disciplines came together in 2023 to bring research to life. Additionally, inspired by the now quite well-known *Fresque du Climat*[29] (*Climate Fresk*) and proposed in Québec since 2023, the *Fresque de l'économie circulaire*[30] is yet another type of outreach initiative that provides an opportunity to understand the challenges of transforming our linear production–consumption system into a more sustainable model. It is presented to a variety of publics, from private to civil society audiences.

4.2 Consolidating a Common Lexicon Centered on Circularity

The initial Québec definition and chart of a *circular economy* were co-constructed as early as 2016 by several key stakeholders from diverse sectors in the *Pôle de concertation en économie circulaire du Québec*. In 2022–2023, the RRECQ and HEC Montréal's *Chaire de gestion du secteur de l'énergie* contributed to defining 120 circularity-related concepts as part of the lexicon (see Fig. 3.5) led by the *Office québécois de la langue française*.[31] From April 2023 to April 2024, this lexicon ranked as the fifth most consulted lexicon of the linguistic office (out of 60). This

[27] *Série Entretiens, Série Ici, Hors-Série, Série qui fait quoi, Série en compagnie de.*

[28] *Association francophone pour le savoir:* https://www.acfas.ca/

[29] The Climate Fresk NGO has been working to bring people and organizations on board with the climate transition since the end of 2018. Its goal is to enable the rapid growth of climate education and shared understanding of the challenge that climate change represents. It began this mission in France and is now active in over 40 countries around the world.

[30] Welcome/The Circular Economy Collage (lafresquedeleconomiecirculaire.com)

[31] Alongside specialists from RECYC–QUÉBEC, Éco Entreprises Québec, and the *Ministère de l'Environnement, de la Lutte contre les changements climatiques, de la Faune et des Parcs*—https://www.oqlf.gouv.qc.ca/ressources/bibliotheque/dictionnaires/vocabulaire-economie-circulaire.aspx

Fig. 3.5 Setting a common language around circular economy. Source: Lancement d'un vocabulaire de l'économie circulaire par l'Office québécois de la langue française (gouv.qc.ca)

lexicon has two main goals: to set a solid scientific foundation for key circularity-related concepts as well as to democratize the circular economy and to foster its diffusion and use by stakeholders and the public alike.

4.3 Reaching Out to Professionals

The *Assises québécoises de l'économie circulaire* organized by RECYC–QUÉBEC resumed in both 2021 and 2023. Academic players were once again consulted on the content selection for the program and were present on-site. Some 600 participants from all sectors of activity attended in 2023, demonstrating Québec's enthusiasm for the transition to this economic model. Along the same vein, new regional events initiated by some regions of the province spearheaded this transition for their territories. For example, in 2022, the *Lanaudière* region launched its Regional Circular Economy Summit (*Sommet régional en économie circulaire Lanaudière*),[32] which was especially geared toward biofood stakeholders. Moreover, in 2023, Victoriaville hosted its first symposium, "Circularity, a Competitive Advantage for Manufacturers" (*La circularité, un avantage compétitif pour les industriels*).

Implementation tools have been published, including several guides. Among those, a toolbox comprising nine tools for socioeconomic organizations to implement circularity was created by the TIESS[33] and released in 2023. These tools result from exchanges with dozens of socioeconomic enterprises whose business models are based on the circular economy throughout Québec.

4.4 In Canada and Abroad

At the Canadian level, interest in circularity is slowly growing. The inaugural Canadian Circular Economy Summit (CCES), co-hosted by Circular Economy Leadership Canada (CELC) and the Circular Innovation Council (CIC; June 2023 in Toronto), gathered 465 delegates from 262 organizations, including more than half from the private sector. Faculty members were consulted by the Planning Committee.

---

[32] https://lanaudiere-economique.org/evenement/sommet-regional-en-economie-circulaire/

[33] *Territoires innovants en économie sociale et solidaire* (TIESS). https://tiess.ca/9-outils-economie-sociale-et-circulaire/

The academic sector represented 10% of those in attendance. The objective was to collaborate on efforts and activities, advance projects, and investments, inform attendees about policies, and develop an action plan to accelerate the implementation of the circular economy in support of Canada's climate change, biodiversity, and innovation agendas.[34] The final Action Plan Framework was released in the fall of 2023.[35] It is intended to chart an implementation framework to accelerate Canada toward circularity. The second edition is scheduled for Montréal in 2025. Last, the CERIEC and other academic actors are also actively participating in international conferences, such as the World Circular Economy Forum (WCEF), organizing or moderating sessions.

4.5 Gaining Traction with the Media

Two approaches to the media have been promoted. The first is the continuation of the gathering and building of a community of interest around circularity in Québec through participation in the *Québec Circulaire* web-based platform. RECYC–QUÉBEC is a major partner of the QuébecCirculaire.org hub's yearly Circular Initiatives Contest, which helps promote the concept of circularity to businesses and NFP organizations alike by showcasing illustrations of the successful implementation of circular strategies. The second constitutes the creation of momentum around the launch in 2021 of the first Circularity Gap Report Québec, which found that the circularity index economy is 3.5%. In other words, most resources used by Québec's society and economy to meet their needs are not recycled.[36] Building on several published studies, this RECYC-QUEBEC-led research was conducted by Circle Economy with the support of several local partners and academics. This report established a baseline and identified key levers and systemic obstacles for a circular transition in Québec. The media coverage, combined with numerous presentations of the report, gave traction to circularity with various audiences.

Third, the Canada-wide public awareness campaign led by the Circular Innovation Council, establishing the Circular Economy Month in October every year, invites Canadians to learn about the circular economy, celebrate individual and collective efforts, embrace circular solutions, and encourage others to act. With themed weeks and resources made available to the public, it contributes to showcasing local initiatives at the provincial level.

Other awareness-raising online or TV campaigns have been designed, targeting specific circular strategies, such as recycling (*Éco Entreprises Québec*) or repair

[34] https://www.circulareconomysummit.ca/_files/ugd/1e0592_9951459863e5465e97f1dc2f43e0ecdc.pdf-p4

[35] https://www.circulareconomysummit.ca/action-plan

[36] "… a high rate of consumption, a hefty material footprint and a low Circularity Metric are typical for an industrialised economy engaged in trade…", The Circularity Gap Report | Quebec 2021, p. 6.

(RECYC–QUÉBEC). Trade or association publications also started creating or structuring content related to circularity.

4.6 *Disseminating Research*

Documentaries have been published, whether on the internet (Savoir.media[37]) or on the radio (Radio Canada). However, awareness-raising efforts probably climaxed with Télé-Québec's acclaimed show "*Y'a du monde à la messe!*" in 2023,[38] with Daniel Normandin, general director of CERIEC. Therefore, the concept is gaining traction with the wider public thanks to the multiplication of outreach and activities geared toward civil society.

Since 2020, more diverse publics have also had access to a greater variety of education and training components. Indeed, education, professional training, and continuing education are in demand.

4.7 *Reaching Out to Professional and Academic Audiences*

The Circular Economy summer school, initiated in 2017 by the EDDEC Institute and CERIUM (*Université de Montréal*), has been hosted at ÉTS with CERIEC since 2023. This highlights the constant interest in this theme. A new international summer school was held in 2023 in Victoriaville. *Vertech city sur l'innovation circulaire*[39] was organized with the *Réseau Vertech.*[40] Over 10 days, students from various countries were asked to propose a circular innovation entrepreneurship project (takeover) linked to the host city or region. The free Massive Open Online Course (MOOC) *Économie circulaire: Une transition incontournable,* initially offered by the EDDEC Institute, was updated in 2022 by CERIEC and RRECQ and has reached 7000 participants since its launch in 2018. Both the summer school and the MOOC bridge the gap between the academic world and the field, as professionals from all horizons can register and participate. Other online training has been offered in English by universities in Canada.

An increasing number of options, whether courses or full programs, are now being offered in diverse post-secondary education institutions. A new CEGEP certification program was proposed in 2024, including an online training program

---

[37] *La société du travail. Repenser le modèle économique en pensant aux limites de la planète* (2023) with the participation of HEC Montréal. https://www.youtube.com/watch?v=8sqFIxIx0Vo

[38] https://www.facebook.com/yadumondeamesse/videos/ya-du-monde-%C3%A0-messe-7-juillet-daniel-normandin/288060327123452/

[39] https://www.vertechcity.com/wp-content/uploads/2023/08/EIE2023_Article-resume.pdf

[40] A network of cities in France, Belgium, and Québec recognized for their original sustainable development initiatives (Lafayette, Namur, Poitiers, and Victoriaville).

developed by the CEGEP of Victoriaville. Professional training is also under development, and academics have been consulted on the continuing education course developed by the CTTEI and offered by *EnviroCompétences*.[41]

4.8 Training Decision-Makers

The Québec Chamber of Commerce Federation (FCCQ), the CERIEC, and the CTTEI co-organized *Accélérer le passage du Québec à l'économie circulaire, Tournée des 17 régions*[42] in 2021–2022. The FCCQ's objective for this tour of all 17 regions of the province was to upgrade the skills of the members of its network of 130 chambers of commerce. Designed primarily to meet the needs of businesses, the program was also open to local economic organizations acting as program and service relays or as circular economy coaches. The program included an introduction to the circular economy model, strategies, and decision-making tools as well as a free coaching clinic. In addition, an electronic toolkit was made available. Out of the approximately 700 participants, 80% of those surveyed confirmed they had been inspired to act by implementing one or more initiatives within their organization. Over the 2021 and 2022 cohorts, 400 organizations and 300 economic players were represented. A total of 43 coaching clinics were organized through the FCCQ.

Geared toward the governmental representation staff, the CERIEC convened in 2022–2023 at the request of the *Ministère des relations internationales et de la francophonie* training, which was dedicated to the staff of Québec's 11 general delegations abroad. Several tools were developed to help them present the concepts and initiatives per key economic sectors of the province and were later made freely available on the Quebeccirculaire.org online platform.[43]

5 Discussion: Beyond 3.5%?

The first decade of circular economy research in Québec has the following four features: (1) a common language, (2) co-construction of knowledge, (3) the wide scope of research, and (4) the view of circular economy as a project for societal transition.

The first feature concerns a common language. The process of structuring around a common language (around a co-constructed Québec-based definition, a visual chart, and, more recently, an official lexicon of the key concepts), a web-based depository of expertise and information (*Québec circulaire*), several communities

[41] https://www.envirocompetences.org/formations/economie-circulaire-des-outils-pour-transformer-votre-modele-d-affaires/

[42] https://tournee-economiecirculaire.fccq.ca/

[43] https://www.quebeccirculaire.org/articles/h/le-quebec-leader-de-l-economie-circulaire-a-l-echelle-canadienne.html

of practice, and the biannual *Assises de l'économie circulaire* organized by RECYC-QUÉBEC have definitely contributed to establishing a "circular economy movement" complementing other sustainability/transition movements in Québec. This movement's key players are well known. New players and newcomers can easily find who and where to go for guidance and scientific support; hence, communication flows easier between actors.[44] The second feature concerns the applied nature of most research conducted. Circular research in Québec has overarchingly relied on processes of co-construction between researchers and organizations in all stages of research; research is conducted with and for organizations around circularity challenges. The research infrastructure, such as CTTEI, ELEC, RRECQ, and, more recently, AdapT for the built infrastructure, has promoted research designs and funding mechanisms to advance partnerships between actors and researchers. The third feature concerns the scope of researcher–practitioner applications of circular economy research. In all, this "movement" encompasses diverse sectors of society, from municipalities to the provincial government, from small and medium-sized enterprises to large industries, and from social and solidarity enterprises to cooperatives. The fourth feature concerns the presence of the relationship between the circular economy—as a quest for efficiency in the production and consumption processes—and the aspiration to shift to what would be a circular society—which would include other flows and concerns for care and justice. In all, these four features are arguably specific markers of the Québec experience in circular economy research, compared to other parts of North America, where a circular economy is promoted in a more restricted, operations-driven fashion.

More recently, the 2021–2024 consolidation period has seen a sharp increase in action–research partnerships and projects developed with practitioners. The simultaneous, joint efforts to build research capacity and outreach have led the circular economy to gain visibility with several audiences interested in addressing sustainability in their organizations, sectors, regions, and value chains. Several researchers who participated in the focus group for the writing of this chapter in early April 2024 emphasized that they are being called by organizations and professionals to engage in new projects. According to them, while in 2020 they needed to explain and engage with practitioners, by 2024, practitioners and decision-makers from diverse organizations and diverse sectors asked them to explore and document circularity-related issues or to support decision-making processes. In all, the circularity movement has definitely contributed to making circularity visible as a way to address sustainability challenges. Overall, these achievements are impressive.

Simultaneously, the rise of circularity in Québec as a research topic and concerns among organizations have not yet translated into tangible results. Quebeckers' consumption of resources per capita is still among the highest in the world, and they have not reduced GHG emissions per capita. This echoes the RRECQ 2023 annual report:

[44] More details about the creation of a common language can be found in the first chapter of the report Transitioning to a Circular economy. Learning from the Québec experience, 2014–2020, (2021).

> There is still a significant gap between theory and practice before the circular economy can not only become the dominant economic model, but also bear fruit in terms of sustainability. Much research remains to be done to perfect the circular economy. As a result, it has become a research theme in its own right, with its own research communities, conferences, symposia, and scientific journals. Publications have grown exponentially in recent years.

The 3.5% circularity rate makes Québec lag far behind leading countries, such as the Netherlands. "*The needs' material footprints originate, to a large extent, from outside of the province through its imports—typical for a developed trade region. Quebec … still relies heavily on imported fossil fuels for transport. The agricultural sector produces unusually large amounts of waste, but little is currently reused or recycled. Also, the goods and services the government invests in… are highly resource-intensive*".[45] As we have seen, the applied nature of research in Québec is a determining factor of an accelerated progression towards a more circular economy. However, funding of schemes such as the ELEC are very lengthy processes. The projects of the Construction Lab were run over 3 years, and have come to fruition only in 2024, while the Food Systems lab started in 2023. The recently launched FRECQ helps provide a framework for governmental bodies to invest increased funds and resources in research and projects in the coming years.

In the words of the *Comité consultatif sur les changements climatiques* in 2023, "*When it comes to the circular economy, the potential is real, but not yet fully realized.*"[46]

Three risks related to "the circular movement in Quebec" need to be highlighted. The first concerns circular washing. The *circular economy*, coined as an umbrella concept that connects circular strategies, such as industrial ecology, recycling, energy recovery, etc., has enabled conversations and actions among practitioners and researchers concerning the circularization of material flows. These conversations have been fruitful in generating novel ways of contemplating sustainability conundrums. Meanwhile, the boundaries of what is and what is not circular are becoming blurry, and circularity may become somewhat of a buzzword. Correspondingly, the risk of *circular washing* could seriously challenge this movement.

The second issue concerns incrementalism. The circularity movement—in which the model of the co-construction of applied research prevails—is contributing to deploying circularity in material-intensive industries, such as agri-food, energy, construction, manufacturing, and metal products, concerning specific issues. This incremental, quick, gains-focused approach engages the onboarding of many players. Unfortunately, this approach may "be dead on the road" if no deeper reflection nor action is invested into the larger questioning of the models. Circularizing a fundamentally linear model may generate increased efficiency; it may not address more efficacy issues if the overall system relies on waste. Focusing only on

---

[45] The Circularity Gap Report | Quebec 2021, p. 6.

[46] "*En matière d'économie circulaire, le potentiel est réel, mais encore peu concrétisé*". Décarbonation du transport lourd de marchandises : construire une voie durable Report, Comité consultatif sur les changements climatiques, 2023, p. 22. Traduction libre.

incremental implementations may prevent radical transformation. As the Council of Canadian Academies stated in their report Turning Point (2021): "There is a risk that incremental actions and a watered-down conception of the circular economy fail to deliver on its promise."[47]

The third risk concerns the depth of change needed. Circularity is one of several perspectives proposed to address the current non-sustainability crises faced by our society. Circularity is reformist and rarely questions broader economic growth, an overconsumption-based model on which the economic system relies. A recent article and research file in *Journal Les Affaires*, one of Québec's most read business newspapers, focused on "The D word",[48] D standing for degrowth. It illustrates how such concepts and perspectives are part of the public and business decision-makers' debates. While decision-makers seem reluctant to promote strategies that would question the premises of the economic system in a more systematic and holistic approach, the research community could reflect on and discuss circularity as opposed to other transition perspectives, such as degrowth, sobriety, and other, to critically reflect on the accomplishments, the limitations, and the road ahead for what a sustainable Québec would be like. This is an intellectual endeavor—and a responsibility for researchers.

Appendix 1: Individuals Consulted and Interviewed

The list of the members of the focus group is as follows:

- Cathy Baptista, Ing., M.Ing, MBA, Scientific Coordinator at RRECQ.
- Sophie Bernard, PhD., Full Professor, Environmental Economics at *Polytechnique Montréal*, Principal Investigator of CIRANO's Sustainable Development theme, Chair for Axis 4 at RRECQ.
- Hélène Gervais, M. Env., Environmental Advisor, RECYC-QUÉBEC.
- Mathias Glaus, Ing, Ph.D., Full Professor, Construction Engineering Department at ÉTS Montréal, Founding Member of CERIEC, Chair for Axis 3 at RRECQ.
- Marc Journeault Ph.D., CPA, CMA, Full Professor, School of Accounting, *Université Laval*, and Head of the *Centre de recherche en Comptabilité et Développement Durable* (CerCeDD), Member of CERIEC, Chair for Axis 2 at RRECQ.
- Hortense Montoux, Project Coordinator, Construction Lab (labCo), ELEC, CERIEC.
- Daniel Normandin, M.Sc., MBA, Director, CERIEC and Executive Director, RRECQ.

[47] Council of Canadian Academies, 2021. Turning Point, Ottawa (ON). The Expert Panel on the Circular Economy in Canada, Council of Canadian Academies, https://cca-reports.ca/wp-content/uploads/2022/01/Turning-Point_digital.pdf

[48] https://www.lesaffaires.com/blogues/marine-thomas/le-mot-en-d/648811

- Chantal Rossignol, M.Sc.A., Coordinator at the Circular Economy Labs Ecosystem (*Ecosystème de laboratoires en économie circulaire*—ELEC).

The leading questions prompted to the members of this focus group aimed at establishing a broad vision of what essential elements should be covered in the context of a book chapter. The objective was therefore to constrain the scope of content to the limits, obstacles, and opportunities those experts believed needed to be put forward. This included but was not limited to—as far as research in the field of circular economy was concerned—milestones, notable policy, regulation or legislative advances, and training developments that have proven to be essential to the actual implementation of the transition to a circular economy over the past 5 years. In this, participants followed the structure of the report Transitioning to a Circular Economy—Learning from Québec experience (2021) while acknowledging that the objective was not the creation of a comprehensive follow-up report of similar scope and reach.

Individual interviews and communications were undertaken with

- Marc Journeault Ph.D., CPA, CMA, Full Professor, School of Accounting, Université Laval, and Head of the *Centre de recherche en Comptabilité et Développement Durable* (CerCeDD), Member of CERIEC, Chair for Axis 2 at RRECQ.
- Annie Levasseur, Full Professor, Canada Research Chair on Measuring the Impact of Human Activities on Climate Change Scientific, Director, CERIEC, and Scientific Director of the AdapT Institute at ETS Montréal.
- Claude Maheux-Picard, Ing., M. Sc. A., General Manager, *Centre de transfert technologique en écologie industrielle* (CTTÉI).
- Mélanie McDonald, Executive Director, *Chemins de Transition*, Université de Montréal.
- Hortense Montoux, Project Coordinator, Construction Lab (labCo), ELEC, CERIEC.
- Marlybell Ochoa Miranda, Project Manager, Roadmap for the transition to a circular economy of the Quebec society, RRECQ.
- Valentina Poch, Operations Manager, *Centre interdisciplinaire de recherche en opérationnalisation du développement durable* (CIRODD), ETS Montréal.

Appendix 2: References

Documentation consulted in the writing of this chapter includes but is not limited to:

- Circle Economy. (2021). The Québec Circularity Gap Report. https://www.recyc-quebec.gouv.qc.ca/sites/default/files/documents/rapport-indice-circularite-en.pdf
- Council of Canadian Academies. (2021). Turning Point, The Expert Panel on the Circular Economy in Canada. https://cca-reports.ca/wp-content/uploads/2022/01/Turning-Point_digital.pdf

- Côté, A., Denoncourt, J. C. (2023). *Pour un droit à la réparation robuste et accessible partout au Québec,* Équiterre. https://cms.equiterre.org/uploads/Fichiers/503_Me%CC%81moire-Pour-un-droit-a%CC%80-la-re%CC%81paration-robuste-et-accessible-partout-au-Que%CC%81bec_A23.pdf
- Girard, A., Thorpe, C., Durif, F., Robinot, E. (2018). *Obsolescence des appareils ménagers et électroniques: Quel rôle pour le consommateur,* Équiterre. https://cms.equiterre.org/uploads/fr_rapportobsolescence_equiterremai2018_0.pdf
- Gouvernement du Québec. (2024). *Accélérer le développement de l'économie circulaire, Feuille de route gouvernementale en économie circulaire* 2024–2028. https://cdn-contenu.quebec.ca/cdn-contenu/adm/min/environnement/publications-adm/developpement-durable/strategie-gouvernementale/feuille-route-economie-circulaire_01.pdf
- Jagou, S. (2021). Transitioning to a circular economy—Learning from the Québec experience 2014–2020, QuébecCirculaire / SPI (2021). https://www.quebeccirculaire.org/data/sources/users/5777/20210519201748-quebec-circulairecereportfinalmay192021tiny.pdf
- Ochoa Miranda, M. (2023). *Feuille de route pour la transition vers l'économie circulaire de la société québécoise, Diagnostic prospectif,* RRECQ. https://rrecq.ca/wp-content/uploads/2023/10/V24.-Diagnostic-prospectif-FREC.-V10-Oct-2023.pdf
- RRECQ. (2023). *Rapport d'activité.* https://rrecq.ca/wp-content/uploads/2024/04/Rapport-dactivite-RRECQ-2023.pdf

Stéphanie Jagou now a Senior Circularity Consultant, she acted as the Training Manager for the circular economy research center CERIEC (Centre d'études et de recherche intersectorielles en économie circulaire until fall 2024). She has been working in the field of sustainable development and circular economy for over 25 years. From NGOs to SMEs, multinationals, and academia, she has managed multiple mandates in Canada, Europe, and New Zealand, and participated in several development projects (Haiti, Gabon, Costa Rica). Ms. Jagou is the author of the reference report Transitioning to a Circular Economy—Learning from the Québec Experience 2014–2020 (2021). As an education specialist since 2012, she has produced tailor-made educational programs, such as the Centre interdisciplinaire de recherche en opérationnalisation du développement durable (CIRODD)'s first summer school on key competencies to implement sustainable development as well as CERIEC's circular economy summer school. The MOOC Économie circulaire, une transition incontournable is one of the most significant projects she co-developed. A lecturer at HEC Montréal, Stéphanie also combines solid experience in management, communications, public relations, and event organization. Stéphanie holds a double B.A. in France and the UK in International Commerce and Marketing, as well as a DESS in Environmental Management from the Université de Sherbrooke.

Emmanuel Raufflet (Ph.D. Management, McGill University) is a Professor of Management at HEC Montréal. His research focuses on social innovation, sustainable development, and circular economy. In 2018–2019, he served as academic director of the IEDDEC (*Institut Environnement, Développement durable et Économie Circulaire*), a joint research center between École Polytechnique, Université de Montréal and HEC Montréal. He is one of the co-heads of the Quebec Research Network on circular economy (260 + researchers/ interdisciplinary) (2021–2026).

Chapter 4
Intervention-Research as a Social-Ecological Transition Belt, Steering Wheel, and Engine

Olivier Riffon and Simon Tremblay

Abstract Social-ecological transition refers to a social movement and a field of research. As a research field, social-ecological transition proposes tools for steering and inciting large-scale transformations toward sustainability. These transformations must take root in, and impact, the social-technical regimes (healthcare infrastructure, transportation system, food regime) that structure our everyday life. Academic research in itself can be seen as a social-technical regime in which changes may occur, so that science contributes to addressing current challenges. This chapter aims to describe and justify a specific research approach rooted in the spirit of transition: intervention-research, an innovative, transdisciplinary, and praxeological approach. Similar to action research, intervention-research implies active and prior involvement of research teams in transition initiatives, with the explicit intention of social transformation. This approach entails the diversification of roles for researchers toward facilitation, management, and mobilization. The proposed chapter examines three intervention-research projects conducted by LAGORA (Laboratoire de Gouvernances Alternatives) in the Saguenay-Lac-Saint-Jean region, Quebec. It illustrates the various stages of intervention-research, the new roles of research actors, as well as the challenges and limitations of such an approach.

Keywords Social-ecological transition · Intervention-research · Transdisciplinarity · Praxeological approach · Social transformation · Roles for researchers

O. Riffon (✉)
Université du Québec à Chicoutimi, Saguenay, QC, Canada
e-mail: oriffon@uqac.ca

S. Tremblay
MRC Domaine du Roy, Roberval, QC, Canada
e-mail: drembl84@etu.uqac.ca

M. Cheriet et al. (eds.), *Accelerating the Socio-Ecological Transition*,
https://doi.org/10.1007/978-3-031-82896-6_4

1 Introduction

Social-ecological transition refers to both an array of social movements and a research field. As a research field, social-ecological transition offers pathways and tools to operate collective and cross-sectoral transformations. These cross-sectoral transformations take root in, and impact, social-technical regimes (healthcare infrastructures, food regimes, transportation systems) that are the foundations of our everyday existence (Geels, 2011). Academic research is in itself a social-technical regime in which transformations can originate and develop, to facilitate and drive social-ecological transitions (Delplancke et al., 2021).

The following chapter describes a specific academic research approach rooted in the spirit of transition. This approach is known as intervention-research, an innovative, transdisciplinary, and praxeological approach. Stemming from action research, intervention-research implies the active involvement of researchers in transition initiatives (Abson et al., 2017), driven by an explicit intention of social transformation. Research involvement beyond the walls of the academic institution enables relevant scientific knowledge to be transferred to social transformation initiatives. In the opposite direction, the production and acquisition of knowledge is enabled through live insights and analysis taking place during these interventions. Moreover, because boundaries between intervention and research become blurred, this particular approach also calls for a diversification of roles. Facilitation, management, and mobilization are the cornerstones of this intervention-research diversification (Wittmayer & Schäpke, 2014).

This chapter is built upon the analysis of three intervention-research projects conducted by LAGORA (Laboratoire de Gouvernances Alternatives) at the University of Quebec in Chicoutimi (UQAC), in the Saguenay-Lac-Saint-Jean region, Quebec. It highlights the various steps of this particular type of research and the different roles expected of research actors. Challenges and limitations of such an approach are also discussed.

2 Context

As the Anthropocene becomes a salient feature of our collective lives, humankind is experiencing a time of uncertainty and social-ecological instability (Bonneuil & Fressoz, 2013; Haché, 2014). An expanding body of scientific literature is informing decision-makers that the rapid acceleration of development has had major impacts on the Earth System (Steffen et al., 2011). Some social and ecological issues have gained recognition in recent years: climate change, increasing socio-economic inequalities, and ecosystem pollution.

In this context, humanity is aware of its responsibilities in shaping the world's ecological decline and the rise of social inequities Huybens, 2011). Historically, this awareness has driven ecodevelopment (Sachs, 1993) and sustainable development.

These two concepts have been prompting humankind to mitigate the detrimental impacts of development on the environment and on communities, by any means. Despite the fact that since the late 1980s, the acknowledgment of humanity's burden has instilled a sense of urgency throughout public institutions and in individual citizens, the pace of economic development is still greater than what is considered to be our planet's capacity (Steffen et al., 2011). This paradox raises questions about systemic locks that hinder necessary social and ecological transitions. While some mechanisms for steering social-ecological transitions have already been documented (Grin et al., 2011), questions remain unanswered about the most effective steering tools that could allow transformative breakthroughs toward transition.

Social-ecological transitions propose a radical, democratic, and peaceful transformation of our relationship to the world, oriented toward new, more resilient, and responsible ways of life. The concept of social-ecological transition was born in the 1970s (Meadows & Delaunay, 1972) and is currently a focal point of research in fundamental, social, and applied sciences. Social-ecological transition studies are driving a shift in the sustainable development framework by integrating several original concepts: Edgar Morin's complexity (Morin, 1999), praxeology (Schön, 1994), collaborative research (Portelance & Giroux, 2009; Anadon, 2007; Callon et al., 2001), transdisciplinarity (Létourneau, 2008; Turcotte & Caron, 2018), ethics of the nature/culture relationship (Descola, 2005; Huybens, 2011), and post-capitalist ecology and economy (Arnsperger, 2009).

Even though transition initiatives are being generated worldwide at various territorial scales (Waridel, 2019), political, economic, and cultural systems are slow to adapt in light of the challenges they face. It seems as though systemic barriers, similar to those faced by sustainable development, are limiting the impact of these territorial initiatives.

Some researchers have proposed to coordinate and unite transition initiatives in what could be termed a transition arena (Grin et al., 2011). Within a given territory, a transition arena is guided by processes of communication and dialogue. Actors discuss transition issues by drawing a map which allows them to pinpoint systemic locks. Actors can then experiment with transition initiatives using relevant mechanisms to address these systemic barriers.

In this spirit, we propose to study the implementation of three transition arenas. The latest are innovative governance structures that mobilize and coordinate the actions of the Saguenay-Lac-Saint-Jean (SLSJ) region to accelerate a social-ecological transition. First is Borée for a sustainable food system. Second is the Grand Regional Dialogue for Transition. Third is Forum Mobilité for a sustainable transportation system. In these projects, actors are engaged in a cross-sectoral manner to coordinate a collective transformation at the regional level, exploring new forms of non-institutional governance.

The proposed chapter aims to analyze these three initiatives through a praxeological approach of research-intervention. In this particular approach, iteration between practice and theory simultaneously improves practice (1) and extracts knowledge (2) (Schön, 1994). Without a doubt, the state of the world has justified deliberate involvement of researchers in transition movements (Wittmayer et al., 2014). The ethical purpose of research on transition underlies this choice (Wittmayer & Schäpke, 2014). Researchers are believed to have a role to play in steering transition initiatives (Grin et al., 2011). The coordination mechanisms with which territorial actors are experimenting are documented. Their potential to overcome system-related barriers is discussed (Grin et al., 2011).

3 Questions and Objectives Addressed by This Study

The current paper is underpinned by two research questions. First, what are the roles of the research sector in social and ecological transitions? And second: how can a research posture that is both rigorous and designed to accelerate transition be implemented through various initiatives?

The objective of this study is to propose responses to these two questions based on three intersectoral initiatives, each working toward social and ecological transition in Saguenay-Lac-Saint-Jean. More specifically, we will describe the evolution, goals, and areas of activity of the three initiatives. We will then analyze and compare the roles played by the research sector in the three approaches, paying particular attention to dimensions such as leadership, action, ownership, and power. We conclude by documenting the limitations and challenges of intersectoral approaches where researcher-practitioners have intervened.

4 Intervention Method

The social-ecological transition concept frames an approach where social-technical regimes progress toward a sustainable state through structural changes. A set of technological, economic, ecological, sociocultural, and institutional innovations reinforce each other in this structural process (Tremblay, 2011). Social-ecological transitions target a general configuration of practices, policies, actors, and/or institutions, but they can also target specific spheres of human activity. Thus, social-ecological transitions propose a large-scale transformation and drive new arrangements and governance modalities, based on increased collaboration among actors at different scales. Transition intervention-researchers seek methods to manage these processes of profound change (Bergman et al., 2008; Loorbach, 2007).

A multi-level perspective allows the study of these multiple forms of transition. Grin et al. (2011) depict transition trajectories using a three-level model: transition niches (1), social-technical regimes (2), and landscapes (3). According to this model, the ethical rationale for transition occurs at the landscape level (climate change, the rise of inequalities, the decline of biodiversity). Over time, these changes put pressure on social-technical regimes (policies, technologies, culture, market, industries). When these regimes are threatened, it is often within them that the greatest resistance is observed. Factors hindering transition are often political, legislative, economic, or cultural.

However, social movements create windows of opportunities for the reconfiguration of social-technical regimes. At a local scale, it is indeed possible to observe a multitude of transition niches. In these niches, alternatives are put into practice while adapting to, or in anticipation of, landscape changes. Pulling together niches is a way to increase coherence and effectiveness of individual transition actions. The interconnection of niches in a transition arena, where interventions are linked and coordinated with specific arrangements, amplifies their effects. Processes of dialogue, learning, and co-construction among transition niches, as well as with actors who are active within regimes, appear to be a key condition for steering transitions.

The regional scale is known to be relevant for the application of sustainable development principles (Gagnon, 2006; Theys, 2002). The same is believed to be true for the coordination of social-ecological transition. Territorial governance as a research field studies the capacities of local actors to mobilize and take charge of regional decision-making processes (Leloup et al., 2005). Territorial governance historically calls for a decision-making process characterized by horizontal decision-making and decentralized power. This is in line with the transition studies approach, in which actors are meant to be enabled through mechanisms that facilitate their collaboration and communication (Riffon, 2016).

Supporting social-ecological transitions with the historically rich perspective of territorial governance allows for an awareness of the multiple natures and statuses of actors associated with transition niches (Riffon, 2016). This perspective also recognizes that societal transformation relies on processes of collaboration and negotiation among networks of actors (Bodin et al., 2006; Leloup et al., 2005). Providing a territorial community access to structures of knowledge and information is thus necessary to overthrow systemic barriers (Granovetter, 1973).

Network studies have also been a field of study since the 1980s, based on two approaches. First, graph theory examines the properties of network structure through several characteristics (connectivity, centrality, density) and the nature of connections (Keeling, 2005). Second, social network analysis focuses on relationships between actors in a given context. By addressing reasons for connections, it serves as an explanatory tool for the evolution and diffusion of ideas, knowledge, and innovations (Leinhardt, 1977). In both frameworks, some actors act as knowledge brokers to close gaps between other actors and enable their latent potential (Bodin et al., 2006).

The facilitation of actor networks can generate social capital, a potential social-ecological catalyst for change. The extent of change depends in part on the ability of local actors to mobilize this social capital. Farinós Dasí (2009) framed it as "territorial intelligence". Mobilizing social capital requires knowledge of both the territorial social system and socio-economic processes. "Territorial intelligence" comes into play when pivotal local actors are able to understand the influence of socio-economic processes on territorial development and institutional functioning. In such conditions, pivotal local actors act as knowledge brokers.

Collaborative and cross-sectoral approaches make good use of territorial intelligence. These approaches are one among many tools to overcome systemic locks and redirect existing territorial governance networks toward social-ecological goals (Beuret & Cadoret, 2010).

Lastly, steering social-ecological transitions involves tools that facilitate mutual understanding, collaboration, and collective action among actors. This will make them feel involved in a common endeavor. Tools for systemic analysis (Villeneuve et al., 2017), analysis of social representations (Riffon, 2016), and tools for ethical dialogue (Segers, 2014) outline the structure of our analysis and discussion.

This chapter explores the role of researchers in social-ecological transition initiatives (Wittmayer et al., 2014). It aims to put into practice and depict three initiatives with an innovative methodological stance that combines explicit ethical engagement and a transdisciplinary praxeological research approach. This ethical engagement is coherent with challenges brought upon the academic social-technical regime by the Anthropocene.

This stance allows, first, the analysis of transition issues without compromising their inherent complexity. Second, it allows for the implementation of concerted action mechanisms tailored to local contexts and informed by scientific knowledge. Understanding the complexity and specificities of territorial dynamics requires considering the interdependence of issues managed by different actors, despite the challenge posed by cross-sectoral endeavors (Morin, 1999). The steering and study of transition processes at the territorial level should, therefore, be transdisciplinary, with transdisciplinarity invoked as both a practice and a research paradigm (Turcotte & Caron, 2018).

Transdisciplinarity involves a respective contribution of a diversity of analytical frameworks to interpret reality (Nicolescu, 1996). This paradigm also enables mobilization of different types of knowledge (Max-Neef, 2005), including non-academic knowledge, which distinguishes transdisciplinarity from interdisciplinarity. In the end, it promotes mutual learning between theorists, practitioners, and researchers-practitioners (Scholz, 2001). An applied transdisciplinary research-intervention approach requires the involvement of different scientific disciplines (social, fundamental, applied, political, economic). Likewise, an array of sectors (mobility, health, urban planning, education) are involved in this undertaking. Over and above, the entire spectrum of actors, from laypeople to experts and from citizens to decision-makers, are called to arms.

The pragmatic nature of transition processes encourages the grounding of research in experience, because researchers are involved in the processes and outcomes they aim to study. LAGORA's approach is rooted in a praxeological methodology of research-intervention. This approach acknowledges that experiential knowledge plays a pivotal role in knowledge production (Schön, 1994).

In the three case studies, practitioner-researchers (Albarello, 2004) are grounded in the practice of a transition arena. Nevertheless, they do engage in analytical thinking in order to extract knowledge and theories (Huybens, 2011). This approach responds to a need for increased involvement of scientific actors in the transition (Wittmayer & Schäpke, 2014). Scientific actors contribute to empowering social-ecological movements toward social-ecological ideals (Lhotellier & St-Arnaud, 1994).

5 Method Employed for the Case Studies

This study was conducted in two parts. In the first part, an exhaustive literature review on transdisciplinary collaboration processes for social-ecological transitions was conducted. Special attention was given to transition arenas involving scientific actors, with the aim of analyzing case studies through a lens of connections, arrangements, and governance modalities for transition. This has been integrated into the analytical framework.

The second part involved a comparative study of three transition arenas that took form in the SLSJ territory over the past 5 years: Borée for a sustainable food system (1), the Grand Regional Dialogue for Transition (2), and Forum Mobilité for a sustainable transportation system (3). Our research-intervention team has been participating in these initiatives since 2018; their respective governance modes are described from an insider viewpoint. Furthermore, governance modes are explained and compared on three bases: participatory mechanisms, dialogic processes, and tools used. The nature of connections and actor dynamics are also underlined. The discussion sheds light on the practice of transdisciplinarity and cross-sectoral dialogue. Positive results obtained to date and the pitfalls that were encountered are presented. In addition, the dynamics between actors and transition niches within social-technical regimes are analyzed. This latest aspect has been underrepresented in the literature.

This comparison of three case studies has been enriched in two ways. An observation journal was kept by the research team during the last 4 years, to document research implications. Observations contained in this research journal were cross-referenced with a dialogue between the two authors of this article, who participated in various roles in each of the three case studies.

6 Theoretical Framework

In this theoretical framework, we will discuss the works of four academics. Barnaud et al. (2016) have highlighted power dynamics in participatory mechanisms. Marcel (2018) theorizes the socio-scientific "third space", a framework to analyze expert and public engagement toward sustainability. Wittmayer and Schäpke (2014) call for a diversification of researchers' roles.

Barnaud et al. (2016) show the existence of power imbalances among stakeholders participating in decision-making processes. These imbalances may arise due to factors such as unequal access to resources, knowledge, or social status. Barnaud et al. (2016) also discuss how actors with greater social, economic, or political power tend to exert more influence over decision-making processes within participatory mechanisms. This influence can shape the outcomes of these processes and may perpetuate existing inequalities. In fact, participation does not always guarantee equal representation or meaningful involvement. Factors such as language barriers, cultural differences, and power dynamics can influence the extent to which diverse voices are heard and integrated into decision-making. To address this, Barnaud et al. (2016) underscore the importance of reflexivity among participants and facilitators of participatory mechanisms. This involves critically examining power dynamics, acknowledging biases, and promoting transparency and accountability to ensure more inclusive and equitable decision-making processes.

According to Cenzano Vilchez et al. (2022), research on social and ecological transition must serve a dual purpose: generating academic knowledge and transforming society. In this vein, Marcel (2010) establishes a typology that goes as follows for practitioner-researchers working within the transition: for, on, with, and by. In other words, Marcel proposes to clearly take a position *for* the transformation of society, *with* the participation of a diversity of actors, *by* the emancipation of mobilized stakeholders, and by producing knowledge *on* this process.

A socio-scientific third space is an interdisciplinary and collaborative environment where scientists, policymakers, stakeholders, and members of the public come together to engage in dialogue, deliberation, and decision-making on socio-scientific issues (Marcel, 2018). This concept emerged from the recognition that complex societal challenges, such as climate change, biodiversity loss, and public health crises, cannot be effectively addressed through traditional disciplinary approaches alone. Instead, they require the integration of scientific knowledge with social, cultural, ethical, and political considerations (Marcel, 2020).

In a socio-scientific third space, diverse perspectives and expertise are valued, and participants work collaboratively to co-produce knowledge, develop solutions, and make informed decisions. This space serves as a bridge between the scientific community and broader society, facilitating mutual learning, understanding, and trust-building. Thus, socio-scientific third spaces are characterized by cross-sectoral endeavors, dialogue, broad public engagement, and reflexive practice (Wittmayer et al., 2014).

Wittmayer and Schäpke (2014) advocate for a more diversified and engaged role for academics in sustainability transitions, emphasizing the importance of engaging directly with stakeholders, communities, and policymakers in collaborative research processes. This can involve co-creation of knowledge, participatory action research, and partnership-building with non-academic actors. Wittmayer and Schäpke (2014) also emphasize capacity building, which involves providing training, resources, and tools to empower stakeholders to participate effectively in various decision-making processes. They also encourage researchers to actively engage with policy processes and contribute evidence-based insights to inform policy development and implementation. Finally, they also call for reflexivity, which is essential to critically reflect on researchers' roles, biases, and assumptions. Diversifying researchers' roles can help bridge the gap between knowledge production and societal action, and contribute to more effective and equitable sustainability transitions. These diverse researchers' roles are: knowledge broker, change agent, facilitator, observer, and knowledge generator.

7 Case Studies

7.1 Borée for a Sustainable Food System

The first project in which researchers were involved was *Borée for a sustainable food system*. Borée was initiated in 2017 by 5 economic and socio-community actors within the food system in order to address prevailing issues. These issues, identified by initiators, include the declining psychological health of agricultural producers, national food insecurity, and the long-lasting problem of food waste. Concerning this last issue, parallels were made with the donut economics model (Raworth, 2017). As a starting point, even though the environmental footprint of food production was estimated to be greater than what is considered to be sustainable, some actors gave evidence demonstrating that many people suffer from hunger in the region.

In 2018, researchers were called upon to facilitate meetings between the five actors. Initially, Borée had not been truly structured, which allowed for quick organizational changes. When the researchers got involved, the group expanded from 5 to 8 organizations, comprising some that were environmentally oriented. The group held broad 3-h meetings every month. There were between 10 and 12 participants, with some organizations represented by two actors. Two research actors were present to facilitate the meetings.

In 2021, a $200,000 grant was received from the Ministry of Economy and Innovation (MEI). Borée's governance structure was then re-examined in view of its expansion. In 2022, the group grew from 8 to 20 organizations. The group continued to hold monthly meetings, with more participants involved. Since then, governance has become more complex with the creation of a management committee and a Human Resources (HR) committee. Researchers no longer have a role in facilitation, as there is a coordinator hired by the group.

7.1.1 Research *For, On, With,* and *By*

In Borée, the notion of social-ecological transition came indirectly, with the establishment of a sustainable food system as a focal point. Bringing together the environmental, socio-community, and economic sectors made it possible to introduce a sustainable development framework, akin to the three dimensions (social, economic, environmental). Territorial development also quickly emerged as a theoretical framework in the quest for a territorial food system. The distinctiveness of the Saguenay food system was determined to be the Boreal identity. The aim, or "For", became to transform the food system to be both boreal and sustainable. Through the concept of system transformation, the idea of ecological transition comes into play.

The "For" aspect is expressed explicitly from the outset with a willingness to transform the food system. This unambiguity is reflected in the Borée's charter and the tools it employs. The "On" aspect involves generating knowledge about the process by which intersectoral work contributes to the transformation of the food regime. Once a coordinator was hired, researchers could then focus on knowledge generation. The "With" aspect is also given prominence by the involvement of actors from three sectors. In Borée, the "By" aspect was made evident, as the Borée group now moderates itself using facilitation and management tools. In fact, from its inception to its expansion to 20 actors, the role of researchers has progressed toward observation, analysis, and knowledge production.

7.1.2 Socio-Scientific Third Space

The third space is Borée's committee of partners itself, comprised of 20 organizations, including four researchers. These researchers represent LAGORA and Agrinova, a food production research institute. The researchers are also involved in a new third space within the management committee. This committee can be considered a subset of the larger third space.

7.1.3 Roles

All practitioner-researchers' roles were involved in Borée. As aforementioned, initially, a need for moderation and facilitation came from a self-formed and autonomous group. Researchers' roles were to facilitate meetings and to mobilize actors in order to build the monthly agenda.

The mediation role came soon after, as actors were involved in an intersectoral context. This context made for tensions and misunderstandings among actors. Researchers were regularly called upon to translate sectoral realities, vocabularies, and perceptions. For instance, an economic actor may not understand a community's culture of consultation. In the private sector, innovations are often kept secret until the launching phase. The translator's role was licensed to researchers due to their impartiality and their ability to defuse and/or attenuate many conflicts before they burst out.

Researchers' roles quickly evolved in 2020 toward being knowledge brokers. Borée had questions and challenges about its governance. Researchers delved into the literature to document existing governance modalities, along with initiatives, strengths, and weaknesses related to each modality. This study was provided to Borée. Researchers were also active in supporting some municipalities aiming to apply the by-products of Borée's work at a local scale, with initiatives and funding.

In parallel, Borée was analyzed for its intersectoral and interdimensional features. These research findings were then integrated into academic courses, presentations, and scientific articles (Villeneuve & Riffon, 2022). This generation of knowledge has overseen much self-reflection. This was especially true because Borée was the first of three intervention-research initiatives.

7.1.4 Challenges and Tools

There were numerous challenges in the intersectoral approach to a food system transformation toward sustainability. Vocabulary issues arose, because of the intersectoral approach. Also, it took 3 years for intersectionality to be established, because some actors were not used to sharing knowledge, resources, and meetings.

The expansion of Borée, in particular, posed a significant challenge. There had to be a pause between the initial phase, with 8 actors, and the current functioning with 20 actors. Financing was also a concern at some moment in time because at first, Borée relied on research funds, whereas now it relies on institutional funding. Ministries participating in financing have requirements, which is to be expected, but this also leads to a loss of freedom, flexibility, and autonomy within Borée. Potential social innovations may therefore have been lost.

Sometimes there were also misunderstandings with ministries, regarding the ways in which different societal sectors work together toward a shared vision of a sustainable food system. Although these differences have been disruptive for ministries, they are now willing to go ahead with this social innovation.

Regarding tools, the mutualization of funds and resources was pivotal in progressing toward a systemic approach to food system transformation. There was very little turnover of personnel, which has proved to be a strength in the making of Borée. Trust took a long time to be established. In the end however, once trust was established, it allowed for the pooling of funds and resources.

7.1.5 Outcomes

We called institutionalization the process by which Borée transitioned from research funding to government funding. Although this positive evolution to the more reliable government funding has had some negative aspects, it has allowed for more recognition and funding. Thus, it serves both as a barrier and as leverage.

Furthermore, with the benefit of hindsight, we can learn from the Borée experience that aligning many niches has gradually led to changes in policies, thanks to the commitment of government ministries. As of today, some ministries involved are exploring how to enhance their sectoral programs to support a sustainable food system, which tries to limit food waste and hunger through structural measures.

To conclude, the grouping of niches has brought interest from ministries. The subsequent institutionalization of Borée has led to emerging change in food policies, most notably in respect to health promotion and prevention. Quebec's Public Health Agency and the Ministry of Municipal Affairs and Housing have contributed to change by their leadership.

7.2 Grand Regional Dialogue for Transition

The Grand Dialogue narrative diverges significantly from that of Borée, as it stems from academic inquiry. Conversely, Borée originated as a grassroots endeavor. As a starting point, Eco-Advisory researchers had long contemplated the mobilization of transition stakeholders in the region to coalesce transition niches within the SLSJ region. The propitious moment to do so became evident when the region experienced a period of social unrest and heightened political polarization surrounding a mega-project for the exportation of liquefied natural gas. Assessment of the political landscape surrounding this issue revealed an absence of platforms and spaces for societal debate.

The concept underlying the Grand Dialogue aimed at instigating a social movement geared toward envisioning the future of the region through innovative and inclusive consultations. These consultations were predicated upon the principles of aspiration, enjoyment, and inclusiveness. Consequently, the overarching objective was to formulate a collective vision involving a broad spectrum of participants, with the explicit aspiration that this vision would align with the spirit of social and ecological transition.

From 2020 to 2023, the Grand Dialogue engaged approximately 9000 individuals, constituting 3% of the regional population, in workshops spanning 1–3 h each. This endeavor culminated in the synthesis of 21 vision statements spanning diverse domains, encompassing the food system, mobility, education, and the arts. These statements were conceived as catalysts for transition processes. Notably, the Mobility Forum and Borée converged around these vision statements. The ultimate aim was to establish dedicated working groups for each of the 21 statements.

An early observation during this process highlighted the fact that the act of inspiring the population through vision workshops served to mitigate disparities in perspectives. This, in turn, facilitated the emergence of aligned visions across economic, social, and environmental sectors, driven by a shared long-term aspiration for territorial development focused on enhancing human well-being qualitatively, while respecting ecological boundaries. The utilization of visionary elements thus engendered a cohesive vision, fostering unity rather than discord. Researchers have played a continuous role in this endeavor.

7.2.1 Research *For, On, With, and By*

Within the framework of the Grand Dialogue, the affirmation of the "For" position is most pronounced. The impetus for transformative action is the most potent among the three initiatives under consideration. The instigators of the Grand Dialogue were driven by a fervent ambition to catalyze social-ecological transition on a regional scale.

This ambition to enact change does not compromise the methodological impartiality of the consultation process. Indeed, rigorous social science methodologies were employed to ensure data saturation and representativeness during the data-gathering phase. The research team maintained methodological neutrality by engaging with a diverse array of stakeholders, notwithstanding potential divergences or conflicts in their perspectives. The overarching objective was to solicit input from the broadest possible spectrum of actors. It is important to note that methodological impartiality does not connote epistemic neutrality, as the ultimate aim remains to facilitate and enact social and ecological transition.

Regarding the "On" dimension of the Grand Dialogue, it is pertinent to highlight the proliferation of research endeavors. While both social science methodologies and fundamental scientific approaches have significantly informed and grounded the development of this initiative, the role of research assumes even greater salience in analyzing the dynamics of the movement. Presently, seven distinct research axes are being pursued. These encompass investigations into the psychosocial impacts of facilitation on youth amidst a backdrop of eco-anxiety, experiments in citizen science methodologies, the regionalization of the planetary boundaries framework, horizontal governance structures within the movement, the ethical design of participatory mechanisms, and the catalytic role of the arts in driving transition processes. These research axes embody a diverse array of scholarly pursuits, which have been disseminated through numerous conferences and scientific publications from 2020 to 2023. It is noteworthy that the delineation of scientific and/or civic roles among participants within the movement is subject to deliberate ambiguity; this facet underscores its innovative nature.

The "With" dimension of the Grand Dialogue assumes considerable significance. A network of 150 regional and national collaborations is actively engaged within this initiative, while socio-scientific third spaces abound. Furthermore, an increasing number of entities are gravitating toward the Grand Dialogue, not only to forge collaborative partnerships in data collection endeavors, but also to gain access to the profusion of accumulated data.

A recent example of this collaborative spirit is embodied by the partnership forged with the provincial (Quebec) Ministry of Municipal Affairs and Housing, tasked with formulating a comprehensive strategy for territorial occupation and vitality. Although the ministry has identified ten priority areas within this strategic framework, it has solicited the assistance of the Grand Dialogue to integrate citizen-centric perspectives into the formulation of actionable strategies addressing these prioritized concerns. Leveraging insights extracted from a meticulous analysis of approximately 27,000 statements relevant to the ministry's identified focal areas, a comprehensive report was produced to guide strategy development.

On the "By" front, both the grassroots social movement embodied by the Grand Dialogue and the broader regional environment share aspirations of empowerment through this research-driven initiative. Firstly, efforts to empower the grassroots field team are still underway. The expansion of the team from two to eight members following the influx of funding in 2022 necessitated a period of organizational transition characterized by staff turnover before reaching stability. Consequently, empowering the field team has been a multifaceted endeavor, involving the transmission of mobilization methodologies and tools to newly recruited personnel.

Secondly, the most recent funding allocation is directed toward empowering the region at large, through the establishment of structured sectoral committees aligned with the 21 thematic visions.

7.2.2 Socio-Scientific Third Space

In this case study, the Grand Dialogue itself serves as the third space. It is a grassroots movement that is not legally constituted and that operates on sociocratic principles, without any hierarchy, in which scientists contribute through attendance at work circles.

In contrast to Borée, which features a single socio-scientific space, the Grand Dialogue consists of multiple subspaces. Each work circle serves as a third space. Thus, there are numerous third spaces operating through semi-formal exchanges, where a significant degree of mobility can be observed between actors. Some circles are permanent, notably those addressing the movement's ethics, research component, and mobilization of political actors. Other circles are associated with specific projects and are dissolved once those projects are completed.

7.2.3 Roles

All the stated roles of transition researchers have been mobilized in the Grand Dialogue initiative. Facilitation has been at the forefront, as 27,000 statements were collected over a period of 3 years. Knowledge brokerage is being carried out, as the movement relies on science for its ideas. The three-level model (Geels, 2011), the planetary boundaries model (Steffen et al., 2011), sociocracy and actor dynamics (Bodin et al., 2006) have been mobilized in this initiative.

While in Borée, the boundaries between roles are very clear, in the Grand Dialogue, the boundaries are excessively blurred. Researchers wear various roles simultaneously: broker, generator, mediator, and facilitator. Moreover, a diversity of actors are involved through these diverse roles: citizens have become research actors by participating in synthesis activities. Some researchers have become practitioners who now prefer to facilitate activities between organizations rather than engage in research.

7.2.4 Challenges, Tools, and Outcomes

Working in a sociocracy poses a significant challenge, as decision-making processes are time-consuming. This reality was even more pronounced in the case of the Grand Dialogue because, at the inception of the initiative, the Why, What, and How of the movement were intended to be defined through sociocratic means. Even in 2023, one-third of employees' work time is dedicated to organizing fieldwork.

This relative organizational slowness makes it more difficult to be responsive and proactive to arising opportunities that need to be seized spontaneously when social movements occur within the regional community. Resources have also been lost over time because not everyone is comfortable with a non-hierarchical organization. Many individuals require more direction rather than less, leading to resource turnover. Additionally, there was a need for willingness to participate in the construction of an innovative initiative, outside the bounds of conventional paradigms.

The approach of Barnaud et al. (2016) to reflexivity is highly evident in the Grand Dialogue. The aspect of social transformation and transition structuring is less prominent. In other words, the team's focus is also on reflecting on work organization and not exclusively on the transition process. Team members sometimes have the impression of taking three steps back to take four steps forward. This is because new ways of doing and acting are explored within this new organization. The "How" is just as important and valued as the "What."

This reality inevitably creates tensions within the team (employees, researchers, and citizens). Tensions arise between transformative aspirations, which demand a certain level of efficiency, and sociocratic functioning, which is more aligned with an ethical approach rather than an outcome-based approach. The multiplicity of roles and the presence of numerous third spaces, as mentioned earlier, also create challenges in terms of clarity and understanding for many individuals and partners. However, this diversity of roles and spaces becomes an asset when it comes to engaging with various societal spheres.

In terms of outcomes, the regional vision articulated across 21 themes is a significant result in itself. The interest of collaborators such as the Ministry of Municipal Affairs and Housing underscores the relevance of this outcome. Furthermore, the establishment of the Mobility Forum based on the vision for sustainable mobility is another substantial outcome. This is just the beginning, as transition committees are expected to take form within each individual transition theme.

It is worth mentioning that, like Borée, the Grand Dialogue has undergone professionalization. As of 2023, 30 individuals are paid at some point to work on different third spaces. However, the movement has not been institutionalized. Thus, the mission is not influenced by the fact that ministries or "regime"-level organizations solicit the team.

This institutionalization is unlikely to occur because the movement's goal is to empower regional actors and to federate transition niches for approximately 10 years, and dissolve itself thereafter. The social transformation aspirations stemming from the Grand Dialogue will hopefully have materialized in the daily routine of the actors and niches.

7.3 "Forum Mobilité" for a Sustainable Transportation System

The Mobility Forum stems from the Grand Dialogue's consultations on sustainable mobility. Early on in the workshops aimed at eliciting a common vision for mobility, a significant number of individuals mentioned the importance of reducing dependence on automobiles. Improving access to public transportation and the need for safer infrastructure for active transportation (walking, cycling) were also raised. This came as a surprise in a region predominantly reliant on car transportation.

Many individuals also expressed the need for a forum to address sustainable mobility issues, which was non-existent at the time. The Regional Council for Environmental and Sustainable Development (CREDD) was pleased to learn about the citizens' interest in such a forum on mobility. Thus, once the vision was synthesized by the research team, CREDD used it to establish a new entity, the Mobility Forum for a sustainable transportation system and sustainable regional mobility in general.

The Forum started with 13 organizations, including ministries, transportation companies, research institutions, and socio-community organizations. Unlike Borée, which chose to work solely based on the actors already present, the Mobility Forum decided to hold annual or biannual events to build networks and gather further input from various stakeholders. This is noteworthy, since the actors would have had ample legitimacy to move on to decision-making. Rather than acting as an autonomous collective, as Borée did, the actors of the Mobility Forum wanted to push further with consultations and networking with stakeholders.

This networking aimed to unite transition niches around sustainable mobility. During the first and second annual events, the actors aimed to reflect and progress with a larger collective of stakeholders. This reflection was based on the synthesis of the population's hopes regarding mobility, as conducted by the Grand Dialogue. During the third event, the Grand Dialogue's vision, refined and enhanced during the first event, became a guiding principle that collectively directed attendees toward accelerating projects and connecting individuals interested in similar initiatives.

The annual events are just the tip of the iceberg for the Mobility Forum. This initiative aims to perpetuate an ecosystem of stakeholders working on mobility. Three key institutional stakeholders, the *Table régionale de l'action climatique* (1), the Table des saines habitudes de vie (2) and the *Conférence administrative régionale en transport et mobilité durable* (3) are part of this ecosystem. Regional transportation's governance has become more cross-sectoral since the inception of the Mobility Forum. Furthermore, the network has densified and become better informed through annual events.

7.3.1 Research *For, On, With, and By*

The "For" aspect of the Mobility Forum is ambiguous. Indeed, the mission of the Forum has never been explicitly formulated. Starting from the premise of sustainable mobility, we can still affirm that this concept is compatible with social and ecological transition. However, the Forum is known for its openness and flexibility rather than for an outcome approach. The same applies to the "On" aspect, as research has not yet been deployed on the functioning of the Forum. Researchers have been more actively involved in providing their practical knowledge contribution to the organization of annual events.

On the other hand, the "With" component involving partners is explicit and well-developed. Intersectoral mobilization has become a reflex and was a goal for the initiators from the outset of the process. The same applies to the "By" aspect, as collaboration aims to empower a large number of stakeholders in the community who are tasked with working on regional mobility.

7.3.2 Roles

During this initiative researchers have played various roles, much like in the other two initiatives. Initially, researchers did not inherit the facilitation role, as it was taken on by CREDD from the outset. However, CREDD employees regularly turn to LAGORA for a certain facilitation expertise during the annual events. Additionally, the LAGORA team facilitates the annual events. Fifteen to 20 members of the research team are present at these events to manage approximately 100 participants during an entire day.

Furthermore, the research team's expertise does not reside specifically in mobility itself, so knowledge brokerage does not occur at this level. Researchers specialize in managing the "container" (structural aspects of facilitation) rather than the content (sustainable mobility). In terms of outcomes, we will see that a strategic entity and an operational entity are part of the Forum's legacy. Considering this, the research team plays a highly relevant role in envisioning and deliberating on governance within and between these two new entities and the previously existing ones. These entities will serve as scientific third spaces with dual links inspired by sociocracy.

The role of change agent is evident, as the leadership for social transformation primarily emanates from the researchers. For example, the idea of implementing dual links inspired by sociocracy within the third spaces originated from the researchers, aiming to enhance the potential for social transformation.

Regarding the roles of observation and analysis, these will eventually enable the generation of knowledge concerning the Mobility Forum process. These roles are too recent for any studies to have been concluded. Two scientific communications have been made public during forums. This knowledge generation will focus on the decision-making process and governance rather than the content of actions. Within the research collective, many lessons have been learned to better support initiatives in the future.

7.3.3 Challenges, Tools, and Outcomes

The positive outcomes of the Forum primarily manifest themselves at the level of governance structure. Two distinct entities will soon be established to facilitate the integration of sustainable mobility in the region. The third annual event has served as a platform to introduce and validate these entities with various stakeholders. Additionally, the networking among stakeholders (1) and the acceleration of projects (2) represent tangible outcomes that transcend the governance framework. The collective intelligence fostered by the Forum facilitates project implementation across the region.

A significant challenge lies in the difficult recruitment of academic expertise in sustainable mobility within the region. This underscores the need to cultivate such expertise in the coming years. Sustainable mobility poses a multifaceted challenge, and the experiential knowledge of researchers in this domain will be essential.

The most significant leverage undoubtedly lies in the overwhelmingly positive response from the stakeholder ecosystem. Over 100 stakeholders participate in the annual events. Organizations have embraced intersectoral collaboration, notwithstanding the inherent challenges associated with this non-traditional approach. The level of engagement among participating stakeholders is remarkably high.

Within the Forum, there is a clear inclination toward collective decision-making. Consequently, there is no undue pressure placed on citizens to abandon their automobile usage, for instance, as the primary focus is systemic change. Structural changes within the mobility system, or the transportation regime, are perceived as pivotal in facilitating the adoption of sustainable mobility practices among individuals.

However, obstacles loom large. These impediments are not exclusive to the Forum but are inherent to sustainable mobility endeavors themselves. The region's extremely low population density (2.9 inhabitants per square kilometer) over vast territories and a deeply ingrained car-centric culture pose significant barriers. Urban sprawl and peri-urbanization are prevalent across the region's few urban centers. Despite the Forum's success, mobility patterns remain largely unchanged. The prevailing cultural norms and reliance on past infrastructure configurations are entrenched features of regimes that are notoriously resistant to change.

Moreover, while the Forum represents a relevant governance innovation, it operates within a broader context characterized by a decline, if not an absence, of concertation possibilities over the past two decades. The demise of Regional Conferences of Elected Officials (CRÉ) in 2015 marked the end of an era where consultation with civil society was common practice.

Presently, there is palpable intent among stakeholders to establish consultation mechanisms. A disparity exists between elected officials and civil society decision-making processes, with communication occurring only sporadically. This dichotomy represents both a challenge and an opportunity to reimagine actor dynamics within the region through innovative means.

8 Discussion

This section discusses potential issues arising from the new roles endorsed by practitioners-researchers. Wittmayer and Schäpke (2014) have played a particular role in showing four potential pitfalls in this particular type of research: leadership, ownership, power, and action.

8.1 Leadership

There was an expectation within the socio-scientific third spaces for the research team to play the role of leader. For instance, in Borée, before the hiring of a coordination resource, if meetings were not planned and structured by the research team, no other actor would undertake this responsibility. In fact, the expressed need within the community was for a neutral actor to assume the coordination of stakeholders, to avoid any one of the stakeholders seizing a larger share of power. A neutral coordination resource was subsequently put in place to take over the leadership role previously held by the research team.

In the Grand Dialogue, the same leadership gap occurred, but at the level of field action; at the time that the grassroots movement grew from 2 to 8 employees, it was the research team who held the major part of experience in the field.

At the Mobility Forum, from the outset, particular attention was paid to the definition of leadership and roles. It was the Regional Environmental Council, a neutral partner organization, that took the lead in mobilizing stakeholders. The Council also had more financial resources to do so. The task of designing activities was attributed to the research team. The design of a mobility governance structure also fell within the realm of research. Leadership was thus planned and clarified from the beginning. This was a lesson learned from the first two case studies.

Through these three case studies, we can see that the main issue at play is not an absence of leadership, but rather that roles and fields of action need to be clarified from the outset. Traditionally, organizations choose someone to delegate to an initiative. Instead, we have observed that it could be useful to identify, within organizations participating in the three initiatives, those individuals already committed to social and ecological transition (Lessard, 2021). These individuals can then initiate a process within their organization to promote the transition. The approach we have observed allows for more leadership expression because the individual's motivation is intrinsic and already established.

8.2 Ownership

Regarding ownership, the main issue is to determine to whom belong the tools and knowledge that are produced. Open data allows for a quick solution to this problem. The clause negotiated with the University allows all parties to disseminate the data. Researchers are currently exploring Creative Commons as a platform.

Furthermore, many activities in the three case studies are not framed within Research Agreements. Activities are collaborative and meant to produce knowledge under the concept of "commons." Indeed, the tools, activity plans, and vision statements within the three initiatives are produced in the form of commons. In turn, this empowers more actors to embark upon a trajectory of social and ecological transition. Producing open knowledge is an alternative to the current model of private ownership and is thus a transition action in itself.

8.3 Power

Some actors will inevitably hold more power in cross-sectoral approaches. We can distinguish two types of power. The first is "real" power or the ability to take action through financial, legal, or organizational means. The second is "soft" power or the ability to influence, often manifested by confidence in speaking and abilities of persuasion. Researchers in the role of facilitators can play a role in rebalancing power dynamics.

Barnaud et al. (2016) distinguish two approaches in this rebalancing of power relations. The dialogical approach emphasizes non-intervention and trusts in dialog to smooth power relations. The critical approach argues that power relations are structural and cannot be smoothed out by rules of dialogue. The dialogical approach seems to work in the Grand Dialogue, but not in the other two case studies.

The critical approach is relevant in helping facilitators to gain awareness of power dynamics. In response to these realities, the facilitation team can create subgroups or distribute speaking opportunities more equitably. In Borée and the Mobility Forum, it is necessary to distribute speaking opportunities so that each actor can express their position. This critical approach is applied intuitively rather than in a structured manner.

The limitation of these interventions by the research-intervention team is that facilitators are not present at all times. Power relations continue when researchers withdraw, for example, when they are not members of the boards of directors of the new governance instances that are created.

8.4 Action

There is an ambient discourse according to which researchers do not act—and that wishes are not compatible with action. However, our approach to research intervention aims to defend the position that reflection is compatible and synonymous with action. In fact, while many stakeholders are reflecting, other actors within these organizations are very active on a day-to-day basis. Of course, there is a difference in timing between action and reflection. The latter occurs much more slowly but still provides a general frame of action over a lasting period.

Nevertheless, actors must keep in mind that the alignment of transition niches comes from reflexivity. Reflection does not hinder action; it optimizes the potential for alignment. Alignment ensures that stakeholders bring together different models that do not initially work as one. Within Borée, organic farming, permaculture, and local agriculture came under the same banner to transform the food system. When they do not participate in the development of a common vision, actors tend to perceive their differences instead. Reflection allows actors to accelerate the transformation of systems by uniting different ideas toward a common vision. Reflective insight does not prevent action.

8.5 Further Challenges

Among the further issues and challenges that have been identified, at least four are noteworthy. First, conflicts can arise. They are transversal to ownership, leadership, action, and power. Practitioners-researchers need to mediate those conflicts or find a way around them before they arise.

Secondly, in the three case studies, we have observed a weak level of involvement by political and private actors. These initiatives involved actors who are ready to move toward a sustainable transition rather than trying to convince those who are not. Civil society is ready to extensively engage in consultation mechanisms, but has somewhat moved forward in the process while elected officials have not been mobilized. The idea of being elected to govern without consulting is cultural. The private sector operates within a slightly different reality. Its absence in these initiatives is related more to the habitual practice of respecting a legal framework rather than contributing to its co-construction in third spaces. Researchers focus on the process, openness, and inclusivity in order to someday mobilize elected officials and private parties.

Thirdly, a huge challenge is overcoming the inertia of systems. Culture itself is a form of inertia. Attempting to plan for social and ecological transition must be a bottoms-up endeavor, because for the transition to succeed, it must be desired by those who will experience it and benefit from it. However, this requires a cultural change. The strategy of the three initiatives we described is original in this respect, because it involves back-casting. The initiators of these initiatives first created a vision, and then mobilized people and actors to find ways to achieve this ideal.

Likewise, urban planning suffers from inertia because urban infrastructures are difficult to change. However, what is possible in that sphere is to avoid repeating the urbanistic pitfalls of the past decades. To this end, a long-term vision will serve as a compass.

Fourthly, aligning innovative social and ecological transition initiatives with existing institutions represents a significant challenge. We can view this issue as a "transition-institution gap." Social-ecological movements must find their niche in the existing space by aligning with institutions. Finding one's niche within existing institutions creates the possibility of not constantly remaining in an oppositional stance.

The Mobility Forum first identified gaps in the ecosystem, in collaboration with the actors. New governance structures were then designed. Conversely, the GD and Borée have no legal existence, which gives them the freedom to find their way through and within existing institutions.

9 Conclusion

In conclusion, as humanity grapples with the realities of the Anthropocene, characterized by uncertainty and social-ecological instability, it is imperative to acknowledge our collective responsibility in shaping the trajectory of our planet and of society. Despite decades of awareness and efforts toward ecodevelopment and sustainable development, the pace of societal development continues to surpass planetary boundaries, resulting in escalating environmental degradation and deepening socio-economic inequalities.

The concept of social-ecological transition offers a radical paradigm shift toward more resilient and responsible ways of life. Drawing from interdisciplinary perspectives and innovative governance structures, transition initiatives aim to catalyze transformative breakthroughs at various territorial scales. However, systemic barriers persist, hindering the full realization of these initiatives.

By studying and implementing transition arenas, such as Borée for a sustainable food system, the Grand Regional Dialogue for Transition, and Forum Mobilité for sustainable transportation, actors in the Saguenay-Lac-Saint-Jean region are pioneering cross-sectoral coordination toward a social-ecological transition. Through a deliberate approach of research-intervention aimed toward social-ecological transformation, these initiatives not only improve practices, but also extract valuable knowledge, contributing to the global discourse on effective transition mechanisms. Clarifying roles, producing common and shared knowledge, redistributing power relations, and aligning niches were observed to be essential tools in the making of sustainability transitions. The weak level of participation by private and political actors, the inertia of systems, and the coexistence of transition initiatives with existing institutions are among the challenges targeted for further investigation.

References

Abson, D. J., Fischer, J., Leventon, J., Newig, J., Schomerus, T., Vilsmaier, U., et al. (2017). Leverage points for sustainability transformation. *Ambio, 46*, 30–39.

Albarello, L. (2004). *Devenir praticien-chercheur: Comment réconcilier la recherche et la pratique sociale* (1th ed.). De Boeck.

Anadón, M., & Association canadienne-française pour l'avancement des sciences. (2007). *La recherche participative: Multiples regards*. Presses de l'Université du Québec.

Arnsperger, C. (2009). *Éthique de l'existence post-capitaliste: Pour un militantisme existentiel*. Cerf.

Barnaud, C., d'Aquino, P., Daré, W. S., & Mathevet, R. (2016). Dispositifs participatifs et asymétries de pouvoir: Expliciter et interroger les positionnements. *Participations, 3*, 137–166.

Bergman, N., Whitmarsh, L., & Kohler, J. (2008). *Transition to sustainable development in the UK housing sector: From a case study to model implementation* (p. 39). Tyndall Center for Climate Change.

Beuret, J. E., & Cadoret, A. (2010). *Gérer ensemble les territoires: Vers une démocratie coopérative* (Vol. 178). ECLM.

Bodin, Ö., Crona, B., & Ernstson, H. (2006). Social networks in natural resource management: What is there to learn from a structural perspective? *Ecology and Society, 11*(2), 1–8.

Bonneuil, C., & Fressoz, J. B. (2013). *L'événement Anthropocène: La Terre, l'histoire et nous*. Média Diffusion.

Callon, M., Lascoumes, P., & Barthe, Y. (2001). *Agir dans un monde incertain: Essai sur la démocratie technique*. Le Seuil.

Cenzano Vilchez, G., Marcel, J. F., & Aussel, L. (2022). La recherche-intervention: Une démarche participative "sur" et "pour" le pilotage intermédiaire des établissements scolaires. *Canadian Journal of Educational Administration and Policy, 199*, 34–46.

Delplancke, M., Picard, S., Patillon, C., Kervarrec, M., & Vimal, R. (2021). Transition écologique: Du défi scientifique au défi pédagogique. *VertigO, 21*(3), 1–30.

Descola, P. (2005). *Par-delà Nature et culture*. Gallimard.

Farinós Dasí, J. (2009). Le défi, le besoin et le mythe de la participation à la planification du développement territorial durable : À la recherche d'une gouvernance territoriale efficace. *L'information Géographique, 73*, 89–111. https://doi.org/10.3917/lig.732.0089

Gagnon, C. (2006). L'Agenda 21 local: Un outil de développement durable et viable, sous-utilisé par les collectivités territoriales québécoises. In M. Robitaille, J.-F. Simard, & G. Chiasson (Eds.), *L'Outaouais au carrefour des modèles de développement* (pp. 133–143). Gatineau, CRDT, CRDC, UQO.

Geels, F. W. (2011). The multi-level perspective on sustainability transitions: Responses to seven criticisms. *Environmental Innovation and Societal Transitions, 1*(1), 24–40.

Granovetter, M. (1973). The strength of weak ties. *American Journal of Sociology, 76*(6), 1360–1380.

Grin, J., Rotmans, J., & Schot, J. (2011). *Transitions to sustainable development: New directions in the study of long term transformative change*. Routledge.

Haché, E. (2014). L'Anthropocène et la destruction de l'image du Globe. In E. Haché (Ed.), *De l'univers clos au monde infini* (pp. 27–54). Éditions Dehors.

Huybens, N. (2011). Comprendre les aspects éthiques et symboliques de la controverse socio-environnementale sur la forêt boréale du Québec. *VertigO, 11*(2), 11119.

Keeling, M. (2005). The implications of network structure for epidemic dynamics. *Theoretical Population Biology, 67*(1), 1–8.

Leinhardt, S. (1977). *Social networks: A developing paradigm*. Academic Press.

Leloup, F., Moyart, L., & Pecqueur, B. (2005). La gouvernance territoriale comme nouveau mode de coordination territoriale ? *Géographie, Économie, Société, 7*, 321–332.

Lessard, C. (2021). Diriger un établissement scolaire: Un leadership plus affirmé, mais multiforme. In *Dans Les directions des établissements au cœur du changement* (p. 285). De Boeck Supérieur.

Létourneau, A. (2008). La transdisciplinarité considérée en général et en sciences de l'environnement. *VertigO—la revue électronique en sciences de l'environnement, 8*, 2.

Lhotellier, A., & St-Arnaud, Y. (1994). Pour une démarche praxéologique. *Nouvelles Pratiques Sociales, 7*(2), 93–109.

Loorbach, D. (2007). *Transition management: New mode of governance for sustainable development* (p. 327). International Books.

Marcel, J. F. (2010). Des tensions entre le "sur" et le "pour" dans la recherche en éducation: Question (s) de posture (s). *Éducation et Socialisation Les Cahiers du CERFEE, 27–28*, 41–64.

Marcel, J.-F. (2018). Esquisse du portrait d'un-e chercheur-e du changement. In *Dans Recherche(s) et changement(s): Dialogues et relations* (pp. 171–188). Cépaduès Éditions.

Marcel, J.-F. (2020). Fonctions de la recherche et participation: Une épistémo-compatibilité dans le cas de la recherche-intervention. *Questions Vives. Recherches en éducation, 33*, 4691. https://doi.org/10.4000/questionsvives.4691

Max-Neef, M. A. (2005). Foundations of transdisciplinarity. *Ecological Economics, 53*, 5–16.

Meadows, D. H., & Delaunay, J. (1972). *Halte à la croissance*. Fayard.

Morin, E. (1999). *Le défi du XXIe siècle, relier les connaissances*. Le Seuil.

Nicolescu, B. (1996). *La transdisciplinarité. Manifeste*. Éditions du Rocher.

Portelance, L., & Giroux, L. (2009). La problématisation dans un processus de recherche collaborative. *Recherches en Education, 6*, 95–108.

Raworth, K. (2017). Why it's time for doughnut economics. *IPPR Progressive Review, 24*(3), 216–222.

Riffon, O. (2016). Une typologie pour l'analyse des représentations du développement durable des instruments de mise en oeuvre à l'échelle territoriale. In J.-C. Némery & F. Thuriot (Eds.), *Les instruments de l'action publique et les dispositifs territoriaux* (pp. 43–58). L'Harmattan, coll. Administration et Aménagement du territoire.

Sachs, I. (1993). *L'écodéveloppement*. Syros.

Scholz, R. W. (2001). *Environmental literacy in science and society: From knowledge to decisions*. Cambridge University Press.

Schön, D. A. (1994). *Le praticien réflexif: À la recherche du savoir caché dans l'agir professionnel*. Logiques.

Segers, I. (2014). *Dialogue, éthique et développement durable pour la pratique de l'éco-conseil*. Université du Québec à Chicoutimi.

Steffen, W., Persson, Å., Deutsch, L., Zalasiewicz, J., Williams, M., Richardson, K., et al. (2011). The Anthropocene: From global change to planetary stewardship. *Ambio, 40*, 739–761.

Theys, J. (2002). Les approches territoriales et sociales du développement durable. *La revue de la CFDT, 48*, 3–13.

Tremblay, L. (2011). *Gouvernance des transitions vers la durabilité*. Centre universitaire de formation en environnement, Université de Sherbrooke.

Turcotte, M.-F., & Caron, M.-A. (2018). *La transdisciplinarité et l'opérationnalisation des connaissances scientifiques*. Éditions JFD.

Villeneuve, F., & Riffon, O. (2022). Mise en place d'un projet d'espaces collaboratifs de transformation alimentaire au Saguenay–Lac-Saint-Jean: Une action du CRRASA-CCLF inscrite dans la planification stratégique de la stratégique de la démarche Borée. *Revue Organisations & Territoires, 31*(3), 25–41.

Villeneuve, C., Tremblay, D., Riffon, O., Lanmafankpotin, G. Y., & Bouchard, S. (2017). A Systemic Tool and Process for Sustainability Assessment. Sustainability, 9(10), 1909. https://doi.org/10.3390/su9101909

Waridel, L. (2019). *La transition, c'est maintenant. Choisir aujourd'hui ce que sera demain* (p. 376). Éditions Écosociété.

Wittmayer, J. M., & Schäpke, N. (2014). Action, research and participation: Roles of researchers in sustainability transitions. *Sustainability Science, 9*(4), 483–496.
Wittmayer, J. M., Schapke, N., Steenbergen, F. V., & Omann, I. (2014). Making sense of sustainability transitions locally: How action research contributes to addressing societal challenges. *Critical Policy Studies, 8*(4), 465–485.

Olivier Riffon is Professor of Eco-Consulting at the Université du Québec à Chicoutimi. A Ph.D. in regional development, he specializes in tools and approaches for implementing sustainable development, social and ecological transition, education for sustainable development, and the integration of participatory and collaborative methods into social and ecological transition processes.

Simon Tremblay has studied stream restoration and participatory mechanisms at Université du Québec à Chicoutimi. He worked for the Grand Dialogue in 2022 and 2023 as a practitioner and researcher. He teaches human geography at Collège d'Alma since 2024.

Chapter 5
Chemins de Transition: An Innovative Method of Knowledge Mobilization to Accelerate the Socio-Ecological Transition in Quebec

Myriam Kayser-Tourigny and Franck Scherrer

Abstract This chapter explores "Chemins de transition," an initiative that leverages its position within a knowledge institution to expedite socio-ecological transitions. Recognizing the potential of universities to contribute beyond their traditional roles, the project addresses pressing ecological challenges such as biodiversity loss, resource scarcity, and climate change. The chapter details the initiative's origins, theoretical foundations, and a foresight-based methodology designed to mobilize over a thousand experts and stakeholders. This approach aims to develop new strategic planning tools for anticipatory governance and adapt them to societal needs. The project's unique method combines academic, professional, and experiential knowledge to create a concrete narrative for transformation. Key aspects include the development of a "transition arena," scenario planning, and participatory workshops that collectively envision desirable futures and outline trajectories to achieve them. The chapter also discusses the project's learning outcomes, challenges, and success factors, emphasizing the critical role of participatory foresight in driving systemic societal change. This innovative approach has informed, trained, and engaged thousands of individuals and organizations across Quebec, demonstrating the importance of interdisciplinary collaboration and long-term strategic planning in achieving sustainable socio-ecological transitions.

Keywords Sustainability transition · Socio-ecological transition · Foresight · Knowledge mobilization · Transition management

M. Kayser-Tourigny · F. Scherrer (✉)
Université de Montréal, Montréal, QC, Canada
e-mail: myriam.kayser-tourigny@umontreal.ca; franck.scherrer@umontreal.ca

M. Cheriet et al. (eds.), *Accelerating the Socio-Ecological Transition*,
https://doi.org/10.1007/978-3-031-82896-6_5

1 Introduction

The *Chemins de transition* initiative was born from the compelling feeling that knowledge institutions, such as universities or museums, could play a direct role in accelerating the socio-ecological transition, beyond their traditional mission of education and research. As with many other projects mentioned in this book, the initial observation is the same: ecological disruptions such as the massive loss of biodiversity, the scarcity of natural resources, and climate change are evolving at an unprecedented pace in the history of humanity. If we do not want to fully endure the impact of these disruptions and bring the development of our societies back within planet boundaries (Steffen et al., 2015), it requires rapid, profound, and systemic societal transformations.

However, these major changes in the ways of eating, housing, producing, consuming, moving, etc., are of such magnitude that they will take at least a generation to unfold. The problem posed to collective action is at once disorienting, unprecedented, and paradoxical: Disorienting because the dizzying nature of the trajectories to follow now, whether to achieve decarbonization targets or slow down the erosion of biodiversity risks to discourage our ability to act; unprecedented, although this is not the first transition that our societies have experienced, it is the first time that the state of our knowledge, particularly scientific, allows us to grasp in advance its full systemic scope; paradoxical because the urgency to act collides with the very uncertain nature of the future to be achieved.

Mobilizing knowledge to create a concrete and intelligible narrative of these forthcoming deep transformations in Quebec society (Bai et al., 2016); developing new strategic planning tools for anticipatory governance of collective action (Muiderman et al., 2020); adapting them to meet the needs of the key forces of society ready to engage in the transition (Gonzalez-Porras et al., 2021): these are the three missions that Chemins de transition has set in response to these challenges. To achieve this, Chemins de transition developed a unique social innovation of its kind in Canada, which made it possible to mobilize the knowledge of more than a thousand academic experts, professionals, citizen activists, gathered to tackle three major challenges of the transition, using a new foresight method adapted for large-scale changes. Since this first phase, developed between 2020 and 2022, several thousand individuals and public and private organizations throughout Quebec have been informed, trained, or have adopted the tools of *Chemins de transition*.

We will first present the context of emergence, the inspirations, and the theoretical framework in which this social innovation was designed. We will then present the nature of the collective needs and expected impacts that guided the implementation of the *Chemins de transition* approach. This approach will then be presented in its main stages, focusing on methodological and organizational aspects. Finally, we will conclude with the main learnings, challenges encountered, and the conditions for success which made it possible to successfully complete the first phase of this unique project.

2 The Context of Emergence and Ideation

Chemins de transition project emerged at the end of the 2010s within a dual context: the collective awareness of the systemic nature of ecological upheavals as well as the importance of trajectories of change to bring human activities within planetary limits was accelerating at the World level. The clearest testimony to this is provided by the grouped publication over a few years of several IPCC reports, such as the one; on land use in 2019 (IPCC, 2019) or on adaptation to climate change in the sixth Report which more or less closely link the climate question to that of biodiversity (IPCC, 2023a); but also a new generation of climate scenarios which model more than before the societal transformations necessary in the long term, such as the common trajectories of socio-economic evolution of the IPCC's sixth report (IPCC, 2023b).

Quebec is part of this global movement recognizing the systemic, long-term, societal, and technological scope of actions to be taken in the face of ecological issues. However, this recognition is still uneven across different actors and layers of society (Romdhani & Audet, 2022). While the term "socio-ecological transition" can serve as a marker of this awareness, its use is then limited to citizen and activist movements, following more or less directly in the footsteps of the Rob Hopkins Transition Network (Hopkins, 2008). It also appears in the titles of certain pioneering research chairs or programs, which are isolated within their academic communities. The dominant frame of reference of public policies, of companies and institutions, including universities, remains largely that of sustainable development, which has been undermined around 2018–2019 by a wave of youth activists advocating for climate emergency action.

The lack of academic, political, discursive, and methodological benchmarks on the socio-ecological transition in Quebec at the end of the 2010s contrasts with other countries, particularly in Europe, where the transition is mobilized by actors, institutionalized, and sometimes instrumentalized by public authorities (Mazeaud, 2021; Semal, 2017). This discrepancy is one of the reasons that led to the creation of Chemins de transition, but it also presented an opportunity to mobilize knowledge on the transition in a more rigorous manner, as well as to transfer it in the most open way possible, under the academic freedom afforded by the university.

More broadly, the research community on sustainable development and climate change shares the observation that it is not so much the lack of knowledge, but the effectiveness of its mobilization that constitutes one of the major obstacles to taking action. For example, in 2019, the Global Sustainability Strategy Forum (GSSF) attempted to address the fact that, after around 40 years, sustainability science has produced many ideas but has not significantly influenced our collective behavior concerning its impact on the environment (Bai et al., 2019). The researchers gathered in this global forum highlighted the significant issue of the lack of communication between scientists and the outside world. Among the eight main recommendations formulated by the forum, three directly inspired the Chemins de transition initiative: "*increase efforts in collecting, unpacking, and supporting narratives for collective behavior change towards sustainability,*" "*initiate and organize processes for co-designing transformations to sustainability with stakeholders,*" and "*place more effort on developing systemic approaches in designing economic and political interventions.*"

The first objective is thus to mobilize knowledge to create a concrete narrative in support of the transition (Jerneck, 2014). However, no single scientific expert can provide definitive evidence on the future of a socio-ecological transition from their expertise alone. This knowledge is not only widely dispersed across various scientific disciplines, but also includes professional knowledge from the field, as well as experiential knowledge to capture weak signals on the evolution of our society.

However, the challenge goes beyond the effective mobilization of knowledge about the future and requires the development of key capacities among society's actors to best understand the complexity of the transition. This includes the ability to imagine the future, learn to navigate a rapidly changing environment, and think more systemically. Finally, it is crucial to share this knowledge effectively: for such complex subjects, conferences are not sufficient. The content and animation approaches should to be adapted to reach each type of actor in society, because everyone is affected by the transition.

Various organizations in different countries, such as foundations, think tanks, and public or parapublic agencies, already carry out all or part of these missions. It is rarer for a university to commit its own financial resources to institutionally support an initiative primarily intended to serve the actors of society, beyond its traditional missions of research and training.

To secure such a commitment from the management of the Université de Montréal, Chemins de transition benefited from four key conditions at its inception: a favorable space for the incubation of new projects, The Innovation Laboratory, whose objective is to change the culture and practices in teaching and research to respond to complex societal challenges and promote transdisciplinarity; the interest of its initial partner, the Espace pour la vie museum complex; a positive experience with the prospective method in a previous knowledge mobilization approach, which led to the development of the Montreal Declaration on Responsible AI (Dilhac et al., 2018); and finally, the experience that the future team had just accumulated by taking up the challenge within the EDDEC institute of making Quebec's active forces understand and adopt an equally disruptive concept, that of the circular economy.

3 A Theoretical Framework: Managing the Socio-Ecological Transition

> Faced with the scale of ecological upheaval, the question is no longer whether we are heading towards a profoundly different society, but whether this transition will be entirely undergone or at least partially chosen.

This statement served as the starting point for Chemins de transition and continues to encapsulate the project's philosophy. It conveys the belief that we can influence the direction of the impending profound transformation, although this requires acting within the framework of a complex societal system and being able to anticipate a long-term future that is difficult to define. The first dimension pertains to the field of transition studies, particularly the theoretical framework of transition management. The second dimension is rooted in the contemporary revival of the prospective approach, incorporating renewed methods and implementation modalities.

Within transition studies, a model stands out and continues, after more than 20 years, to be a reference for transition researchers. This is the model of Geels (Geels, 2002, 2011; Loorbach et al., 2017) on the multilevel perspective. This model allows for the complexity of socio-ecological systems to be modeled by demonstrating that transition is driven by a set of simultaneous transformations at different levels involving a multitude of actors. Geels identifies three levels: the sociotechnical landscape, the sociotechnical regime, and niche-innovations. The first concerns variables that are beyond the control of society, such as demography or ecological upheavals. The second represents the set of norms (economic, cultural, political, etc.) that govern society at a given moment. Finally, niche-innovations represent fringe ideas that could become societal norms if they gain momentum. This concept can also be linked to that of weak signals or emerging variables in the foresight approach.

In this model, the transition represents the passage from one sociotechnical regime to another. This passage, or transition path, is caused by landscape changes that create opportunities for transformation, which can be fueled by innovations. An amalgamation of innovations can thus gain importance in shaping the new sociotechnical regime.

This model has faced repeated criticism, but it persists because no alternative has yet replaced it. Among these criticisms is the emphasis on innovations as the primary driver of transition, to the detriment of cultural and social changes. Additionally, it fails to account for the agency of the various actors involved (Audet, 2015). Despite these shortcomings, it serves as a significant, albeit distant, theoretical basis for the Chemins de transition approach.

The connection is more direct with the application framework of transition management developed by Kemp and Rotmans (Rotmans et al., 2003) to create management tools aimed at guiding the transition toward sustainability and fostering collective action. The objective of transition management is to promote innovation to accelerate the transformation of the sociotechnical regime. It proposes an iterative cycle in four phases. First, it involves creating and developing a "transition arena." This arena comprises researchers and diverse stakeholders, including pioneers in their respective sectors. The role of the arena is to build medium- and long-term visions of the future (25–50 years) and then develop transition paths, summarized in a transition agenda. This agenda leads to a third phase where transition experiments are launched and developed. Finally, this process is continuously evaluated to allow for adjustments along the way.

This framework remains very general, but it is useful for linking the theory of societal system transitions to numerous field initiatives. Chemins de transition utilized it as an intelligibility framework to identify which phases were most useful for developing new tools to serve potential transition arenas in Quebec. The downstream phases of experimentation and continuous evaluation are perhaps better documented and supported today, particularly in Quebec (Van Neste et al., 2024), while the upstream phases are much less so, both in academic literature and in terms of best practices. This is especially true for the phase where the transition arena develops a vision of the desirable future and outlines the transition paths to achieve it. Our objective was to invest in this phase, where the methodological shortcomings in foresight were particularly glaring in Quebec.

During the first phase of developing and testing the Chemins de transition method across three major challenges (digital, food system, territory), we essentially simulated a transition arena, albeit with important nuances compared to the traditional transition management model. On the one hand, from the perspective of knowledge mobilization, our primary objective was the diversity of knowledge rather than the diversity of stakeholders per se. However, we realized that achieving one necessitated the other. Furthermore, we broadened the diversity of participants beyond researchers and professionals to include other audiences, notably citizens, from this early stage, not only during the downstream experimentation phase. This is a learning experience that we were able to maintain during the transfer phase of our tools and content to real transition arenas, such as Collectivités ZéN (Projet Québec, 2020). This learning also reflects the parallel evolution of foresight, which has increasingly become a more participatory approach.

3.1 Foresight at the Service of the Transition

It was in the twentieth century that the major methods of anticipating the collective future were developed, along with what we can call the principles of anticipatory governance (Muiderman et al., 2020). The methods that have been most successful with governments and large organizations are those that anticipate foreseeable or probable futures. These methods are based on mathematical models that simulate the future evolution of a given situation in advance. Such models are pervasive in our societies, particularly in guiding the fight against climate change, from the scenarios established by the IPCC to the various decarbonization trajectories developed by public agencies worldwide. These methods rely on the scientific measurement of evolution determined by precise initial conditions, essentially predicting the future based on the present.

Simultaneously, other approaches to anticipation emerged, such as foresight, which explore the realm of possible futures. What these approaches have in common is their reliance on the richness of human imagination, enabling the creation of alternative visions of possible futures, which can be more or less radically different from the present. Ultimately, the foresight approach aims to illuminate present actions based on a vision of the future (de Jouvenel, 1999) (Fig. 5.1).

In a context of economic growth and stability of the socio-technical system, probabilistic methods, which the foresight theorist Riel Miller describes as "anticipation for the future," where the planned future is a goal on which a bet is made, meet most societal needs. However, in a context of profound transformation of the socio-technical system, "anticipation for emergence," which allows us to perceive and think about emerging novelties that would otherwise remain invisible, becomes more essential (Miller, 2018). Other authors emphasize the complementarity between these two major models of anticipation in collectively addressing the complexity of the socio-ecological transition (Muiderman et al., 2020). They particularly highlight the value of "qualitative" foresight to reveal the limitations of solutions based on projecting the present. This approach also shifts the focus from a purely technical perspective to exploring the social and behavioral nature of the transition. Furthermore, qualitative foresight helps to build a common language and make the transition more tangible for all stakeholders involved (Saujot & Waisman, 2020). Finally, it encourages the strengthening of knowledge and breaking out of disciplinary silos by promoting an interdisciplinary and collaborative approach (Saujot et al., 2020).

Another movement in our societies promoting alternative modes of anticipating the future comes from the growing demands of activist circles, citizens, and others to unleash the power of future imagination and reappropriate it through collective narratives. These narratives, developed in a highly participatory and inclusive manner, express an accessible and desirable vision of the transition at the local level. This approach is one of the manifestations of future literacy (Miller, 2015), which UNESCO defines as "the skill that allows people to better anticipate the future to understand the role it plays in what they see and do today." For UNESCO, developing this skill in the most democratic way possible is essential in the twenty-first century, when the future is particularly uncertain (Jennische & Sörbom, 2023).

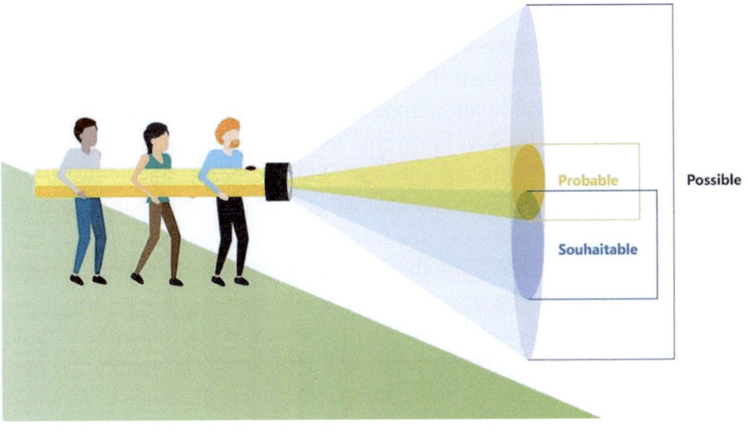

Fig. 5.1 The futures Cone of Chemins de transition, according to Voros (2017)

Finally, the concept of transition significantly enriches the role and importance of foresight. Beyond its traditional purposes of anticipating risks associated with an uncertain future or providing upstream vision for strategic planning, transition foresight aims to liberate imagination from a social and economic system locked in its routines, initiate collective reflection on future potential to make it tangible, and restore the capacity to act. It is from this perspective that Chemins de transition sought to evaluate how to mobilize the foresight approach in Quebec to support collective action.

3.2 A Social Innovation, for Whom and for What?

To develop and continuously evaluate this social innovation, we relied on a development evaluation tool: The Theory of Change (Anderson, 2006). This conceptual and methodological tool is used in development, strategic planning, program evaluation, and leadership. Its purpose is to describe and visualize how and why change should occur in each system, as well as the steps and conditions necessary for this change to happen.

In practice, a Theory of Change articulates the assumptions about how a program or social innovation is expected to produce desired outcomes, whether they are social, economic, political, or environmental. It identifies drivers of change, key interventions, and causal mechanisms that link these interventions to the expected outcomes.

We had to define how this social innovation could respond to specific needs identified within society. We started from the following postulates: First, Quebec society, like other Western societies, finds itself needing to increase the number of its actors involved in the transition. The very notion of socio-ecological transition remains an elitist concept, lacking a collective understanding by the public. Additionally, a certain weariness and pessimism can be perceived among "transitioners" who go through phases of doubt about their ability to shape a desirable future. Therefore, it was necessary to identify ways to overcome this inertia.

To do this we first established that, the development of skills such as systemic thinking and long-term anticipation appeared to be crucial, especially in a culture that favors short-term actions and prevents us from grasping the transformational scope needed to adapt to tomorrow's world. On the other hand, the distressing perception of the urgency to act in the face of ecological upheavals can hinder the necessary perspective. Furthermore, the tools required to guide long-term collective action were poorly available, and there were sometimes excessive expectations placed on science to make the transition affordable and to guide priority actions.

A third need also became apparent: the promotion of interdisciplinarity and collaboration. Although the knowledge was available, it was scattered and not easily accessible to transition actors on the ground. Structures for inter-stakeholder consultation on the transition are also insufficiently present, even though they are essential for establishing a common vision of the transition and the measures to be adopted. This context also does not facilitate discussion around difficult yet crucial subjects for collective progress. Instead, we observe actors passing on the responsibility and waiting for others to take the first step.

Finally, it was necessary to embrace uncertainty, which requires implementing mechanisms for continuous adaptation. The complexity of the transition makes any prediction impossible, thus necessitating the development of consultation and knowledge-sharing tools to respond effectively to future societal transformations.

To meet these needs, Chemins de transition has developed a participatory foresight method in several stages, which we tested on three major transition challenges across the province of Quebec: "The Digital Challenge", "The Food Challenge", and "The Territory Challenge".

4 Choosing the Challenges

We were fortunate to be able to experiment with our method without constraints imposed by public authorities, which allowed us great freedom in our approach and enabled us to determine the three major challenges we focused on. These challenges were formulated as questions, with a horizon of 20–25 years to consider the transformations they required. These challenges were considered on the scale of Quebec society, whose national coherence is strong enough to trace a transition trajectory that is both autonomous and socially situated.

The first challenge we addressed was to integrate the digital transition with the socio-ecological transition (Deron & McDonald, 2022). At the time, this was by far the least documented area in Quebec. While some work had been done in Europe on the environmental footprint or responsibility of digital technology, we not only had to adapt this knowledge to our context but also go further. Public opinion, lacking this knowledge, swung between an almost absolute belief in technological solutions (which would save the planet) and activist calls for more or less enforced digital sobriety, with a large segment of the population remaining ignorant in between. The collective contributions gathered by Chemins de transition illustrated that the potential of digital innovation can indeed be aligned with greater attention to carbon and resource sobriety in digital technology. This sector has the potential to accelerate the ecological transition, provided that the environmental gains enabled by digital technology are not offset by its impacts in terms of GHG emissions, resource use, and energy consumption. This requires collective decisions on essential uses, the direction of innovation, and the economic models guiding its development.

The second challenge sought to answer the question: How can we accelerate the socio-ecological transition of Quebec's food system by 2040 (Henry, 2023)? Unlike the first challenge, this question was already well documented: the measures to be taken, both in terms of individual behaviors and collective actions, as well as the organization of the food system, to reduce our environmental footprint while maintaining social justice objectives, are generally well known. The main challenge lies in the implementation of these measures and the complexity of the food system itself. The added value here was making this systemic nature more accessible so that all stakeholders could find their place in the proposed trajectory.

The third challenge focused on ways of inhabiting the territory in a more sober and resilient manner, a challenge that was the most complex to address (Verdun & McDonald, 2023). The occupation of a territory is not a "system" in itself; it is the result of the projection of other socio-economic systems of production, consumption, housing, transport, etc. Therefore, removing barriers to the use of the territory to make it more compatible with planetary limits requires embracing the entire socio-economic system. Another problem is that changes in a spatial system (locations, spatial forms, infrastructures…) occur in the long term, while the impacts of climate change and the transformation of ecosystems will increasingly and rapidly affect it. The scale of the challenge did not prevent the 500 participants in the process from demonstrating extraordinary collective intelligence to tackle it.

5 The Foresight Method Designed by Chemins de Transition

The methodology implemented by Chemins de transition uses the foresight approach to revitalize strategic planning tools in light of the challenges of the transition. It is the traditional role of foresight to guide strategic action (Godet & Durance, 1997). Chemins de transition did not innovate through the choice of tools; instead, the tools were selected for their robustness and prior validation. The innovation lies in the combination of two essential aspects of the approach.

Firstly, the foresight approach is explicitly placed at the service of knowledge mobilization (SSHRC, 2019), a now essential activity in scientific research that involves effectively sharing knowledge with recipient communities. The particularity of mobilizing knowledge about the future is that no expertise can claim to hold "the truth" about what will happen. Therefore, it is crucial to move away from a traditional, top-down vision of scientific knowledge transfer toward a non-hierarchical sharing approach between academic knowledge, often inaccessible to field actors, and the professional knowledge essential for capturing transitional innovations and their feedback, as well as citizens' experiential knowledge. Since it is not realistic to summon all this knowledge simultaneously, the approach was designed as a "passing of the ball" between stakeholders, in a climate of trust and transparency.

Then, it is a complete foresight prospective approach (Hines & Bishop, 2013). This may seem standard, as it involves following the steps recommended by foresight manuals, which lead from exploring possible futures to an action plan, through establishing a strategic vision of a desirable future and determining a trajectory to achieve it. However, it is often observed that this expected succession of exploratory and normative phases is rarely present in most documented examples. This may be due to differences in national cultures of foresight use—normative foresight by backcasting (Dreborg, 1996) is more frequent in Northern Europe, while exploratory foresight by scenarios dominates in France (Durance & Cordobes, 2007)—, a lack of time or resources, or, as is often the case in North America, a strict distinction in mandates between experts who provide visions of the future and decision-makers who determine the strategy to achieve them. Chemins de transition had the academic freedom necessary to fully deploy these two phases, exploratory and normative, by subdividing them into several distinct stages, which we will describe here. The entire process took between 18 months and 2 years for our three challenges, but it can now, in our transfer phase, be completed in about a year.

6 Horizon Scanning

The exploration of possible futures is a classic first step in exploratory foresight, characterized by two distinct phases: **Horizon Scanning** and the Scenario development technique through morphological analysis. We systematically repeated this process for each of our challenges.

The Horizon Scanning begins with identifying the "ingredients of the future" (Gabilliet, 1999). This involves, but is not limited to, observations related to the current situation in Quebec concerning relevant issues such as land use planning methods, production methods, consumption habits, and the use of digital technologies. The most important ingredients are major societal trends and weak signals, which are emerging practices or elements currently marginal but potentially significant in the future. These also include key implementation issues or future points of tension, referred to as "nodes of the future." To complete this initial step, we enlisted the help of students from the Université de Montréal to conduct comprehensive literature reviews for each of our three challenges. Based on this, we identified the nodes of the future and developed notebooks listing these various foresight ingredients (Fig. 5.2).

The notebooks were then submitted to discussion groups (ranging from 3 to 6 depending on the challenges). These groups comprised multidisciplinary experts

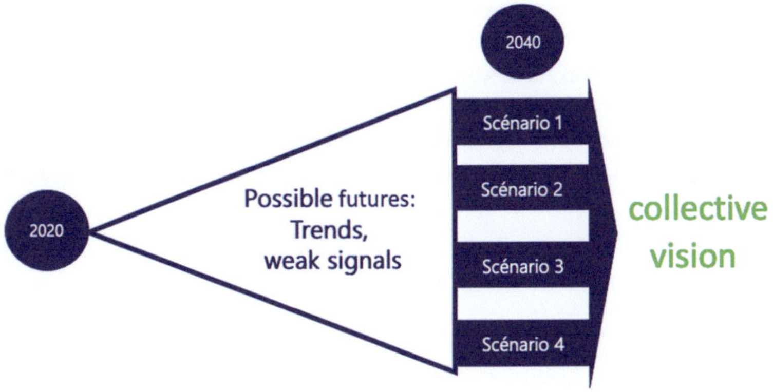

Fig. 5.2 The stage of possible futures

representing the trend areas identified in the literature, as well as professionals from local innovative networks in the transition field. They were selected for their ability to capture weak signals from various forms of emerging innovations on a global scale, thanks to their active participation in these networks. The objective was to enrich the information collected during the Horizon Scanning.

This approach allowed us to gather a broader range of trends, including more marginal ones. By avoiding a direct approach based solely on expert opinion, such as the Delphi method (Linstone & Turoff, 1975), we prevented limiting ourselves to dominant trends in the literature, instead encouraging experts to consider weak signals. We then ensured the collection of knowledge covered all aspects of the challenge. For example, for the challenge related to the territory, around 55 separate areas of expertise were consulted. The Horizon Scanning represents a completely new deliverable in Quebec.

6.1 The Scenario Development Technique

Once the Horizon Scanning has been completed and submitted to the focus groups, we compile the data into a morphological analysis table (Lamblin, 2018; Zwicky, 1969). This table aggregates various societal variables, independent of each other and specific to the challenge studied, such as the economic system, the evolution of values, digital culture, the food production system, and so on. It is best to limit these variables to 4–6 to facilitate understanding for collective intelligence, which generally leads us to group sub-variables into broader categories.

The table integrates four evolution hypotheses for each of these variables, informed by the trends and weak signals identified in the Horizon Scanning projected over the desired time horizon (e.g., 25 years). The next step involves combining evolution hypothesis from each variable. We ensure that all combinations are used only once across four scenarios. By not excluding any possibilities, we ensure that our scenarios reflect the full range of potential futures (Fig. 5.3).

Each scenario describes a possible future in which the socio-ecological transition has taken place, but with different directions each time. We also ensure that these scenarios do not lean too far toward optimism or pessimism but embrace the benefits and tensions of each possible future. There is no trend scenario, to avoid introducing probability bias. The scenarios, being nuanced, illustrate the difficult choices that society had to face, allowing us to depict the critical junctures that had to be resolved.

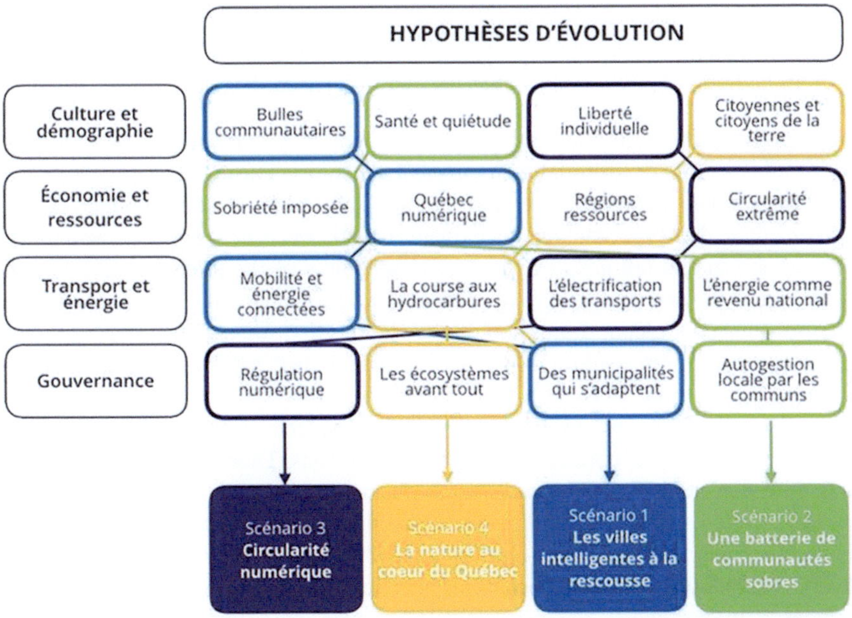

Fig. 5.3 Construction of foresight scenarios of the territorial challenge through morphological analysis

7 Co-construction of the Vision of the Desirable Future

If these first two stages are classic in foresight, the next one is more innovative in its participatory aspects. The scenario frameworks are necessarily dense and complex, and cannot easily be debated, even by experts familiar with the language of foresight. However, debating possible futures is an essential step in arriving at the most collective vision possible of a desirable future. Following a well-established practice of design fiction from the foresight approach (Dunne & Raby, 2024), the combinations brought together in scenario frameworks are translated into short stories. These stories, embodied by characters living anecdotes of everyday or professional life, are much more accessible for debate by a non-expert audience.

Therefore, the next step in the approach was to organize between 15 and 30 future exploration workshops conceived by the Foresight City Lab (Lavoie et al., 2021). Each workshop was dedicated to examining one of the four scenarios, allowing each scenario to benefit from the enriched debate by a diverse group of people from Quebec society. For the digital challenge, we had 150 participants, while this number reached 250 for the food system challenge, and 350 for the territory challenge.

It is at this stage that the approach truly takes on an essential participatory character. Having already benefited from a diversity of expertise during the first phase, we now ensure the collection of experiential knowledge from citizens, ensuring geographic, generational, and professional diversity, among others.

The workshops offer participants the opportunity to free themselves from the constraints of the present and project themselves into the future. They follow a specific formula where the facilitator begins by inviting participants to express what they find desirable or formidable in the presented story. Participants quickly realize they share many points of view and identify consensus. However, points of disagreement also emerge, on which we dwell further. We seek to collect a wide range of testimonies and opinions on the critical junctures of the future, aiming to construct a realistic and desirable vision rather than a utopian dream. In summary, these are moments of in-depth discussion on the choices that society must make to progress toward a desirable future for a Quebec that has successfully achieved its socio-ecological transition.

At the end of these workshops, our project managers are left with a multitude of post-its and comments collected during the discussions. Their task is then to synthesize these elements. This mission is complex, as it is essential to consider all points of view while establishing a relatively concise common vision. This vision is presented in the form of the main achievements of Quebec society of tomorrow, effectively representing the transformations that have become, in a certain way, the standards. In our reports, the vision is detailed over approximately ten pages, but it can be summarized in a few major achievements. These are not simple abstract objectives but rather a portrait of the future state of Quebec society, reflecting the choices made during the future exploration workshops. This vision is then submitted to the participants for feedback. It forms the foundation for the next major phase: the development of trajectories.

8 The Development of Trajectories

Just like the construction of scenarios through morphological analysis, the backcasting approach (Quist, 2016)—which involves tracing a trajectory from a desirable future back to the present—is a classic method in foresight. From a transition management perspective, it involves addressing two key hurdles. Firstly, since the goal is to reach a state of post-transition society, the trajectory must be truly transformational. Secondly, this transformation must be mapped out on the scale of the societal system, while remaining understandable and transferable to various transitional arenas to provide them with useful tools. Finally, developing this trajectory, which is more demonstrative than prescriptive, must respect the trust-based approach of "passing the ball" between stakeholders.

For each challenge, an expert committee of around 10–15 members is tasked with starting from the current situation observations made during the horizon scanning to reach the vision of the desirable future. Submitting the vision to experts who did not participate in its development can be difficult, as they must take ownership of it to engage in the validation process without significantly altering it, to avoid losing key elements that were chosen collectively. To do this, we ask them two questions: is the vision achievable, and is the formulation of future achievements accurate?

The main task of the expert committee is to identify milestones, which are intermediate steps linking the present to the desired future. In addition to the expert committee members, discussion groups are organized to make this list as exhaustive and relevant as possible.

We then challenge our expert committee to interrelate these steps: which milestones are prerequisites for others? Which ones could facilitate another step without being necessary to achieve it? Finally, the expert committee must position these milestones in time, considering not only the steps prior to others but also the degree of difficulty that achieving each milestone represents (Fig. 5.4).

Once the trajectory is established, it undergoes a robustness test. This test assesses how the trajectory withstands potential future events such as accelerated climate effects, water shortages, or waves of climate migration. This step ensures the resilience of the trajectory and allows for adjusting milestones if necessary.

To ensure the relevance and transferability of the trajectory, detailed documentation is maintained throughout the process. This documentation includes

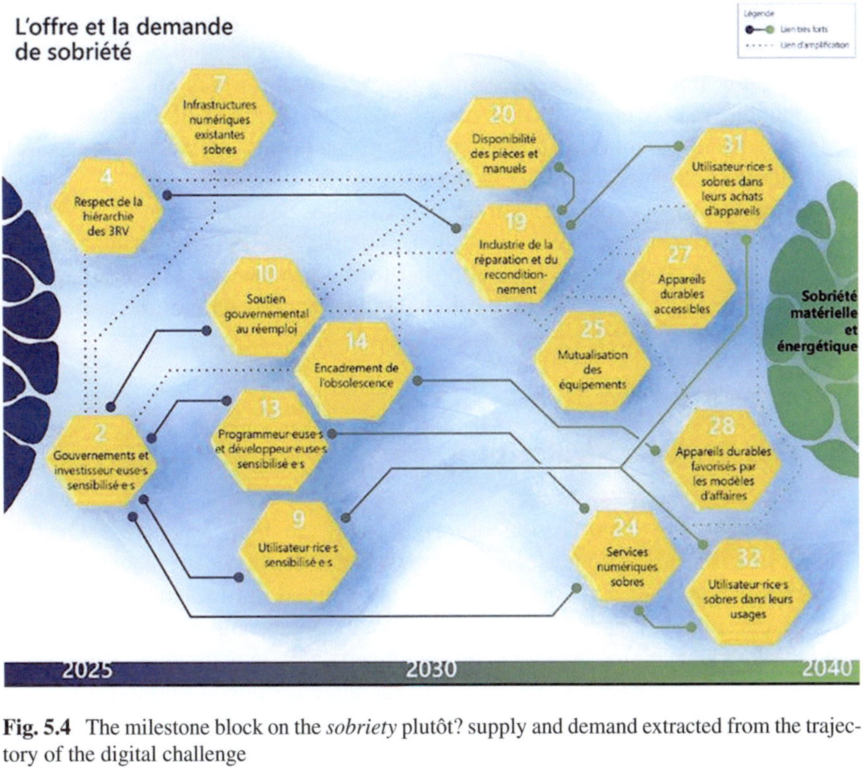

Fig. 5.4 The milestone block on the *sobriety* plutôt? supply and demand extracted from the trajectory of the digital challenge

descriptions of milestones, reasons for their inclusion, and examples of possible solutions. This ensures that all critical information is centralized and easily accessible.

9 The Result

For each of the challenges, we produced a report detailing a trajectory to manage the transition in the area concerned across Quebec. These reports identify between 35 and 60 transformational changes deemed priorities, which we have designated as "milestones," and position them over time. These milestones illustrate not only the scale of the changes needed but also our capacity to act.

Trajectories illuminate the interrelationships between milestones, with some being prerequisites for others, or their achievement influencing the progress of other milestones. Thus, an overall picture emerges, providing a better understanding of the intermediate steps needed to achieve seemingly distant goals, such as fundamental changes in perception. In our trajectories and reports, we also identify the specific roles of different stakeholder groups, such as governments, businesses, investors, and citizens.

Finally, our trajectories identify what we call future nodes. These nodes represent significant tensions between collective priorities, requiring decisions to be made at a given moment. The future nodes may also raise unresolved questions regarding the implementation of certain milestones, for which solutions remain uncertain due to existing societal obstacles. We consider it crucial to identify these nodes in advance, to initiate a collective dialogue on how to approach them, as well as a dialogue with researchers to resolve some thorny issues.

This first phase of experimentation led by Chemins de transition mobilized more than 1,000 people through the three challenges across Quebec. As a result, a significant number of people have been exposed to the process and are familiar with Chemins de transition. This continues today in the second phase of the project, the transfer phase.

10 Our Transfer Strategies

As we saw in the identification of needs, they are immense across Quebec. Considering time and resource constraints, we had to make choices about the audiences for our transfer activities. Early on, we decided not to address the public directly, considering that our partner, Espace pour la Vie, would be able to fulfill this need through educational activities for young people or an upcoming exhibition at the Biosphere (the museum of the environment and transition in Montreal).

Our priority has been to target partners capable of maximizing the impact of Chemins de transition by integrating our advancements into their own initiatives and reaching a wider audience through them. We have designated them as "bridging partners." A typical example is the partnership established with the "Front Commun pour la Transition Énergétique", an alliance between citizen groups, environmental NGOs and other civil society organizations, notably unions and community organizations, to support the "Collectivité ZéN" project in establishing their roadmap toward carbon-neutral and resilient territories. The desired impact was to strengthen the effectiveness of knowledge transfer, increase the capacity for future thinking and systemic thinking, and promote inter-actor collaboration. The main objective of this partnership is to learn how to adapt the Chemins de Transition method to the specific needs of field actors with the progressive aim of granting them total autonomy.

The second strategy targets large public and private organizations. This proved particularly relevant to the digital challenge, where everything needed to be done, especially in terms of relevant information and awareness of digital environmental issues. This involves a wide variety of interventions, from simple awareness conferences to personalized support for organizations wishing to integrate this challenge into their strategic priorities. More and more organizations or networks wish to go further and chart their own path by adopting the entire foresight method, as the "Réseau de recherche en économie circulaire du Québec" is currently doing.

Finally, we chose to also engage with the Université de Montréal ecosystem, particularly by participating in numerous courses across fields as varied as our challenges, culminating in an annual summer school to train future professionals in transition arenas.

11 Learning, Conditions for Success, and Limits

Several conditions played an essential role in the implementation of Chemins de transition. First, the ongoing funding provided by the University of Montreal gave us the necessary means to take the time to develop and test all the stages of social innovation. This allowed us to subsequently seek additional resources for transfer activities. Not being dependent on project-specific funding, we had great freedom in choosing relay partners, enabling us to favor those who could have the most impact on the ground or in their community but who could not afford the services of a foresight agency. Thus, we were able to function as a public service offered by the university to Quebec society. Furthermore, the legitimacy of the university institution provided the necessary credibility to convince stakeholders to participate in this pan-Quebec dialogue on these major transition challenges.

Second, the stability of the team over time was a key factor in fostering a solid appropriation of the project's knowledge and complex methodology. This continuity facilitated coherent actions and reinforced the robustness of the initiative.

We also implemented an impact measurement strategy that promoted continuous improvement of the process. We deliberately chose to delay the start of each challenge by several months, one after the other, and proceeded partly by trial and error to immediately integrate the learning from one challenge to the next. This iterative approach contributed to refining animation techniques and strengthening our capacity to manage multi-actor contexts.

The diversity of people participating in the different phases, in order to mobilize a wide variety of knowledge and points of view throughout the process, is an essential variable. The first phase of the three challenges allowed the mobilization of nearly 100 experts from around 40 different disciplines and half of whom come from the professional world. The future exploration workshops phase brought together 750 people who were citizens or from civil society, respecting the diversity of age and geographical origin, which were the only ones that we could provide. The backcasting phase made it possible to mobilize the varied expertise of 350 people. This participatory dimension also leads to a certain length of the process, almost 18 months, which can prove to be a hindrance. In the method transfer phase we were able to reduce this duration to 1 year.

The transfer phase, from which we have not yet been able to learn all the lessons since it is still ongoing, constituted a new and different challenge. For organizations and collectives that have adopted the entire Chemins de transition method to plan their roadmap to transition, adaptation to needs, context, and scale has generally gone well, allowing for new learning. However, the transfer of capacities to operate the process autonomously remains uneven. It is easier for the downstream backcasting phase and the mobilization of stakeholders, but less so for the construction of foresight scenarios, which remains complex to learn. Furthermore, the organizational fragility of the relay partners, whom we chose primarily for their social relevance or proximity to communities, poses a risk or even an obstacle to the long-term appropriation of the method and content of Chemins de transition.

Mobilization and awareness tools, which take the form of more ad hoc interventions with different audiences, respond to a clear appetite for anything that can help collective action in the face of the socio-ecological transition. However, beyond meeting the needs of how and in what direction to act, the impact of the social innovation represented by Chemins de transition will be measured above all by the capacity of these strategic planning tools for transformational trajectories to ultimately generate action plans that stakeholders will implement.

In conclusion, the ultimate medium-term impact we have formulated for Chemins de transition is to contribute significantly to the maturity of the socio-ecological transition in Quebec. This assumes that enough individuals and organizations will have made profound changes in their values, choices, and ways of operating. The contribution of Chemins de transition could be that these actors mobilize long-term thinking, a capacity for continuous adaptation to a changing context, and resilience in the face of multiple shocks caused by ecological upheavals. Our hope is that approaches like ours will serve to converge the strategies and initiatives of various actors around common objectives and visions, particularly regarding the most complex challenges of the transition.

References

Anderson, A. A. (2006). *The ommunity builder's approach to theory of change: A practical guide to theory development*. The Aspen Institute Roundtable on Community Change.

Audet, R. (2015). Le champ des sustainability transitions: Origines, analyses et pratiques de recherche. *Cahiers de Recherche Sociologique, 58*, 73–93.

Bai, X., Van Der Leeuw, S., O'Brien, K., Berkhout, F., Biermann, F., Brondizio, E. S., Cudennec, C., Dearing, J., Duraiappah, A., Glaser, M., Revkin, A., Steffen, W., & Syvitski, J. (2016). Plausible and desirable futures in the Anthropocene: A new research agenda. *Global Environmental Change, 39*, 351–362.

Bai, X., Begashaw, B., Bursztyn, M., Chabay, I., Droy, S., Folke, C., Fukushi, K., Gupta, J., Hackmann, H., Hege, E., Jaeger, C., Patwardhan, A., Renn, O., Safonov, G., Schlosser, P., Skaloud, P., Vogel, C., van der Leeuw, S., & Zhang, Y. (2019). *Changing the scientific approach to fast transitions to a sustainable world. Improving knowledge production for sustainable policy and practice*. IASS Discussion Paper. https://doi.org/10.2312/iass.2019.018.

De Jouvenel, H. (1999). La démarche prospective. Un bref guide méthodologique. *Futuribles, 247*, 47–68.

Deron, M., & McDonald, M. (2022). *Comment faire converger la transition numérique et la transition écologique au Québec dans un horizon de 20 ans*?. Chemins de transition. https://cheminsdetransition.org/download/7/defi-numerique/1849/rapport-final-defi-numerique-chemins-de-transition.pdf

Dilhac, M. A., Abrassart, C., & Voarino, N. (2018). *Rapport de la Déclaration de Montréal pour un développement responsable de l'intelligence artificielle*.

Dreborg, K. H. (1996). Essence of backcasting. *Futures, 28*(9), 813–828.

Dunne, A., & Raby, F. (2024). *Speculative everything, with a new preface by the authors: Design, fiction, and social dreaming*. MIT Press.

Durance, P., & Cordobes, S. (2007). *Attitudes prospectives: Éléments d'une histoire de la prospective en France après 1945*. L'Harmattan.

Gabilliet, P. (1999). *Savoir anticiper: Les outils pour maîtriser son futur*. ESF.

Geels, F. W. (2002). Technological transitions as evolutionary reconfiguration processes: A multi-level perspective and a case-study. *Research Policy, 31*(8–9), 1257–1274. https://doi.org/10.1016/S0048-7333(02)00062-8

Geels, F. W. (2011). The multi-level perspective on sustainability transitions: Responses to seven criticisms. *Environmental Innovation and Societal Transitions, 1*(1), 24–40. https://doi.org/10.1016/j.eist.2011.02.002

Godet, M., & Durance, P. (1997). *La prospective stratégique*. Futuribles.

Gonzalez-Porras, L., Heikkinen, A., Kujala, J., & Tapaninaho, R. (2021). Stakeholder engagement in sustainability transitions. In *Research handbook of sustainability agency* (pp. 214–229). Edward Elgar Publishing.

Henry, P. (2023). *Comment accélérer ensemble la transition socio-écologique du système alimentaire québécois d'ici 2040?: Rapport final du défi alimentaire*. Chemins de transition. https://cheminsdetransition.org/download/17/defi-alimentaire/2716/rapport-final-defi-alimentaire-chemins-de-transition.pdf

Hines, A., & Bishop, P. C. (2013). Framework foresight: Exploring futures the Houston way. *Futures, 51*, 31–49. https://doi.org/10.1016/j.futures.2013.05.002

Hopkins, R. (2008). *The transition handbook: From oil dependency to local resilience*. Bloomsbury Publishing.

Intergovernmental Panel on Climate Change (IPCC). (2023a). *Climate change 2022—Impacts, adaptation and vulnerability: Working Group II Contribution to the Sixth Assessment Report of the Intergovernmental Panel on Climate Change*. Cambridge University Press.

Intergovernmental Panel on Climate Change (IPCC). (2023b). Annex III: Scenarios and modelling methods. In *IPCC, Climate change 2022—Mitigation of climate change: Working Group*

III Contribution to the Sixth Assessment Report of the Intergovernmental Panel on Climate Change (pp. 1841–1908). Cambridge University Press.

Jennische, U., & Sörbom, A. (2023). Governing anticipation: UNESCO making humankind futures literate. *Journal of Organizational Ethnography, 12*(1), 105–119.

Jerneck, A. (2014). Searching for a mobilizing narrative on climate change. *The Journal of Environment & Development, 23*(1), 15–40. https://doi.org/10.1177/1070496513507259

Lamblin, V. (2018). *L'analyse morphologique, une méthode pour construire des scénarios prospectifs*. Prospective and Strategic Foresight Toolbox, Futuribles . https://www.futuribles.com/wp-json/futuribles/v1/pdf/27976

Lavoie, N., Abrassart, C., & Scherrer, F. (2021). *Imagining the city of tomorrow through foresight and innovative design: Towards the regeneration of urban planning routines?* (pp. 40–54). Transactions of the Association of European Schools of Planning.

Linstone, H. A., & Turoff, M. (Eds.). (1975). *The delphi method* (pp. 3–12). Addison-Wesley.

Loorbach, D., Frantzeskaki, N., & Avelino, F. (2017). Sustainability transitions research: Transforming science and practice for societal change. *Annual Review of Environment and Resources, 42*(1), 599–626.

Mazeaud, A. (2021). Gouverner la transition écologique plutôt que renforcer la démocratie environnementale: Une institutionnalisation en trompe-l'œil de la participation citoyenne. *Revue française d'administration publique, 3*, 621–637.

Miller, R. (2015). Learning, the future, and complexity. An essay on the emergence of futures literacy. *European Journal of Education, 50*(4), 513–523.

Miller, R. (2018). *Transforming the future: Anticipation in the 21st century*. Taylor & Francis.

Muiderman, K., Gupta, A., Vervoort, J., & Biermann, F. (2020). Four approaches to anticipatory climate governance: Different conceptions of the future and implications for the present. *Wiley Interdisciplinary Reviews: Climate Change, 11*(6), 673.

Projet Québec ZéN (2020). *Feuille de route pour la transition du Québec vers la carboneutralité*. https://collectivitezenquebec.org/about/

Quist, J. (2016). Backcasting. In *Foresight in organizations* (pp. 125–144). Routledge.

Romdhani, A., & Audet, R. (2022). Quatre discours de la transition écologique pour la région métropolitaine de Montréal. *Les Contributions de la Chaire de recherche UQAM sur la transition écologique, 21*.

Rotmans, J., Kemp, R., van Asselt, M., Geels, F., Verbong, G., Molendijk, K., & van Notten, P. (2003). Transition management. In *Key to a sustainable society* (p. 243). Koninklijke Van Gorcum.

Saujot, M., & Waisman, H. (2020, February). *For a better representation of lifestyles in energy-climate foresight studies* (Study No. 2).

Saujot, M., Le Gallic, T., & Waisman, H. (2020). Lifestyle changes in mitigation pathways: Policy and scientific insights. *Environmental Research Letters, 16*(1), 015005. https://doi.org/10.1088/1748-9326/abd0a9

Semal, L. (2017). Une mosaïque de transitions en catastrophe. Réflexions sur les marges de manœuvre décroissantes de la transition écologique. *La pensée écologique, 1*(1). https://doi.org/10.3917/lpe.001.0145

Social Sciences and Humanities Research Council. (2019). *Guidelines for effective knowledge mobilization*. https://www.sshrc-crsh.gc.ca/funding-financement/policies-politiques/knowledge_mobilisation-mobilisation_des_connaissances-eng.aspx

Steffen, W., Richardson, K., Rockström, J., Cornell, S. E., Fetzer, I., Bennett, E. M., et al. (2015). Planetary boundaries: Guiding human development on a changing planet. *Science, 347*(6223), 1259855.

Van Neste, S. L., Mele, P., & Larrue, C. (2024). *Transitions socioécologiques et milieux de vie: Entre expérimentation, politisation et institutionnalisation*. Les Presses de l'Université de Montréal.

Verdun, J., & McDonald, M. (2023). *Comment habiter le territoire québécois de façon sobre et résiliente d'ici 2042?: Rapport final du Défi territoire*. Chemins de transition. https://chemins-detransition.org/download/24/defi-territoire/2875/rapport-final-defi-territoire.pdf

Voros, J. (2017). Big history and anticipation: Using big history as a framework for global foresight. In *Handbook of anticipation: Theoretical and applied aspects of the use of future in decision making* (pp. 1–40).

Zwicky, F. (1969). *Discovery, invention, research through the morphological approach*. Macmillan.

Myriam Kayser-Tourigny has a degree in International Studies and has completed her Masters in Environment and Sustainable Development at the Université de Montréal. As part of the latter, she joined the Chemins de transition team and supported the team in a variety of projects. She is particularly interested in the social and democratic challenges of socio-ecological transition.

Franck Scherrer, alumnus of ENS Paris, PhD in urban planning (Institut d'urbanisme de Paris), is a full professor in the Faculty of Urban Planning at the Université de Montréal. He is a former director of the School of Urban Planning and Landscape Architecture and of the Institut Environnement—développement durable—économie circulaire (IEDDEC). His research focuses on urban infrastructure (water, transport), metropolitan collective action, the circular city, foresight, and urban planning for the socio-ecological transition. He is currently Associate Vice-Rector for Research, Discovery, Creation and Innovation at UdeM, where he directs the Chemins de transition initiative.

Chapter 6
Methodological Proposal for Graphic Design Education Aimed at Fostering Social and Environmental Responsibility

Geneviève Raîche-Savoie and Claudia Déméné

Abstract According to the United Nations and its Sustainable Development Goals (SDGs), educational systems must adapt their learning objectives and methods to develop skills that promote sustainable development (Rieckmann et al., 2017). This chapter explores an ongoing doctoral research that utilizes a methodology enabling education to shape the role of graphic designers within the socio-ecological transition in Quebec. In class, graphic design students typically follow "creative briefs" that simplify problem complexity (Haylock, 2020), which limits their autonomy and sometimes prevents them from addressing social and environmental concerns. When they undertake projects independently, they often prioritize personal expression over the needs of users and stakeholders (Valencia et al., 2021). To address this, the research employs Learning Experience Design (LXD), an emerging transdisciplinary discipline focused on creating effective and engaging learning experiences. The primary objective is to implement LXD to tailor learning for graphic design students at Cégep de Sainte-Foy. Specifically, it adopts responsible entrepreneurship as a pedagogical approach, empowering students to initiate their own projects while fostering social and environmental responsibility. Employing a Mixed Methods Research (MMR) approach, the study combines quantitative and qualitative data collection through questionnaires, interviews, co-design workshops, and checklists. Notably, the research stands out for its collaborative nature, involving stakeholders such as students, faculty, administrative bodies of Cégep de Sainte-Foy, and external experts. This chapter presents initial results demonstrating the potential of education to prepare learners for meaningful contributions to sustainable development, emphasizing the transformative role of tailored pedagogical approaches like LXD in fostering responsible design practices.

Keywords Learning experience design · Human-centered design · Responsible entrepreneurship · Graphic design education · Mixed Methods Research

---

G. Raîche-Savoie (✉) · C. Déméné
Université Laval, Quebec City, QC, Canada
e-mail: Genevieve.Raiche-Savoie@design.ulaval.ca; Claudia.Demene@design.ulaval.ca

M. Cheriet et al. (eds.), *Accelerating the Socio-Ecological Transition*,
https://doi.org/10.1007/978-3-031-82896-6_6

1 Introduction

According to the United Nations Educational, Scientific and Cultural Organization (UNESCO), "education is both a goal in itself and a means for attaining all the other sustainable development goals[1] (SDGs)" (Rieckmann et al., 2017). The SDGs set the path toward a better and more sustainable future for all humanity by 2030. Since 1992, UNESCO has been promoting Education for Sustainable Development (ESD) so that educational systems can meet the urgent need for developing new skills that lead to more viable societies (UNESCO, 2024). This viability aims for greater social justice, poverty reduction, and, above all, better resilience to climate change. ESD sets learning objectives and content relevant to sustainable development while implementing teaching methods that promote learners' autonomy (Rieß et al., 2022). It is crucial to consider this approach as a fundamental strategy for promoting socio-ecological transition in Quebec. This transition refers to a transformation process aiming at profound changes in production and consumption patterns, as well as in Quebec's political and institutional structures, to address current challenges such as climate change, biodiversity loss, social inequalities, and environmental degradation (Markard et al., 2012). ESD has the potential to promote Quebec's socio-ecological transition by raising awareness among individuals about social and environmental issues, equipping them with necessary skills to contribute to sustainable solutions, and fostering creativity and collaboration in the search for practices that are more respectful of communities and the environment.

The main objective of this chapter is to present, to teachers from higher education institutions and universities, an example of an ongoing doctoral research that stands out for its approach combining education, design, and entrepreneurship sciences. This research utilizes the methodology of Learning Experience Design (LXD) to support the design of a learning experience for graphic design students at Cégep de Sainte-Foy (simple case study). According to Floor (2023), "Learning Experience Design (LXD) is the process of creating learning experiences that enable the learner to achieve the desired learning outcome in a human-centered and goal-oriented way." (p. 54). In this study, the LXD process is utilized to create a learning experience that will empower future social and environmental responsible graphic designer to successfully complete their entrepreneurial projects. Furthermore, this study employs the LXD methodology for the collection, analysis, and integration of quantitative and qualitative data using various methods such as questionnaires, interviews, co-design workshops, and checklists. The contribution to the advancement of knowledge in this doctoral research lies in proposing a

[1] (1) No poverty, (2) Zero hunger, (3) Good health and well-being, (4) Quality education, (5) Gender equality, (6) Clean water and sanitation, (7) Affordable and clean energy, (8) Decent work and economic growth, (9) Industry, innovation and infrastructure, (10) Reduced inequalities, (11) Sustainable cities and communities, (12) Responsible consumption and production, (13) Climate action, (14) Life below water, (15) Life on land, (16) Peace, justice, and strong institutions, (17) Partnerships for the goals.

methodological approach to design a pedagogical approach adapted to the field of graphic design with the aim of fostering social and environmental responsibility.

The chapter will be structured into seven sections following the steps of a research protocol. The next section will introduce the problem statement and research objectives within the context of a Graphic Design program at a higher education institution in Québec City, Québec, Canada. Subsequently, the chapter will outline the conceptual framework of the research, providing a structured approach to understanding and investigating the research problem while clarifying key concepts such as Learning Experience Design (LXD), Sustainable Entrepreneurship Education (SEE), and Design-led Entrepreneurship. Additionally, it will explain the use of Mixed Method Research (MMR) to provide a more comprehensive understanding of the research problem. This section will also highlight the proposed methodological structure, particularly the Learning Experience Design (LXD) process. The following section will detail the various methods used to collect both quantitative and qualitative data, followed by the presentation of preliminary results and data analysis as the research is still ongoing. Finally, the last section will discuss the anticipated impact of the research.

2 Problem Statement and Research Objectives

The various stakeholders involved in the fields of scientific, professional, and educational design call for an evolution in education to prepare future designers to solve complex problems, emphasizing collaboration between stakeholders and considering social and environmental impacts (ICoD, 2020; Meyer & Norman, 2020; Noël, 2020a, 2020b). During their studies at the college level, students enrolled in the Graphic Design program at Cégep de Sainte-Foy are assigned a series of design briefs developed by their teachers or real clients. Design briefs serve as guides for designers by describing the project's objectives, functional and aesthetic requirements, target audience, technical constraints, deadlines, and any other relevant elements. These documents have the effect of simplifying or significantly reducing the complexity of a problem and limiting students' ability to act by providing them with directives that already suggest certain types of solutions to consider (e.g., designing a poster, a website, a brochure, etc.) (Haylock, 2020). Yet, according to Haylock (2020), problem formulation is a vitally important part of the design process, but one that has historically been under-recognized in the teaching and traditional practice of graphic design. It becomes essential to promote problematization through projects initiated by the students themselves. This offers them the opportunity to develop their autonomy and agency, that is, to influence the circumstances of their lives by enabling them to become active agents who affect the course of events through their actions (Bandura, 2018).

However, students who decide to initiate projects themselves do not always succeed in realizing their ideas into actions that create value for society. In this perspective, Valencia et al. (2021) demonstrate that in the initial phase of a designer's entrepreneurial mindset development, the focus is on personal expression, at the expense of the requirements of users and stakeholders involved, resulting in a lack of consideration for social and environmental dimensions. This type of project holder is primarily motivated by exploring various materials, technologies, shapes, textures, or aesthetic properties, and the desire to undertake is based more on its passion and quest for meaning (Valencia et al., 2021). In this regard, Haylock (2020) explains that human-centered design is not necessarily a concept emphasized in graphic design education compared to that of experience design (UX/UI).

Despite the various opportunities offered to graphic design students at Cégep de Sainte-Foy to explore entrepreneurship within their study program, the majority of these initiatives target students nearing the end of their program (fifth and sixth semesters). Additionally, they do not receive guidance from faculty in developing their own design projects. Furthermore, despite the promotion of *Entrepreneurship Sainte-Foy*, among graphic design students, to date, a very small percentage of them have utilized their services to launch their projects (less than 5 students in 5 years). A general approach based on management program models that targets all students in the institution does not necessarily consider the actual practices of certain types of project holders (Bureau & Fendt, 2012). Pedagogical approaches that are not adapted may have a counterproductive effect on students with a creative profile who do not identify as entrepreneurs (Roberts, 2012), despite their aspirations to develop projects for which entrepreneurial skills are essential (Damásio et al., 2017).

Considering the elements of this problem statement, this doctoral research focuses on the following research question: *How can graphic design students be encouraged to develop their agency by integrating social and environmental aspects into their graphic design projects?* The main objective of this research is to design, using the Learning Experience Design (LXD) process, a responsible entrepreneurial learning experience tailored to students enrolled in the Graphic Design program at Cégep de Sainte-Foy.

To address the main objective, five successive sub-objectives have been defined:

Depict an entrepreneurial profile of graphic design students at Cégep de Sainte-Foy.
Identify the desired learning outcomes of graphic design students at Cégep de Sainte-Foy who are project holders.
Design various concepts for responsible entrepreneurship-focused learning experiences.
Enable graphic design students at Cégep de Sainte-Foy to choose the most suitable responsible entrepreneurial learning experience for their reality.
Evaluate whether the responsible entrepreneurial learning experience achieves the desired objectives and learning outcomes and adheres to the theories, practices, and design principles associated with Learning Experience Design (LXD).

The upcoming section introduces the key concepts underlying the research problem. The conceptual framework outlined here anchors the research in established theories and literature, illustrating its reliance on interdisciplinary methodologies. A conceptual framework is a brief explanation based on the logical arrangement of a set of interrelated concepts and sub-concepts brought together due to their affinity with the research problem (Fortin & Gagnon, 2022).

3 Conceptual Framework

This research draws on the foundations of three scientific disciplines: education, design, and entrepreneurship. The intersection of these three disciplines allows for the creation of a conceptual framework (Fortin & Gagnon, 2022) based on the arrangement of three emerging concepts: (1) Learning Experience Design (intersection of education and design); (2) Sustainable Entrepreneurship Education (intersection of entrepreneurship and education); (3) Design-led Entrepreneurship (intersection of design and entrepreneurship). These three concepts are brought together due to their relevance to the problem statement (Fig. 6.1).

3.1 Learning Experience Design (LXD)

To develop a responsible entrepreneurial learning experience tailored to students enrolled in the Graphic Design program at Cégep de Sainte-Foy, the research follows the process of Learning Experience Design (LXD) as its methodology. LXD is transdisciplinary and that draws from various fields, including educational

Fig. 6.1 The research conceptual framework

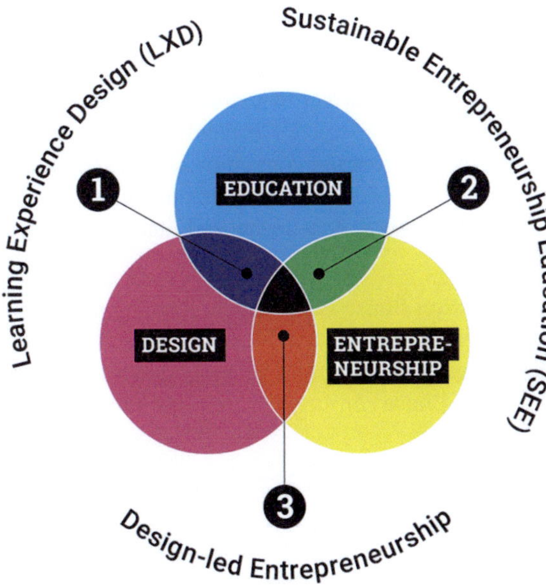

technology, informatics, human-computer interaction (HCI), information technology, education, instructional design, and psychology (Schmift, 2021).

First, the use of the term "learning" demonstrates that this discipline focuses on learning rather than teaching, instructions, or training (Floor, 2023). Teaching involves the transmission of knowledge, skills, and values from a teacher to learners, whereas learning refers to the process by which learners acquire these elements. The learner and their learning process are placed at the center of the experience design process. The designer must show empathy to determine the learners' needs, what motivates them, and how to keep them engaged (ELM Learning, n.d.). Meeting the needs and desires of the learner will help achieve their learning outcomes (Berthet, 2016). Learners' needs concern the basic requirements necessary to achieve a specific learning objective, while desires encompass wishes, preferences, or aspirations that go beyond these fundamental needs.

The use of the term "experience" refers to "any encountered situation that takes a certain amount of time and leaves some kind of impression" (Floor, 2016). A learning experience designed according to the LXD methodology should aim to evoke positive emotions in the learner, thereby enabling them to take an active role in their own learning process (Berthet, 2016). Positive emotions notably enhance learner performance by affecting cognitive processes such as attention, memory, and decision-making (Ibid, 2016). This type of experience can take place in a variety of settings, including schools, museums, workplaces, homes, or even in a natural environment. Furthermore, learning can occur in a physical world, a virtual environment, or a combination of both, thus hybrid (LXD, n.d.; Wasson & Kirschner, 2020).

Finally, like other design disciplines, it is an activity focused on problem-solving where the complexity of all influencing factors must be considered, and where feedback, evaluation, and iterations are part of the process (Galaykova, 2021). The main difference from other design disciplines (e.g., Experience Design) is that the theories, methods, and design tools have been adapted to focus on creating learning experiences. In addition to being an emerging design process in practice, Learning Experience Design is a growing research topic in academia in recent years (Schmidt & Huang, 2021; Wasson & Kirschner, 2020). In this regard, design is no longer only concerned with creating content or a single technological tool, but rather with the entirety of the learning environment and all its actors (Wasson & Kirschner, 2020).

LXD follows a non-linear and iterative process consisting of six phases[2] (Fig. 6.2):

The Question Phase stems from a learning need among the learners.
The Research Phase focuses on the learners and the desired learning outcomes.
The Design Phase involves generating ideas and transforming them into concepts.
The Build Phase converts the concepts into prototypes or learning artifacts.

[2] The phases bear resemblance to the Design Thinking process, a more general approach to problem-solving and user-centered design.

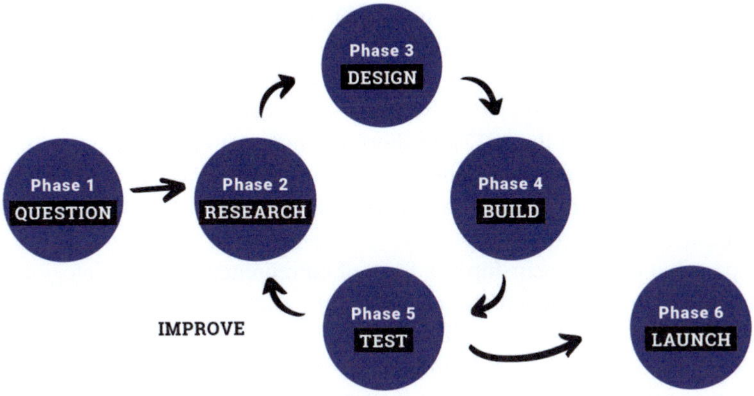

Fig. 6.2 Learning experience design process (Floor, 2023)

The Test Phase assesses whether the prototype achieves the desired objectives and learning outcomes. The findings from the evaluation lead to iterations, revisiting some of the previous phases or repeating the entire cycle.

The Launch Phase involves putting the final product, the learning experience, into practice with the learners.

3.2 Sustainable Entrepreneurship Education (SEE)

Sustainable Entrepreneurship Education (SEE) aims to foster an entrepreneurial mindset among learners. This pedagogical approach is compared to two other approaches to entrepreneurial education: one where entrepreneurship is a subject of study (*Education about enterprise*) and the other where it is a professional skill to develop (*Education for enterprise*) (Fayolle & Gailly, 2008; Hägg & Gabrielsson, 2019; Kakouris & Liargovas, 2020; Pepin, 2015). Sustainable Entrepreneurship Education (SEE), as a third approach, is a learning process where students themselves identify a problem, a need, or an opportunity and take action through the development of a plan to undertake (e.g., a business plan, project proposal, etc.) (Rae, 2003). In other words, "undertaking involves establishing actionable goals and subjecting them to the scrutiny of real-life experience within a specific context" (Translation based on Pepin, 2015, p. 62). To this end, the entrepreneurial activity or project is guided by a reflective structure so that learners can reflect in action throughout the process (Pepin, 2015; Schön, 1983).

SEE is a process of personal development that can go beyond the creation of a new business or job (Verzat, 2015). Students become "project holders" rather than "entrepreneurs," and entrepreneurship education becomes a means rather than an end (Pepin et al., 2017). Teaching entrepreneurship does not consist of acquiring a set of knowledge, but rather using entrepreneurship as a pedagogical tool to organize learning experiences (Pepin, 2015). This pedagogical tool moves away from the strictly economic and utilitarian vision of entrepreneurship in education (Pepin et al., 2017). Alternatively, it emphasizes on "value creation," wherein the entrepreneurial endeavor pursued by the learner offers benefits to users or the intended audience (Fayolle, 2017). In short, teaching entrepreneurship, particularly responsible entrepreneurship, can help make a positive and responsible contribution to society and the environment (Ballereau et al., 2020; Ben-Hafaïedh, 2020; Pepin et al., 2021).

Responsible entrepreneurship is a form of entrepreneurship that integrates ethical, social, and environmental considerations into value creation (e.g., business creation, one-off projects, citizen initiatives, etc.) (Pepin et al., 2017). According to Pepin et al. (2017), individuals who hold projects take the initiative and complete various types of projects, developing their ability to act and assuming the role of agents of change. In a pedagogical context, responsible entrepreneurship involves individual, social, and ecological awareness (reflection) and responsibility (action) of the project holder (Ibid, 2017).

3.3 Design-Led Entrepreneurship

Design-led entrepreneurship, derived from the field of design sciences and practice, is often associated with Design Thinking. Design Thinking is a method of reasoning and acting in practice to elaborate solutions to specific problems, such as products, services, interactions, organizations, strategies, etc. (Baran, 2018).

The design approach in entrepreneurship offers much more than just design methods or tools. It also involves thinking through empathy, integrative thinking, experimentation, opportunity pursuit, and collaboration (Baran, 2018; Gamba, 2017). Firstly, empathy enables understanding and interpreting the viewpoints and emotions of the target audience or users. Integrative thinking generates a holistic approach by considering multiple perspectives and influencing factors of a problem. Experimentation aims to exploit the constraints of a problem to generate various creative solutions. Opportunity pursuit transforms problems and constraints into favorable change opportunities. Lastly, collaboration allows for teamwork and incorporating the perspectives of other disciplines. Consequently, the design approach becomes a preferred ally with the field of entrepreneurship for solving complex problems (Maher et al., 2018).

In this regard, research by Fridman et al. (2022) suggests that the principles of social responsibility, environmental sustainability, ethics, critical thinking, and accountability can be combined with entrepreneurship principles (viability, feasibility, and desirability) to create a new model of responsible design thinking in education. Responsible design thinking enables students to critically consider and anticipate the social and environmental consequences of their designs as part of their learning process. This model allows them to reflect on various current issues (e.g., crime, education, government, health, fair trade, ecology, social inclusion, and economic policy) to make more ethical and responsible choices (Fridman et al., 2022).

3.4 *The Arrangement of the Three Concepts*

In the context of this research, the discipline of Learning Experience Design (LXD) places graphic design students at the center of the design process. Sustainable Entrepreneurship Education (SEE) empowers students to identify issues autonomously as project holders and suggest solutions that address social and environmental concerns. Finally, Design-led Entrepreneurship, through the processes of Design Thinking, helps students to reflect, become aware, and act through the development of their design project. In other words, each concept will guide the development of the learning experience through this ongoing doctoral research.

Education + Design = Learning Experience Design (LXD) is a design methodology that will be used to create a responsible entrepreneurial learning experience tailored for graphic design students, guiding them to initiate their own design projects in a conscientious manner.

Entrepreneurship + Education = Sustainable Entrepreneurship Education (SEE) is a pedagogical approach that will be used within the responsible entrepreneurial learning experience to develop new skills that can contribute to Quebec's socio-environmental transition.

Design + Entrepreneurship = Design-led Entrepreneurship, through the use of responsible design thinking, **is a methodology for reflection and problem-solving** that will be used by graphic design students to understand the impact of their design project and take responsibility for their actions.

After clarifying these three concepts and their interrelations, the following section explains the use of Mixed Method Research (MMR) to offer a more comprehensive understanding of the research problem.

4 Methodology

This doctoral project utilizes Mixed Methods Research (MMR) due to its potential to provide rigorous methodologies for addressing and examining issues in a more holistic manner (Nagels, 2022). In Mixed Methods Research (MMR), the door is opened to the use of multiple methods, different perspectives, and various forms of data collection and analysis (Creswell & Creswell, 2018) to address the research question: *How can graphic design students be encouraged to develop their agency by integrating social and environmental aspects into their graphic design projects?* Furthermore, MMR contributes to achieving the main objective of designing a responsible entrepreneurial learning experience tailored to the students of the

Graphic Design program at Cégep de Sainte-Foy, as well as the five successive sub-objectives (see Sect. 2).

According to Creswell et al. (2018), mixed methods can be defined as rigorous methods of collecting, analyzing, and integrating two types of data: quantitative (closed-ended) data and qualitative (open-ended) data. The integration of quantitative and qualitative data occurs within a broader framework that may include one of the three fundamental core designs of MMR: the Explanatory Sequential Design, the Exploratory Sequential Design, or the Convergent Design (Ibid., pp. 217–233). The first type of design, known as the Explanatory Sequential Design, begins with a quantitative phase followed by a qualitative phase. The quantitative results obtained are used to define the research questions and guide the qualitative phase, which deepens the understanding of the phenomena under study. In contrast, the Exploratory Sequential Design starts with a qualitative phase followed by a quantitative phase. The qualitative phase allows for exploration and hypothesis formulation, which are then tested and validated using quantitative methods in the subsequent phase. Finally, the Convergent Design involves the independent collection and analysis of quantitative and qualitative data. The results are then compared and combined to obtain a comprehensive understanding of the phenomena under study.

MMR can also be conducted using a Complex Design (Creswell, 2022) which embeds the fundamental designs (explanatory sequential, exploratory sequential, and convergent) (Creswell, 2022, p. 62) within a specific framework or process that involves a greater number of steps. The Complex Design of this doctoral research is based on the six cyclical and iterative phases of the Learning Experience Design (LXD) process (Fig. 6.2): (1) Question, (2) Research, (3) Design, (4) Build, (5) Test, and (6) Launch. The LXD process within this Mixed Method Research has been chosen for its ability to offer a thorough, reliable, and nuanced understanding of complex research problems. It integrates the strengths of quantitative and qualitative approaches, focusing on data collection as well as the creation of the learning experience.

4.1 LXD—Phase 1: Question

The LXD process begins with a question to be answered or a problem to be solved. In the context of the research presented, the researcher aimed to understand: *How can graphic design students be encouraged to develop their agency by integrating social and environmental aspects into their graphic design projects?* Through the responsible entrepreneurial learning experience that will be designed as part of this research, graphic design students will become project holders by initiating design projects by themselves. The development of this design project served as a pretext to expose students to current social and environmental issues. Student-initiated projects could include creating self-published, small-circulation, handmade publications. Additionally, students could design merchandise such as t-shirts or buttons.

4.2 LXD—Phase 2: Research

The second phase, LXD Research, focuses on two elements: the learners and their desired learning outcomes (Floor, 2023). The objective of the LXD Research Phase is to develop empathy toward the learners and gain a deep understanding of what they wish to gain from their learning experience, namely competences.

In the context of the doctoral research, this phase initially aimed to acquire an in-depth understanding of graphic design students that were project holders. This understanding focused on their interests, passions, and values as well as the type of projects they enjoyed undertaking. Subsequently, the LXD Research Phase identified how the responsible entrepreneurial learning experience could positively impact these students' lives through its relevance and value. This phase aimed to understand the challenges they faced in developing their entrepreneurial project as well as what they wanted to learn. The desired learning outcomes of the students influenced the formulation of the different learning objectives of the responsible entrepreneurial experience (Fig. 6.3). These learning objectives were inspired by the learning objectives of various stakeholders: (1) Design Education (Noel, 2020a), (2) Responsible Entrepreneurship (International and Entrepreneurship Office of Cégep de Sainte-Foy (BIE, 2022)), (3) Sustainable Development Goals (UNESCO (Rieckmann et al., 2017)).

4.3 LXD—Phase 3: Design

The LXD Design Phase includes four sub-steps: divergence ideation, convergence ideation, draft concept, and detailed concept (Fig. 6.4). The LXD Design Phase involves an ecosystemic approach by ensuring consultation with all stakeholders involved in the design and pedagogical implementation. In the research presented, this involved students and teachers from the Graphic Design program, along with members of the International and Entrepreneurship Office (BIE), who took part in a co-design workshop to generate ideas.

The LXD Design Phase begins with a period of divergence ideation aimed at generating multiple ideas using various tools (e.g., mental maps or brainstorming). Divergence ideation encourages exploration and diversification of ideas without imposing initial constraints. In the context of the doctoral research, participants

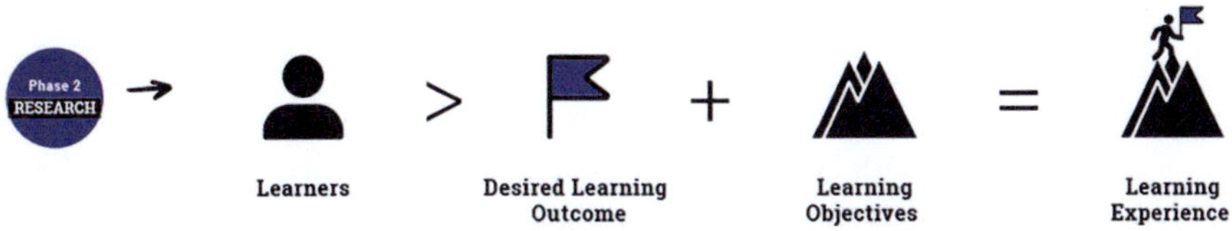

Fig. 6.3 Phase 2: LXD research phase

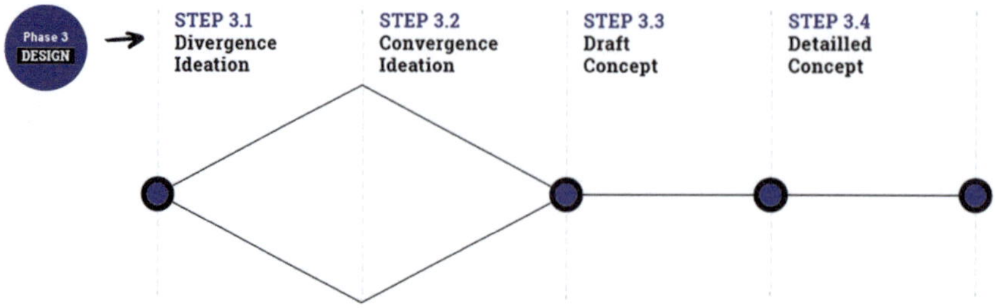

Fig. 6.4 Phase 3: LXD design phase

freely proposed ideas regarding the form that the responsible entrepreneurial learning experience could take. Next, the convergence ideation step allows for the selection of the most suitable idea for the context. The selected idea is transformed into a draft concept to validate its viability. This draft includes the structure of the experience (plan), the main activities, and the resources that will be used. Finally, the detailed concept allows for the presentation of the main elements of the experience in tangible and visual forms (e.g., educational tools, etc.).

4.4 LXD—Phase 4: Build

The next phase, LXD Build, involves transforming the selected concept into a series of prototypes that will evolve into the final design of the experience. Prototypes come in various forms, including visual mock-ups to represent aspects of the learning experience such as tools or the virtual environment, interactive mock-ups integrating interactive elements such as quizzes or simulations, as well as detailed scenarios describing activity sequences and interactions between learners and content. The main objective of a learning experience design prototype is to bring a concept to life to test it, gather feedback from stakeholders, identify potential issues, and make improvements before moving to implementation (phase 6). In this research, prototypes of the responsible learning experience will be developed in Fall 2024.

4.5 LXD—Phase 5: Test

The LXD Test Phase allows to verify whether the different prototypes achieve the desired learning outcomes (from the learners) as well as the learning objectives (of the learning experience). Additionally, this phase ensures that the design of the learning experience aligns with the theories, practices, and design foundations associated with Learning Experience Design (LXD). Finally, the LXD Test Phase involves consulting all stakeholders. The conclusions drawn from the evaluation

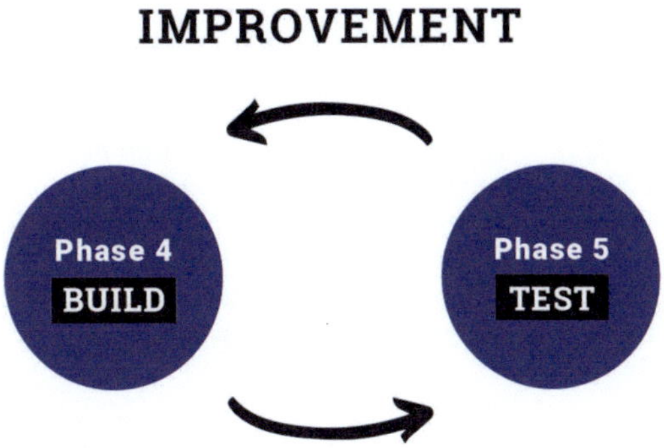

Fig. 6.5 The build and test phases of LXD (Phase 4–5)

enable iterations by revisiting some of the previous phases (Fig. 6.5) or by repeating the entire cycle (Fig. 6.2).

4.6 LXD—Phase 6: Launch

Finally, the LXD Launch Phase involves putting the final product into action. In the context of the doctoral research, it involved implementing the responsible entrepreneurial learning experience for all students in the Graphic Design program at Cégep de Sainte-Foy.

After establishing the methodological structure, the subsequent section outlines the various methods employed for qualitative and quantitative data collection and their alignment with the research questions and sub-objectives. This includes details on sampling, recruitment, delivery, and consent. The section also specifies the type and number of participants, the activities involved (data collection methods), the frequency and duration of these activities, the locations where they occur, and the measurement instruments used.

5 Data Collection

The data collection for this doctoral research relied on the use of multiple methods, both quantitative and qualitative, throughout phases 2, 3, and 5 of the LXD process. Table 6.1 illustrates how the research sub-objectives determined the type of data that were collected and the chosen methods as well as the timeline of the ongoing research. Figure 6.6 illustrates the integration of each method within each phase.

Table 6.2 provides a summary of: (1) individuals targeted for recruitment and the selection criteria (sampling); (2) methods by which they were approached to participate in the project (recruitment); (3) delivery methods; and (4) consent. Recruitment and obtaining consent were treated as distinct steps (involving separate documents), allowing for a possible period of reflection before providing informed and voluntary

Table 6.1 The types of data to be collected according to the sub-objectives of the research phases

Research phases	Research question + sub-objectives	Data collection methods	Schedule
Phase 1: LXD Question	*How can graphic design students be encouraged to develop their agency by integrating social and environmental aspects into their graphic design projects?*	N/A	N/A
Phase 2: LXD Research	Depict an entrepreneurial profile of graphic design students at Cégep de Sainte-Foy	Open-Ended Questionnaire (Quantitative data)	February–March 2024
	Identify the desired learning outcomes of graphic design students at Cégep de Sainte-Foy who are project holders	Semi-Structured Interviews (Qualitative data)	
Phase 3: LXD Design	Design various concepts for responsible entrepreneurship-focused learning experiences	Co-design Workshop (Qualitative data)	April–May 2024
	Enable graphic design students at Cégep de Sainte-Foy to choose the most suitable responsible entrepreneurial learning experience for their reality	Close-Ended Questionnaire (Quantitative data)	
Phase 4: LXD Build	Transforming the concept chosen by the students into a series of prototypes that will evolve into the final design of the experience.	N/A	June–August 2024
Phase 5:LXD Test	Evaluate whether the responsible entrepreneurial learning experience achieves the desired objectives and learning outcomes and adheres to the theories, practices, and design principles associated with Learning Experience Design (LXD)	Control List (Quantitative data)	September–December 2024
		Open-Ended Interviews (Qualitative data)	
Phase 6:LXD Launch	Implementing the final product, namely the responsible entrepreneurial learning experience for all graphic design students at Cégep de Sainte-Foy	N/A	2025

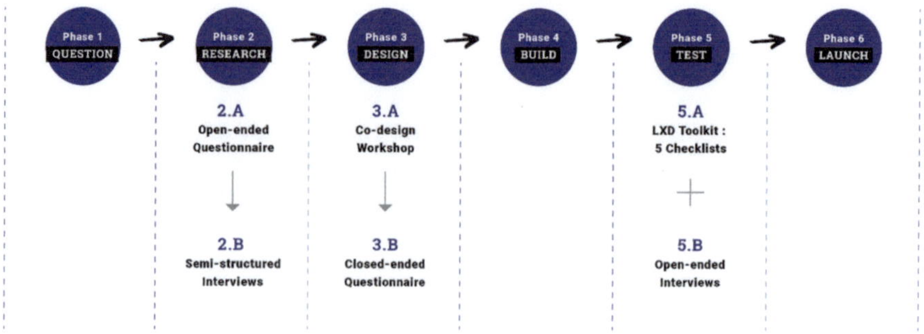

Fig. 6.6 The integration of methods in each phase of the research

Table 6.2 Sampling, recruitment, delivery methods, and consent

Phases + objectives	Methods	Sampling	Recruitment	(3) Delivery	(4) Consent
Phase 2: LXD Research	**2.A Open-ended Questionnaire**	183 students enrolled in the Graphic Design program at Cégep de Sainte-Foy. *(N = 45)*	Email + Video	An email was sent by the program coordinator This message contained a link to a video explaining the research project by the first author researcher. It also included a link to the online questionnaire	Before accessing the online questionnaire, participants reviewed the consent form. They then selected a checkbox to indicate that they had read the information and consented to participate in the research
	2.B Semi-structured Interviews	183 students enrolled in the Graphic Design program at Cégep de Sainte-Foy who are project holders. *(N = 5)*	Flyer (PDF)	The PDF flyer was attached to the email from the previous phase (2.A). Additionally, a question was added at the end of the questionnaire so that interested students could provide their name and matricula number for participating in semi-structured interviews	A written consent form with signature was sent to interested individuals Semi-structured interviews were only scheduled with individuals who had completed the consent form

(continued)

Table 6.2 (continued)

Phases + objectives	Methods	Sampling	Recruitment	(3) Delivery	(4) Consent
Phase 3: LXD Design	**3.A Co-design Workshop**	10 students who participated in semi-structured interviews (project holders) *(N = 5)*	Email + Flyer (PDF)	The PDF flyer was attached to an email that was sent to students that participated in the previous phase (2.B)	A written consent form with signature was sent in the Email The co-design workshop was only conducted with individuals who had completed the form
		12 teachers from the Graphic Design program at Cégep de Sainte-Foy. *(N = 3)*			
		7 members of the International and Entrepreneurship Office from Cégep de Sainte-Foy. *(N = 2)*	Email	An email was sent by the Ethic Committee at Cégep de Sainte-Foy	
	3.B Close-ended Questionnaire	183 students enrolled in the Graphic Design program at Cégep de Sainte-Foy. *(N = 26)*	Email + Video	An email was sent by the program coordinator This message contained a link to a video explaining the different concepts. It also included a link to the online questionnaire	Before accessing the online questionnaire, participants reviewed the consent form. They then selected a checkbox to indicate that they had read the information and consented to participate in the research

Phase 5: LXD Test	**5.A Checklists**	N/A	N/A	LXD Toolkit: 5 Checklists (O'Neal, Hindman, & Donaldson, 2023)	
	5.B Open-ended Interviews	183 students enrolled in the Graphic Design program at Cégep de Sainte-Foy who are project holders. *(Objective of 1–2 student participants)*	Email + Flyer (PDF)	An email with the flyer was sent by the program coordinator	A written consent form with signature was sent in the Email Documents pertaining to the learning experience design was exclusively sent to as well as interviews were conducted only individuals who had filled out the form
		12 teachers from the Graphic Design program at Cégep de Sainte-Foy. *(Objective of 1–2 teacher participants)*			
		7 members of the International and Entrepreneurship Office from Cégep de Sainte-Foy. *(Objective of 1–2 participants)*			
		40 pedagogical advisors from Cégep de Sainte-Foy.*(Objective of 1–2 participants)*	Email	An email was sent by the Ethic Committee at Cégep de Sainte-Foy.	
		20 experts in design and/or responsible entrepreneurship (professionals, teachers, researchers) *(Objective of 1–2 participants)*	Email or private message	The contact details were obtained through the LinkedIn platform or institutional websites, and individuals were contacted via their professional page or email address.	

consent. It should be noted that each of these research methods involved the use of non-probability sampling (Fortin & Gagnon, 2022), meaning that participant selection was not random, and samples were composed based on specific criteria, such as participant characteristics (e.g., graphic design students who were project holders) and their environment (e.g., Graphic Design program at Cégep de Sainte-Foy). This intentional sampling resulted from the selection of specific criteria (project holders in graphic design) by the main author to choose participants likely to meet the study's objectives.

Finally, Table 6.3 provides a summary of (1) types of participants, (2) number of participants, (3) activities (data collection methods), (4) frequency of activities, (5) duration of activities, (6) location of activities, and (7) measurement instruments.

6 Preliminary Results and Analysis

As this doctoral research is still ongoing at the time of this chapter's publication, this section provides a preview of some preliminary results and analysis, with an emphasis on findings related specifically to fostering social and environmental responsibility. As illustrated in Fig. 6.7, each fundamental design of the MMR (explanatory sequential, exploratory sequential, and convergent) was integrated into some of the LXD design processes (Phase 2, 3, and 5).

6.1 Phase 2: LXD Research Phase > Explanatory Sequential Design

The second phase, the **LXD Research Phase**, followed an explanatory sequential design to outline the entrepreneurial profile and understand the desired learning outcomes of graphic design students at Cégep de Sainte-Foy. It began in February 2024 with the collection of quantitative data through an open-ended questionnaire (2.A), followed by the collection of qualitative data through semi-structured interviews (2.B) in March 2024. According to Aguilera and Chevalier (2021), the integration of data from the explanatory sequential design aims to confirm, enrich, and complexify the results (Fig. 6.8). First, qualitative data confirm the quantitative results. Then, enrichment adds or reveals other response elements. Finally, complexification nuances the quantitative analysis by revealing previously unobserved or observable results. While the questionnaire quantified the entrepreneurial profile of students by presenting numbers and percentages, the interviews provided a more detailed explanation, through the words and expressions of the students, of their desired learning outcomes (Savoie-Zajc, 2009).

Table 6.3 Types of participants, number of participants, activities, frequencies, durations, locations, and measurement instruments

(1) Types of participants	(2) Number of participants	(3) Activities (methods)	(4) Frequencies	(5) Durations	(6) Locations	(7) Measurement instruments
Students enrolled in the Graphic Design program at Cégep de Sainte-Foy	Minimum objective of 100 participants	**2.A** Open-ended Questionnaire	One questionnaire	10–15 min	Online	*LimeSurvey*
Students enrolled in the Graphic Design program at Cégep de Sainte-Foy who are project holders	Objective of 10 participants	**2.B** Semi-structured Interviews	One interview per participant	15–20 min	In-person (Cégep de Sainte-Foy)	Interviews were recorded with Zoom and stored on the Cégep de Sainte-Foy's server
Students enrolled in the Graphic Design program at Cégep de Sainte-Foy who are project holders	Objective of 10 participants	**3.A** Co-design workshop	One co-design workshop	3 h	In-person (Cégep de Sainte-Foy)	Collection of written records + photographs and videos
Teachers from the Graphic Design program at Cégep de Sainte-Foy						
Members of the International and Entrepreneurship Office from Cégep de Sainte-Foy						Files were recorded and stored on the Cégep de Sainte-Foy's server
Students enrolled in the Graphic Design program at Cégep de Sainte-Foy	Minimum objective of 100 participants	**3.B** Close-ended Questionnaire	One questionnaire	10–15 min	Online	*Microsoft Forms*
N/A	N/A	**5.A** Checklists	N/A	N/A	N/A	LXD Toolkit: 5 Checklists (O'Neal, Hindman & Donaldson, 2023)

(continued)

Table 6.3 (continued)

(1) Types of participants	(2) Number of participants	(3) Activities (methods)	(4) Frequencies	(5) Durations	(6) Locations	(7) Measurement instruments
Students enrolled in the Graphic Design program at Cégep de Sainte-Foy who are project holders Teachers from the Graphic Design program at Cégep de Sainte-Foy Members of the International and Entrepreneurship Office from Cégep de Sainte-Foy Pedagogical advisors from Cégep de Sainte-Foy Experts in design and/or responsible entrepreneurship (professionals, teachers, researchers)	Objective of 10 participants	**5.A** Open-ended interviews	One interview per participant	15–20 min	In-person (Cégep de Sainte-Foy) Online (*Zoom*)	Interviews were recorded with Zoom and stored on the Cégep de Sainte-Foy's server

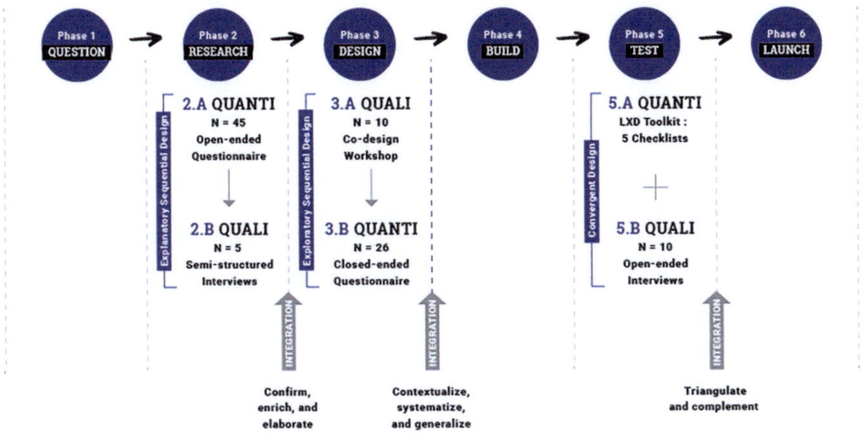

Fig. 6.7 The integration of MMR fundamental designs in each phase of the research

To start, one of the questions from the open-ended questionnaire ($N = 45$) was: "Regarding desired learning outcomes, what would you like to gain from this experience? (You may select multiple responses)." The results were as follows:

- 57.78%: Undertake a project independently to gain autonomy
- 53.33%: Monetize their activities (ex. creating their own job)
- 51.11%: Learn about entrepreneurship or topics not covered in graphic design courses
- 44.44%: Develop a project that showcases their strengths and interests (ex. Adding projects in their portfolio)
- 42.22%: Increase their self-confidence
- 40%: Identify problems, opportunities, or needs around them
- 24.44%: Create a project that aligns with their personal values, including environmental, inclusion, and diversity concerns.

The results indicated that students were twice as likely to want to undertake a project showcasing their strengths and interests compared to creating a project aligned with social or environmental issues. This finding aligns with Valencia et al. (2021), who demonstrated that in the initial phase of developing a designer-entrepreneur mindset, referred to as "the artisan," the emphasis is on personal expression rather than the needs of users and stakeholders involved.

Following the open-ended questionnaire, the semi-structured interviews revealed that the personal values expressed by the students are often directly integrated into their projects as themes. For instance, one student is illustrating a volume of poetry that "revolves around nature and the Québec territory". Another student explains that one of their projects serves as a true expression of their cultural identity sharing (Québécois and Brazilian): "I was going through an existential crisis. Because I was born in Brazil, and for about a year and a half, I've realized that I know nothing about [it]. I went in December, and it had been 10 years since I last went, so I wanted to shake things up and put it all into a [magazine]". Furthermore, these values influence how students structure their projects. A participant explains that they work on a collaborative magazine: "I gravitate a lot towards things that come from the community. I like feeling the presence of people in things, and things that bring people together".

Although only 24.44% of the questionnaire participants expressed a desire to create a project aligned with their personal values, all interview participants demonstrated that these values were already integrated into the development of their projects. The integration of data from Phase 2 provided meaning and direction during the Design and Build Phases of the LXD process (Phase 3 and 4).

6.2 Phase 3: LXD Design Phase > Exploratory Sequential Design

The third phase, the **LXD Design Phase**, followed an exploratory sequential design. First, aimed to collect qualitative data through a co-design workshop (3.A) in April 2024 to develop various ideas related to designing a responsible entrepreneurial learning experience. Then, the closed-ended questionnaire (3.B) in May 2024 aimed to collect quantitative data to identify the most suitable responsible entrepreneurial learning experience for graphic design students at Cégep de Sainte-Foy. The qualitative results served as the basis for designing a quantitative assessment considering the specific context of the project holders (Creswell, 2022, p. 55). The co-design workshop ($N = 10$) allowed the development of elements, including a list of concepts intended for the design of a responsible entrepreneurial learning experience. This list was then used to develop the closed-ended questionnaire intended for a larger sample ($N = 26$), namely graphic design students at Cégep de Sainte-Foy (Greene et al., 1989, p. 251; Schrauf, 2016). The integration of data from the exploratory sequential design provided contextualization, systematization, and generalization of the results (Aguilera & Chevalier, 2021) (Fig. 6.8). First, in terms of contextualization, the qualitative data collected justified the selection of components used during quantitative data collection. Next, systematization validated hypotheses while considering the sampling (internal validity). Finally, generalization aimed to verify validity with respect to a broader population (external validity).

Initially, participants in the co-design workshop generated a total of ninety-two ideas pertaining to the development of a responsible entrepreneurial learning experience during the divergent ideation activity. Following this, they initiated a convergent ideation activity with the goal of organizing, assessing, and refining the ideas, as well as selecting the most appropriate solutions for graphic design students at Cégep de Sainte-Foy. Nine categories were identified by the participants:

1. Action
2. Activities/Workshops
3. Outside
4. Collaboration

5. Courses/Specialization
6. Program Integration
7. Internships
8. Tools
9. Program Life

After discussing the categories and various constraints involved in creating a new learning experience within the Graphic Design program, three ideas were chosen as the most suitable solutions for the graphic design students at Cégep de Sainte-Foy:

1. Promoting internal and free resources
2. The creative retreat
3. The boutique

Following the analysis, the first option, "Promoting internal and free resources," was excluded from the choices included in the closed-ended questionnaire. This idea was not deemed as a learning experience but rather as a tool that could be integrated into the other two options. The collected responses from the close-ended questionnaire were conclusive and helped determine the form the learning experience should take. There was a clear distinction between the two options: "The creative retreat" received 65.38% votes, while "The boutique" received 34.62% votes. The choice of "The creative retreat" thus represented two-thirds of the total votes.

The concept of "creative retreat" or "residency" is well-known in the fields of arts and literature (Conseil des arts et des lettres du Québec, 2009). Creative retreats are dedicated periods that create environments conducive to creativity and provide participants with access to resources (TransArtists, n.d.). These retreats typically provide a quiet and inspiring setting, often in natural surroundings or culturally rich locations, where participants can focus on their work away from the distractions of everyday life. These periods may also include workshops, mentorship sessions, group activities, and free time for rejuvenation and reconnecting with one's practice. In the context of this doctoral research, this is a space where graphic design project holders can retreat to an external environment away from Cégep to concentrate intensively on their own entrepreneurial projects. Available to all program students, this occasion would occur in either a natural or urban environment, planned for the winter term, ideally positioned before or after to avoid interfering with students' coursework and hectic schedules. Lasting from 3 to 5 days, it would involve individual work sessions, theoretical aspects of responsible entrepreneurship and design-led entrepreneurship, and various collaborative activities. The retreat would not only provide a workspace for students to develop their entrepreneurial projects but also assist them in building new competencies. Because of the external venue, students may face expenses for space rental and equipment, but financial options could be explored to mitigate these costs. The main aim of this retreat would be for participants to advance their personal projects while deepening their awareness of their impact as graphic designers and motivating them to contribute to Quebec's socio-ecological transition. The integration of data from Phase 3 led to preparing the ground for the LXD Build Phase (Phase 4) of the LXD process.

6.3 Phase 5: LXD Test Phase > Convergent Design

The fifth phase, the **LXD Test Phase**, will take place in the Fall 2024. It will involve collecting mixed data through checklists (5.A) and open-ended interviews (5.B). The objective of this phase will be to verify if the responsible entrepreneurial learning experience achieves the desired learning objectives and outcomes and adheres to the theories, practices, and design principles associated with Learning Experience Design (LXD). The five checklists come from the Learning Experience Design (LXD) Toolkit created by Stephanie Perry O'Neal et al. (2023). The LXD Toolkit includes five advanced diagnostic tools specifically designed to enhance the development of learning designs: (1) *Elements of Learning Experience Design*, (2) *UX/UI for Experience Design*, (3) *Lifelong Learning and Engagement Strategies*, (4) *Advanced Tools and eLearning Trends*, (5) *Foundation for a Solid Course Design*. These tools enable designers to assess their learning design using a detailed rating scale that covers 12 distinct design theories and practices, as well as 17 foundational aspects of learning design. The primary aim of the LXD Toolkit is to facilitate the creation of engaging and impactful learning experiences.

During this phase, quantitative and qualitative data will be collected independently and in parallel. The collected data will aim to show similar results. In other words, the results of both methods will corroborate each other (Greene et al., 1989, p. 251; Schrauf, 2016). The integration of mixed data from the convergent design will allow for triangulation and complementarity of results (Aguilera & Chevalier, 2021). Triangulating data involved collecting and analyzing data from different sources or methods to strengthen the validity and reliability of the results (Van der Maren, 1996). By using two evaluation methods (checklists and open-ended interviews), it will be possible to ensure the robustness of the results (confirmation or invalidation). Finally, complementarity will aim to articulate different results that were mutually exclusive to provide clarification (e.g., quotes from project holders). The integration of qualitative and quantitative data in the LXD Test Phase (Phase 5) will provide the opportunity to proceed with revisions by reexamining certain previous steps of the LXD process. In particular, the researcher will alternate between the LXD Build and Test Phases (Phases 4–5) to craft a sequence of prototypes, which will iteratively refine into the final design of the experience intended for implementation in 2025.

7 Discussion and Conclusion

Education plays a crucial role in achieving the sustainable development goals (SDGs) set by the United Nations by 2030, preparing individuals to contribute to creating a more sustainable future. In this regard, educational systems must adjust their learning objectives and methods to develop skills conducive to sustainable development. This chapter examined a methodology to prioritize so that Education for Sustainable Development (ESD) supports socio-ecological transition in Quebec. Learning Experience Design (LXD) is a creative pedagogical design process centered on the learner, focused on objectives, guided by learning theories, and oriented toward real-life situations. This methodology not only strives to meet the needs of today's learners but also aims to deliver a high-quality learning experience. In the context of this research, LXD guided choices that suited the graphic design students and the context. This process is helping the researcher shape the learning experience to focus on raising awareness about social and environmental issues, equipping students with the necessary skills to make better decisions for communities and the environment.

To support this research, a Mixed Methods Research (MMR) approach was adopted, combining different forms of data collection and analysis. Furthermore, this research was conducted in collaboration with stakeholders, including the student and faculty body of the graphic design program, the administrative body of the Cégep de Sainte-Foy (Office of International and Entrepreneurship and pedagogical counselors), as well as other specialists (university professors, graphic designers, and learning experience designers). The level of engagement of the actors in this educational ecosystem is important as the research aims to assist the graphic design practice community (Mattessich & Monsey, 1992).

In the context of this MMR, three fundamental designs were used: the explanatory sequential design, the exploratory sequential design, and the convergent design. Some of the phases of the LXD process were associated with a specific design, with data collection methods adapted to the sub-objectives of each phase. Firstly, the LXD Research Phase (Phase 2), implemented through an explanatory sequential design, aimed to analyze the learners and their needs. Then, the LXD Design Phase (Phase 3), through the exploratory sequential design, was deployed to develop a learning experience specifically tailored to these learners. Finally, the LXD Test Phase (Phase 5), using the convergent design, strived to confirm that this experience not only meets the needs of learners (desired learning outcomes), but was also in line with current societal and environmental challenges (learning objectives) as well as the principles of LXD. The analysis of the results of each method was carried out gradually as the research progressed, with integration of mixed data at each phase. This approach strengthened the validity and reliability of the results while providing the opportunity to make revisions by re-examining certain previous phases of the LXD design process.

Although this example is grounded in the context of higher education institutions, the LXD methodology can be adapted to other contexts. We can think about how to enhance the skills and awareness of citizens, decision-makers, and local stakeholders to promote a better understanding and greater adherence to socio-ecological transition. What these individuals need to learn is important; however, it is equally essential to consider their emotions and reflections throughout their learning journey. Placing individuals at the heart of initiatives is crucial to achieving established sustainable development goals.

References

Aguilera, T., & Chevalier, T. (2021). Les méthodes mixtes pour la science politique: Apports, limites et propositions de stratégies de recherche. *Revue française de science politique, 71*, 365–389. https://doi.org/10.3917/rfsp.713.0365

Ballereau, V., Pepin, M., Toutain, O., & Tremblay, M. (2020). La formation à l'entrepreneuriat durable et responsable: Un champ scientifique et pédagogique en émergence. *Entreprendre & Innover, 45*(2), 5–9. https://doi.org/10.3917/entin.045.0005

Bandura, A. (2018). Toward a psychology of human agency: Pathways and reflections. *Perspectives on Psychological Science, 13*(2), 130–136. https://doi.org/10.1177/1745691617699280

Baran, G. (2018). Design-led approach to entrepreneurship. International Journal of Contemporary Management, 14(4), 7–26. https://doi.org/10.4467/24498939IJCM.18.034.10020

Ben-Hafaïedh, C. (2020). Se former en rendant service à la communauté: Quand des étudiant·e·s se font consultants pour des organisations sociales. *Entreprendre & Innover, 45*(2), 69–80. https://doi.org/10.3917/entin.045.0069

Berthet, S. (2016). *Design d'expérience apprenant: Placer les émotions au cœur de l'apprentissage*. Medium. Retrieved June 28, 2024, from https://medium.com/@SoleneBerthet/design-dexp%C3%A9rience-apprenant-placer-les-%C3%A9motions-au-coeur-de-l-apprentissage-af809154396d

BIE—Bureau de l'international et de l'entrepreneuriat. (2022). *De l'idée à l'action: Contribuer au développement des habiletés essentielles par le biais de l'esprit d'entreprendre*. Document d'orientation institutionnel Cégep de Sainte-Foy. 1er novembre 2022. https://bie.csfoy.ca/fileadmin/documents/bie/Document_d_orientation_institutionnel-VF-A2022.pdf

Bureau, S., & Fendt, J. (2012). La dérive situationniste. Le plus court chemin pour apprendre à entreprendre? Revue française de gestion (4), 181–200.

Conseil des arts et des lettres du Québec. (2009). *Conseil des arts et des lettres du Québec creative residencies: Paris, Rome, Buenos Aires, London, Tokyo, Berlin, Mexico City, Lyon, Barcelona, Montréal, Inukjuak, Gatineau, Québec City*. Conseil des arts et des lettres du Québec. http://collections.banq.qc.ca/ark:/52327/1972546.

Creswell, J. W., Creswell, J. D., & Creswell, J. D. (2018). *Research design: Qualitative, quantitative, and mixed methods approaches* (5th ed.). SAGE.

Creswell, J. W. (2022). A concise introduction to mixed methods research. Second Edition. SAGE.

Damásio Manuel José, & Bicacro, J. (2017). Entrepreneurship education for film and media arts: how can we teach entrepreneurship to students in the creative disciplines? Industry and Higher Education, 31(4), 253–266.

ELM Learning. (n.d.). *What is learning experience design?* Retrieved June 28, 2024, from https://elmlearning.com/learning-experience-design-everything-you-need-to-know/

Fayolle, A. (2017). Chapitre 4. Création de valeur nouvelle et innovation. In *Entrepreneuriat* (pp. 107–130). Dunod. https://www.cairn.info/entrepreneuriat%2D%2D9782100765072-page-107.htm

Fayolle, A., & Gailly, B. (2008). From craft to science: Teaching models and learning processes in entrepreneurship education. *Journal of European Industrial Training, 32*(7), 569–593. https://doi.org/10.1108/03090590810899838

Floor, N. (2016, September 28). *This is learning experience design*. Retrieved June 28, 2024, from https://www.linkedin.com/pulse/learning-experience-design-niels-floor

Floor, N. (2023). *This is learning experience design: What it is, how it works, and why it matters* (1st ed.). New Riders.

Fortin, M.-F., & Gagnon, J. (2022). *Fondements et étapes du processus de recherche: Méthodes quantitatives et qualitatives* (4th ed.). Chenelière éducation.

Fridman, I., Meron, Y., & Roberts, J. (2022). Responsible design thinking: Informing future models of cross-disciplinary design education. *Journal of Design, Business and Society, 8*(2), 145–166. https://doi.org/10.1386/dbs_00037_1

Galaykova, M. (2021). *Learning experience design—get to know the elephant*. Retrieved June 28, 2024, from https://galayketti.medium.com/learning-experience-design-get-to-know-the-elephant-385c5e9c259d

Gamba, T. (2017). D'où vient la « pensée design » ?. I2D - Information, données & documents, 54, 30-32. https://doi.org/10.3917/i2d.171.0030

Greene, J. C., Caracelli, V. J., & Graham, W. F. (1989). Toward a conceptual framework for mixed-method evaluation designs. *Educational Evaluation and Policy Analysis, 11*, 255–274.

Hägg, G., & Gabrielsson, J. (2019). A systematic literature review of the evolution of pedagogy in entrepreneurial education research. *International Journal of Entrepreneurial Behavior & Research, 26*(5), 829–861. https://doi.org/10.1108/IJEBR-04-2018-0272

Haylock, B. (2020). "Problem formulation is the problem" dans Haylock, B. et Wood, L. (Eds.) One and many mirrors: perspectives on graphic design education. Occasional Papers. ISBN 978-0-9954730-1-0

IcoD—International Design Council. (2020). *Disruption is an opportunity for change: What role can designers have in creating the new normal?* Retrieved June 28, 2024, from https://www.theicod.org/en/resources/news-archive/disruption-is-an-opportunity-for-change

Kakouris, A., & Liargovas, P. (2020). On the about/for/through framework of entrepreneurship education: A critical analysis. *Entrepreneurship Education and Pedagogy, 4*, 396–421. https://doi.org/10.1177/2515127420916740

LXD—Learning Experience Design. (n.d.). *Learning experience design basics*. https://lxd.org/fundamentals-of-learning-experience-design/

Maher, Raymond & Maher, Melanie & Mann, Samuel & Mcalpine, Clive. (2018). Integrating design thinking with sustainability science: a Research through Design approach. Sustainability Science. 13. https://doi.org/10.1007/s11625-018-0618-6.Markard, J., Raven, R., & Truffer, B. (2012). Sustainability transitions: An emerging field of research and its prospects. *Research Policy, 41*(6), 955–967. https://doi.org/10.1016/j.respol.2012.02.013

Mattessich, P. W. Monsey, B. R. (1992). Collaboration: What Makes It Work- A Review of Research Literature on Factors Influencing Successful Collaboration. Amherst H Wilder Foundation. ISBN : 978-0940069022.Meyer, M. W., & Norman, D. (2020). Changing design education for the 21st century. *She Ji: The Journal of Design, Economics and Innovation, 6*(1), 13–49.

Nagels, M. (2022). Les méthodes mixtes, une perspective pragmatique en recherche. In *Traité de méthodologie de la recherche en Sciences de l'éducation et de la formation*. Enquêter dans les métiers de l'humain. https://hal.science/hal-03857724

Noël, G. (2020a) We All Want High-Quality Design Education: But What Might That Mean? She Ji: The Journal of Design, Economics, and Innovation, 6(1), 5-12. https://doi.org/10.1016/j.sheji.2020.02.003

Noël, G. (2020b). Fostering design learning in the era of humanism. *She Ji: The Journal of Design, Economics, and Innovation, 6*(2), 119–128. https://doi.org/10.1016/j.sheji.2020.05.001

Pepin, M. (2015). *Apprendre à s'entreprendre en milieu scolaire: Une étude de cas collaborative à l'école primaire*. Université Laval.

Pepin, M., Tremblay, M., & Audebrand, L. K. (2017). *L'entrepreneuriat responsable: Cadre conceptuel et implications pour la formation*. Document de travail 2017–008, Faculté des sciences de l'administration, Université Laval. ISBN 978-2-89524-451-6.

Pepin, M., Audebrand, L. K., Tremblay, M., & Keita, N. B. (2021). Evolving students' conceptions about responsible entrepreneurship: A classroom experiment. *Journal of Small Business and Enterprise Development, 28*(4), 570–585. https://www.emerald.com/insight/content/doi/10.1108/JSBED-02-2020-0035/full/html

Perry O'Neal. S. L. Hindman, L. and Donaldson, J. (2023). Learning Experience Design (LXD) Toolkit. https://workbookwoman.s3.us-west-1.amazonaws.com/LXD+Toolkit+6-6-23/index.html#/

Rae, D. (2003). Opportunity centred learning: An innovation in enterprise education? *Education Training, 45*(8/9), 542–549.

Rieckmann, M., Mindt, L., & Gardiner, S. (2017). *L'éducation en vue des objectifs de développement durable. Objectifs d'apprentissage*. https://unesdoc.unesco.org/ark:/48223/pf0000247507
Rieß, W., Martin, M., Mischo, C., Kotthoff, H.-G., & Waltner, E.-M. (2022). How can education for sustainable development (ESD) be effectively implemented in teaching and learning? An analysis of educational science recommendations of methods and procedures to promote ESD goals. *Sustainability, 14*, 1–16. https://doi.org/10.3390/su14073708
Roberts, J. (2012). Infusing Entrepreneurship within Non-Business Disciplines: Preparing Artists and Others for Self-Employment and Entrepreneurship. Artivate 1 (2), 53–63.
Savoie-Zajc, L. (2009). L'entrevue semi-dirigée. In Dans B. Gauthier (Dir.), Recherche sociale: De la problématique à la collecte de données 5th ed. : Presses de l'Université du Québec.
Schmidt, M., & Huang, R. (2021). Defining learning experience design: Voices from the field of learning design & technology. *TechTrends, 66*, 141–158. https://doi.org/10.1007/s11528-021-00656-y
Schmift, M. (2021). *Understanding the complexity of learning experience design*. Retrieved June 28, 2024, from https://medium.com/ux-of-edtech/understanding-the-complexity-of-learning-experience-design-a5010086c6ee
Schön, D. A. (1983). *The reflective practitioner: How professionals think in action*. Basic Books.
Schrauf, R. (2016). *Mixed methods: Interviews, surveys, and cross-cultural comparisons*. Cambridge University Press. https://www.cambridge.org/core/books/mixed-methods/540C45FDBF5A2E7FB67C6B5F24E1DB13
TransArtists. (n.d.). *What are residencies*? Retrieved June 28, 2024, from https://www.transartists.org/en/what-are-residencies
UNESCO. (2024). *Greening curriculum guidance: Teaching and learning for climate action*. https://doi.org/10.54675/AOOZ1758
Van der Maren, J. M. (1996). *Méthodes de recherche pour l'éducation*. De Boeck Université.
Valencia, A., Lievesley, M., & Vaugh, T. (2021). Four mindsets of designer-entrepreneurs. The Design Journal, 24(5), 705–726. https://doi.org/10.1080/14606925.2021.1958601
Verzat, C. (2015). "Esprit d'entreprendre, es-tu là?" Mais de quoi parle-t-on ? *Entreprendre & Innover, 27*(4), 81–92. https://doi.org/10.3917/entin.027.0081
Wasson, B., & Kirschner, P. A. (2020). Learning design: European approaches. *TechTrends: Linking Research and Practice to Improve Learning A publication of the Association for Educational Communications & Technology, 64*(6), 815–827. https://doi.org/10.1007/s11528-020-00498-0

Geneviève Raîche-Savoie , currently pursuing her doctoral studies at Université Laval, brings over a decade and a half of experience teaching design at both college and university levels across various Canadian institutions. Professionally, Geneviève is also engaged in an independent communication design practice focused on promoting environmental awareness and fostering positive social change. This blend of teaching and practice has led her to embrace the role of a "practitioner-researcher," bridging the realms of Education, Design, and Entrepreneurship. Her research is dedicated to nurturing an entrepreneurial mindset among graphic designers, aiming to foster a more reflective, socially, and environmental conscious approach to their profession.

Claudia Déméné is a professor of sustainable design at the School of Design, Université Laval, Canada. Her expertise focuses on the social and environmental issues surrounding the life cycle assessment. Her main research aims to develop an environmental label of the lifespan of electronic devices in order to defy different obsolescence forms. She is also a collaborative member of the Interdisciplinary Research Center for Operationalization of Sustainable Development (CIRODD) and the Québec Circular Economy Research Network (RRECQ). These research networks aim at combining research in sustainable development and in circular economy to solve complex problem in the socio-ecological transition in Québec.

Chapter 7
Exploring Emerging NLP and Machine Learning Methods in Climate Change Discourse Analysis on Social Media: A Systematic Literature Review

Hana Ghiloufi, Nicolas Merveille, and Sehl Mellouli

Abstract This study systematically examines emerging methods, particularly NLP and ML, for analyzing climate change discourse on social media platforms. Within this framework, sub-objectives encompass presenting methodological approaches and identifying prevalent climate change themes, and data sources. As climate change communication has evolved rapidly in the digital age, with social media becoming a pivotal arena for public discourse, opinion dissemination, and information exchange. The intersection of ML and NLP techniques offers unprecedented opportunities to transform vast amounts of unstructured data into valuable information, ready to be consumed by climate policymakers and different stakeholders. Drawing upon a comprehensive review of 56 articles, this study identifies and synthesizes six different methods that are further divided into sub-approaches and techniques, addressing climate change themes and platforms used. This research contributes to the literature by presenting the most used and effective methods and identifying potential areas needing more investigation in the future. It also provides insight into trending themes and overlooked ones, offering best practices and future research directions.

Keywords Climate change · Computational methods · Natural language processing · Machine learning · Social media

H. Ghiloufi · N. Merveille (✉)
Université du Québec à Montréal, Montreal, QC, Canada
e-mail: ghiloufi.hana@courrier.uqam.ca; merveille.nicolas@uqam.ca

S. Mellouli
Université Laval, Quebec City, QC, Canada
e-mail: sehl.mellouli@fsa.ulaval.ca

M. Cheriet et al. (eds.), *Accelerating the Socio-Ecological Transition*,
https://doi.org/10.1007/978-3-031-82896-6_7

1 Introduction

Climate change (CC) is a global concern, with far-reaching implications for our environment, societies, and economies (Chen et al., 2019; Shen & Wang, 2023). The adoption of the 2015 Paris Agreement and the integration of combatting CC as an independent goal within the Sustainable Development Goals (SDGs) have significantly elevated global discussions, debates, and actions across local, national, and international platforms (Kaushal et al., 2022).

As temperatures rise and extreme weather events become more frequent, the urgency of addressing this critical issue has never been clearer. As CC directly impacts our environment, individual and collective safety, and health, it necessitates collective discussion and resolution by both the public and policymakers (Ibrohim et al., 2023). Understanding public attitudes is an imperative component when it comes to tailoring effective climate policies and it should be taken into consideration (Khatibi et al., 2021; Hwang et al., 2021). Given the controversial nature of CC as a topic, a wide spectrum of opinions and positions has emerged (Howe et al., 2015). People tend to adopt specific viewpoints, often leading to the circulation of misinformation and fake information, especially in online platforms. These perspectives significantly impact climate action (Corner et al., 2012), either propelling or hindering its progress. Indeed, there is substantive evidence confirming that attitudes expressed on Social Media (SM) have a significant impact on individuals' inclinations toward embracing climate action (Mavrodieva et al., 2019). Consequently, policymakers must remain current with the ongoing discourse surrounding this critical issue. Therefore, it becomes essential to gather comprehensive information about public perceptions, attitudes, and sentiments toward CC and maximize the utility of available raw data by transforming it into actionable insights.

Simultaneously, there is a notable surge in the proliferation of SM platforms that facilitate discussions on various topics (Lee & Kwak, 2012). SM has undergone a paradigm shift in the communication area, significantly impacting the discourse on CC (Falkenberg et al., 2022). The immediacy and widespread reach of these digital platforms have altered the dynamics of information dissemination. This dialogue exposes researchers and stakeholders to a wealth of user-generated content (UGC), offering invaluable insights into this pressing concern and a massive data source that has not yet been used to the extent that matches its severity (Upadhyaya et al., 2023). The challenge lies in comprehensively analyzing and categorizing the diverse opinions expressed by users regarding CC within this massive volume of available data. Therefore, the demand for automatic processing of such data is increasing significantly alongside the rise in volumes of this type of data (Alturayeif et al., 2023). Fortunately, technological advancements have provided us with the means to address this challenge. Natural language processing (NLP) and text mining tools now offer powerful instruments for extracting maximum information from these platforms.

According to IBM Solutions,[1] NLP refers to the branch of artificial intelligence (AI) concerned with the interaction between computers and human language. It involves the development of statistical and machine learning (ML) models that enable computers to understand, interpret, and generate human language in a meaningful way. It encompasses various methods, including text mining, which involves extracting valuable insights from vast textual datasets (Usai et al., 2018) along with others such as sentiment analysis (SA), topic modeling (TM), and stance detection, which play pivotal roles in understanding and analyzing vast amounts of textual data. SA, for instance, enables the classification of text into positive, negative, or neutral sentiments (Liu, 2010). TM, on the other hand, uncovers underlying themes or topics present in a collection of documents, helping in the organization and summarization of textual data (Blei et al., 2003). Additionally, stance detection identifies the viewpoints expressed toward specific topics or issues in textual data, providing valuable insights into public opinion and discourse (Alturayeif et al., 2023). ML, a fundamental component of NLP, enhances the effectiveness and scalability of these techniques by providing algorithms and models that can learn patterns and relationships from large amounts of textual data (Glaz et al., 2021). By employing ML algorithms, NLP tasks can be automated and scaled to handle massive datasets efficiently (Glaz et al., 2021). Therefore, while NLP, ML, and text mining are distinct fields with their own methodologies and objectives, they are inherently interconnected, with ML serving as a critical enabler for advancing NLP techniques within the broader context of text mining and analysis (Nagarhalli et al., 2021).

These emerging methods are crucial in the decision-making process for businesses (Kasztelnik, 2020), policymakers, and government entities (Ibrohim et al., 2023), as they have demonstrated high accuracy and precision in detecting valuable insights. This practical utility prompts researchers to explore the various applications of these tools across different domains, including environmental concerns and the topic of CC, which forms the focus of this investigation. Indeed, they have garnered widespread adoption among researchers who seek to make use of publicly available data on CC through NLP from gauging public opinion concerns (Shen & Wang, 2023) to detecting climate hazards (Yigitcanlar et al., 2022).

However, there is a lack of holistic reviews that cover the intersection of CC discourse as an independent environmental issue, SM as a primary source of data, and NLP and ML methodologies. Previous reviews have typically focused on one or two dimensions. For example, Balaji et al. (2021) conducted a survey on ML algorithms used in SM analysis, discussing 17 supervised and unsupervised algorithms, while Ghani et al. (2019) concentrated on big data analytics for SM. Some have been undertaken to explore the literature regarding the efficacy and applications of specific methodologies, with SA emerging as a predominant focus (Ibrohim et al., 2023; Stede & Patz, 2021; Du et al., 2020; Tan et al., 2023; Wankhade et al., 2022; Tedmori & Awajan, 2019; Devi & Somasundaram, 2020) positioning SA at the

[1] What is Natural Language Processing? | IBM.

forefront of research inquiries. Other tasks have been targeted too, such as fake news detection (Agrawal et al., 2021), sarcasm detection (Shah & Shah, 2021), hate speech detection (Mullah & Zainon, 2021), and TM (Jelodar et al., 2019; Laureate et al., 2023; Asgari-Chenaghlu et al., 2021). Other works have focused on specific climate topics such as large language models LLMs for climate technology innovation (Toetzke et al., 2023). In a broader methodological aspect, Glaz et al. (2021) investigated the usage of ML and natural language processing techniques for mental health research, while patient experience feedback was explored in the study by Khanbhai et al. (2021).

Throughout the range of studies mentioned and numerous others, CC remained unaddressed. Bias was consistently limited to techniques or platforms across various themes, yet the intersection of the three was overlooked. These shortcomings serve as motivation to specifically explore this uncovered intersection. This study systematically examines emerging methodologies, particularly NLP and ML, for analyzing climate change discourse on social media platforms. To achieve this goal, we set different sub-objectives. These sub-objectives include outlining various emerging techniques, categorizing specific methods undertaken, and extracting different climate change themes discussed in the literature. Additionally, we seek to identify primary platforms commonly used for sourcing data and explore potential platforms suitable for gathering unstructured text data for climate change discourse.

Through these objectives, we look forward to contributing to the advancement of knowledge in both computer science and sustainability and environment studies fields by providing a deeper understanding of the public discourse on CC and guiding researchers by identifying useful and effective methods and data sources employed in the literature.

The following sections of this article are structured as follows: Sect. 2 outlines the methodology employed for conducting the SLR. Section 3 presents and discusses the findings of this SLR in relation to the research questions. Section 4 presents the gaps found and the possible future research avenues. Finally, Sect. 5 provides the conclusions of this survey.

2 Methodology

In this study, we undertake the compilation, categorization, and presentation of a current survey on the application of ML and NLP techniques for understanding CC discourse on SM platforms. To achieve this objective, we adopt the (Kitchenham, 2004), Systematic Literature Review (SLR) framework, which is particularly suited for computer science surveys and aligns to investigate state-of-the-art methodologies. Following this framework, we establish a review protocol encompassing four key stages: definition of research questions, design of search strategies, data collection, and data extraction.

2.1 Research Questions

RQ1: What is the current state of research on emerging methodologies for studying CC discourse on SM?
RQ2: What are the most effective methods recommended?
RQ3: Which taxonomy could be used to represent these works?
RQ4: What are the predominant CC-related topics and themes discussed on SM?
RQ5: What are the platforms for which discourse analysis models have been proposed?

2.2 Search Strategy

Preliminary searches were performed to determine the number of possibly relevant studies in our investigated research area. We conducted the search based on title, abstract, and keywords since searching based on full texts generated a large volume of irrelevant papers where their filtering is time-consuming and hence it's not an effective search strategy. This multidisciplinary research topic spans three main domains: emerging methods, the discourse surrounding CC, and online platforms. The selection of keywords used for the search query then is based on the relevance to the research objectives and the need to encompass various aspects of the topic. The query comprises three main categories: methodologies (e.g., SA, ML), CC-related terms (e.g., global warming, GHG emissions), and SM platforms (e.g., Twitter, online platforms).

The primary search query was then:

("sentiment analysis" OR "opinion mining" OR "machine learning" OR "nlp" OR "data mining" OR "computational linguistics" OR "text mining" OR "stance detection") AND ("climate change" OR "global warming" OR "GHG emissions" OR "carbon" OR "emissions" OR "sdg13") AND ("social media" OR "social network" OR "Twitter" OR "online platforms").

Four primary electronic databases, well known for their extensive coverage for major journals in the field of social computing and ML (Alturayeif et al., 2023), were employed for data collection namely: Scopus, Web of Science (WoS), IEEE Xplore, and the Association for Computing Machinery (ACM). All identified records were saved in a Zotero Library to facilitate the manual selection process.

2.3 Study Selection

This systematic literature review adopted Kitchenham (2004) framework for software engineering as a methodological backbone. Following their guidelines, adapted and adjusted by Ibrohim et al. (2023), this survey implements the paper selection as a four-step process, as shown in Table 7.1.

The paper selection process initiates with an identification and screening phase, where metadata from the four specified databases is gathered through direct search queries, resulting in a total of 1931 articles. To streamline the process, both automatic and manual selection procedures are employed. Initially, automatic procedures are executed, encompassing identification, screening, and initialization phases, during which duplicates are eliminated. For initialization, papers published within the last 10 years are filtered, aligning with the focus on emerging technologies and advancements. Subsequently, a relevance assessment is conducted based on predefined criteria detailed in Table 7.1, involving a thorough review of titles and abstracts to ascertain their relevance. As inclusion and exclusion criteria should be based on the research question (Kitchenham, 2004), they should be effectively set to ensure reliable interpretability and correctly classify studies. This evaluation is

Table 7.1 Inclusion and exclusion criteria

Inclusion	Exclusion
Stage 0: Identification and screening: automatically	
	• More than 10 years passed from publication
	• Duplicates
Stage 1: Initialization: automatically	
Research paper	• Not a research paper: Excludes papers that are not original research, such as commentaries, technical reports, or opinion pieces
	• Conference paper
Stage 2: Title and abstract: manually	
Meet relevance criteria:	• Does not meet the relevance criteria
By answering the following questions:	• Survey paper
1. Does the paper explicitly mention and discuss CC? (mandatory)	• Propose basic text mining techniques/qualitative content analysis
2. Does the paper propose an NLP or ML approach? (mandatory)	
3. Is SM the primary source of text data? (mandatory)	
Stage 3: Full-text selection: manually	
Meet the paper quality criteria by answering the following questions	• Fails one or more mandatory questions
1. Did the paper clearly describe the research goals/problems? (mandatory)	
2. Did the paper present relevant literature or previous studies? (mandatory)	
3. Did the paper clearly explain the research methodology? (mandatory)	
4. Did the paper come from a top journal? (optional)	

guided by three specific questions designed to assess papers' alignment with the research questions. Papers failing to meet any of these criteria are promptly excluded from further consideration. Lastly, we undertake a thorough assessment of paper quality through in-depth analysis of full-text content assessment. Here, we pose three mandatory questions and one optional question, each evaluated on a binary scale of fulfillment or not. Papers failing to satisfy one or more of the mandatory criteria are consequently excluded from the review process. After the filtering process, we had a final set of 56 papers to be included in the review. The following PRISMA diagram illustrates the filtering process (Fig. 7.1).

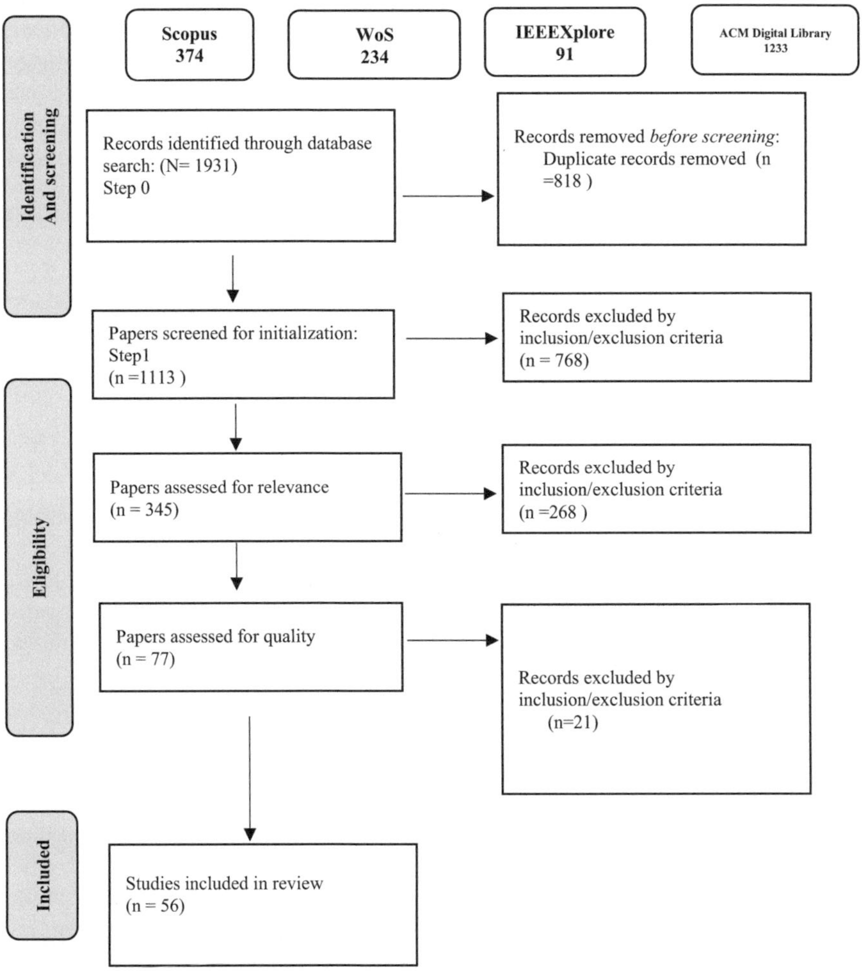

Fig. 7.1 PRISMA flow diagram

2.4 Data Extraction

This step involves gathering pertinent data from collected papers for analysis and transformation into meaningful insights, holding significant importance due to its direct impact on the conclusions drawn. Hence, we thoroughly extracted both quantitative and qualitative data. To synthesize the findings from the collected studies and address the research questions (RQs), we employed a combination of manual and automatic techniques. For this step, we compiled the necessary information into an Excel spreadsheet to address the proposed research question, dividing the process into two distinct phases.

The initial phase concentrates on extracting global information, encompassing elements that fulfill the RQs and metadata such as paper title, authors' names, journal name, abstract, keywords, techniques utilized, dataset volume (if applicable), labels (if applicable), performance accuracy (if applicable), SM platform, manually labeled CC topic, country of study (if applicable), and dataset language. Subsequently, the second phase zooms in on the methodologies, with a separate sheet dedicated to breaking down the global methodology into specific methods and approaches. This sheet includes the paper's identifier by title and authors' names, NLP or ML method (e.g., method1, method2…), independent approaches employed (e.g., approach 1, approach 2…), as well as the lexicon, embedding, or feature extraction techniques used.

3 Results and Discussion

3.1 Methodological Dimension

The first section will answer the first three research questions about the emerging methods observed in the literature. The review uncovered six methods employed by researchers for analyzing CC discourse on SM. While some studies chose to utilize independent approaches, others integrated various methods. Tables 7.2 and 7.3 present an outline of these methods and summarize the diverse combinations identified and the corresponding works, respectively.

3.1.1 Sentiment Analysis

SA emerged as the prevailing method in the reviewed literature, and featured prominently in 40 out of 56 papers, either as a standalone method or in tandem with other approaches. Among these, 16 papers exclusively focused on SA. Within this subset, lexicon-based SA was adopted in ten works, classic ML in four, and deep learning in two. Notably, SA frequently complemented other techniques, with TM being its most common counterpart, found in 14 papers. Additionally, seven papers explored

Table 7.2 Detected methods

Method	Definition	#works
Sentiment analysis	Assessing the sentiment or emotional tone expressed within a piece of text, typically to determine whether it is positive, negative, or neutral	41
Topic modeling	A statistical technique used to identify abstract topics or themes within a collection of documents to discover hidden semantic structures	22
Text classification	Categorizing text documents into predefined classes based on their content or features, often using ML algorithms	7
Cluster analysis	Grouping similar documents into clusters or categories based on their content, without predefined classes, to discover natural groupings within the data	6
Named entity recognition (NER)	Identifying and categorizing named entities (such as names of people, organizations, locations, etc.) within a text corpus, often used for information extraction	2
Stance detection	The process of determining the perspective or stance expressed toward a particular topic or claim within a given text, typically categorized as supporting, opposing, or neutral	1

Table 7.3 Summary of existing works

	Method		Combination of methods	Papers	#works
1	SA	1.a	SA alone	Ballestar et al. (2022), Yao et al. (2022), Sham and Mohamed (2022), Cheng et al. (2023), Schonfeld et al. (2021), Yigitcanlar et al. (2022), Abdar et al. (2020), Barrios-O'Neill (2021), Noviello et al. (2023), Xiang et al. (2021), Klingenberger et al. (2022), Jeong et al. (2023), Kim et al. (2021), Kirelli and Arslankaya (2020), Meenar et al. (2023), Lydiri et al. (2022)	17
		1.b	SA/Emotion analysis	El Barachi et al. (2021), Smirnov and Hsieh (2022), Chen (2020), Thukral et al. (2021), Arce-García et al. (2023), Bergstedt et al. (2018), Loureiro and Alló (2020)	7
		1.c	SA+ TM	Mulyani et al. (2024), Zeng (2022), Lee et al. (2023), Shen and Wang (2023), Pani et al. (2023), Bui et al. (2023), Shen and Li (2023), Zhang et al. (2022), Gjorshoska et al. (2023), Wu et al. (2023), Dahal et al. (2019), Qiao and Williams (2022), Zander et al. (2023b), Mouronte-López and Subirán (2022)	14
		1.d	SA+ text classification	Karimiziarani et al. (2023)	1
		1.e	SA+ cluster analysis	Zhang et al. (2021), Li et al. (2023)	2

(continued)

Table 7.3 (continued)

	Method		Combination of methods	Papers	#works
2	Text classification	2.a	Text classification	Ji et al. (2024), Rizzoli (2023), Styve et al. (2022)	3
		2.b	Text classification + SA	1.d	
		2.c	TM + Text classification	Camarillo et al. (2021)	1
3	Cluster analysis	3.a	Cluster analysis	Li et al. (2023), Han and Sun (2024), Kaushal et al. (2022)	3
		3.b	Cluster analysis + SA	1.e	
		3.c	Cluster analysis + NER	Chen et al. (2023)	1
		3.d	TM + cluster analysis	Yang et al. (2021)	1
4	NER	4.a	NER + TM + SA	Navarro and Tapiador (2023)	1
		4.b	Cluster analysis + NER	3.c	
5	TM	5.a	TM + text classification	Camarillo et al. (2021), Bennett et al. (2021)	2
		5.b	TM	Zander et al. (2023a), Moghadas et al. (2023)	2
		5.c	TM + SA	1.c	
		5.d	TM + text classification	2.c	
		5.e	TM + cluster analysis	3.d	
6	Stance detection		Stance detection	Tyagi et al. (2021)	1

SA with a more nuanced approach of emotion analysis, while one paper incorporated it into text classification. Table 7.4 presents the summary of observed techniques.

Four SA approaches were spotted: classic ML, deep learning (DL), Lexicon-based, and hybrid approach.

Studies ranged from simple implementation to complex ones and comparative models, Klingenberger et al. (2022) opted for a simpler approach where they deployed a Naïve Bayes classifier for SA across 5284 posts and comments sourced from three diverse SM platforms: Twitter, YouTube, and LinkedIn, all discussing the low-carbon agenda. Their analysis revealed significant levels of optimism toward sustainability, particularly evident on LinkedIn. Moreover, the study highlighted the versatility of ML approaches in capturing sentiment across varied SM platforms. Naïve Bayes was also used by Kirelli and Arslankaya (2020) alongside with K-Nearest-Neighbor (KNN) and Support Vector Machine (SVM) on 32,848 Turkish tweets on the general topic of CC Among these methods, KNN exhibited the highest

Table 7.4 SA approaches summary

SA	Library/algorithm	Papers
Lexicon-based	VADER	Ballestar et al. (2022), Smirnov and Hsieh (2022), Pani et al. (2023), Xiang et al. (2021), Gjorshoska et al. (2023), Meenar et al. (2023), Dahal et al. (2019), Karimiziarani et al. (2023), Navarro and Tapiador (2023), Mouronte-López and Subirán (2022)
	NRC Lexicon	Smirnov and Hsieh (2022), Barrios-O'Neill (2021), Thukral et al. (2021), Arce-García et al. (2023), Bergstedt et al. (2018), Loureiro and Alló (2020), Qiao and Williams (2022)
	Textblob	Sham and Mohamed (2022), Zander et al. (2023b), Mouronte-López and Subirán (2022)
	SnowNLP	Zeng (2022), Cheng et al. (2023), Shen and Wang (2023)
	LIWC	Lee et al. (2023), Jeong et al. (2023)
	SentiStrength	Yao et al. (2022), Sham and Mohamed (2022)
	Customized lexicon	
	SentiWordNet	Sham and Mohamed (2022)
	Hu and Liu	Sham and Mohamed (2022)
	MPQA	Sham and Mohamed (2022)
	Tidytext	Chen (2020)
ML-based	SVM	Sham and Mohamed (2022), Schonfeld et al. (2021)
	Naive Bayes	Sham and Mohamed (2022), Klingenberger et al. (2022), Kirelli and Arslankaya (2020)
	Logistic regression	Sham and Mohamed (2022)
	KNN	Kirelli and Arslankaya (2020)
Deep learning-based	LSTM	Noviello et al. (2023)
	RNN	Shen and Wang (2023)
	BiLSTM	El Barachi et al. (2021), Thukral et al. (2021), Wu et al. (2023)
	BERT	Mulyani et al. (2024), Shen and Wang (2023)
	EmoRoBERTa	Bui et al. (2023)
	GRU	Noviello et al. (2023)
	DistilBert transformer	Noviello et al. (2023)
	RoBERTa	Kim et al. (2021)
	CNN	Lydiri et al. (2022)
Ensemble learning	SVM + RNN-GRU + RNN-LSTM + Gradient Boosting + Logistic regression	Schonfeld et al. (2021)
Instant end user tools	WEKA	Yigitcanlar et al. (2022)
	NLPIR Parser	Shen and Li (2023)
	NetMapper	Tyagi et al. (2021)

performance, achieving an accuracy of 74.63%, which is regarded as excellent within the context of a Turkish corpus.

Works like Schonfeld et al. (2021) tried more complex combinations but on one platform evaluated two ML approaches for SA. They compared a basic SVM with an ensemble classifier, which combined Recurrent Neural Network with Gated Recurrent Unit (RNN-GRU), Recurrent Neural Network with Long Short-Term Memory (RNN-LSTM), a logistic regression classifier, and gradient-boosted decision Tree. This evaluation was conducted on a dataset comprising 2.71 million tweets categorized into anti/pro and neutral sentiments toward CC. The results indicated that the ensemble learning approach outperformed the simple SVM, achieving an accuracy of 78.5% recommending the use of DL and ensemble approaches. This is evidenced by Kim et al. (2021) who applied RoBERTa—transformer-based model—to analyze 266,686 tweets related to Solar energy as a CC mitigation strategy achieving 80.2% accuracy and confirming the utility of DL models in tackling such problems. Drawing from this recommendation, Noviello et al. (2023) conducted experiments employing various deep learning (DL) models, including GRU, two variants of LSTM, and DistilBERT. These experiments were conducted on a substantial dataset comprising 35 million climate-related tweets, 300,000 news articles, and an additional 130 million tweets related to natural disasters in the US spanning a decade. The results indicated that a single LSTM model achieved superior performance, reaching an accuracy of 96.1%, coupled with high scalability.

The second approach of lexicon-based SA was widely adopted, in total 58% of SA-scanned papers used this approach in both general SA and emotion analysis. In their study, Ballestar et al. (2022) employed the VADER lexicon to analyze the sentiments expressed in tweets by CC influencers, specifically Greta Thunberg and Bill Gates, along with their replies during COP25, categorizing them into three polarities. Their findings revealed a notably more positive sentiment toward Gates compared to Thunberg.

Similarly, Xiang et al. (2021) and Meenar et al. (2023) used the VADER lexicon to investigate sentiments regarding China's carbon policy globally and perspectives on heat waves, respectively. Xiang et al. (2021) examined the sentiment across a vast dataset of 1,590,143 tweets, while Meenar et al. (2023) supplemented their analysis of heatwaves with survey responses, amalgamating 107 survey answers with 367 tweets.

Other lexicons and libraries were employed including SentiStrength by Yao et al. (2022) to contrast the sentiments of French and English users toward carbon neutrality over 46,756 tweets. Their findings highlighted the prevalence of negative sentiments in both groups, particularly among the French population. Similarly, Barrios-O'Neill (2021) leveraged the NRC sentiment lexicon to examine the social contagion of environmental organization advocacy on Twitter, with a primary emphasis on CC and overexploitation. They argued the amplifying effect of emotional negativity on social contagion dynamics, this finding emphasizes the role that SA plays in understanding complex social phenomena.

The DJLex of TextBlob and the Linguistic Inquiry and Word Count (LIWC) lexicons were employed respectively by Abdar et al. (2020) in their analysis of Reddit users and by Jeong et al. (2023) in their study focusing on Twitter users from Alaska. Both investigations centered on renewable energy as a strategy for mitigating CC.

In addition, Xi and Xidian developed customized lexicons for conducting SA on global warming, demonstrating the adaptability of lexicon-based approaches to various environmental discourse contexts and languages too as demonstrated by Cheng et al. (2023) who used Baidu Dictionary containing the sentiment scores of Chinese words with SnowNLP on 4,850,000 Sina Weibo posts to study users' perceptions on heat waves.

Other studies focused on a specific augmented SA with a more nuanced approach namely emotion analysis which is a computational technique used to identify and analyze the emotional content present in textual data. Emotion mining goes beyond traditional SA to more nuanced granularities to detect specific emotions expressed in text, such as joy, anger, sadness, fear, and disgust (Yadollahi et al., 2017). In this category, we observed the dominance of the usage of NRC Lexicon (Smirnov & Hsieh, 2022; Arce-García et al., 2023; Bergstedt et al., 2018; Loureiro & Alló, 2020).

In examining CC discussions amidst the finite pool of societal concerns and the COVID-19 pandemic on Twitter, Smirnov and Hsieh (2022) employed lexicon-based SA techniques, namely the NRC emotion lexicon and VADER. This analysis focused on discerning sentiments into negative and positive categories, along with seven basic emotions of fear, anger, sadness, disgust, trust, joy, and surprise with both lexicons performing well achieving similar sentiment and emotion prediction. The same set was used by Arce-García et al. (2023), to examine SM responses to climate activists during CC conferences, drawing on Pulchnik's basic emotions across 1,395,054 tweets. Furthermore, the same lexicon was applied in the analysis of emotions related to energy policy by Loureiro and Alló (2020). Their study, encompassing 1,700,000 tweets from the UK and Spain, identified eight basic emotions. Interestingly, they found that messages regarding CC in the UK tended to be less negative compared to those in Spain. Additionally, anticipation emerged as the predominant feeling in the UK, while fear was more prevalent in Spain. Using the same lexicon and set of emotions along with Hu and Liu's (2004) lexicon, Bergstedt et al. (2018) analyzed Alaskans' perceptions toward CC over 1,900,000 tweets.

Finally, two papers employed a hybrid approach, combining multiple methodologies to enhance their analyses. A hybrid approach typically involves the integration of different techniques (lexicon-based and ML-based) or models to leverage their respective strengths and compensate for their weaknesses. In their study, Thukral et al. (2021) employed such an approach, integrating lexicons and deep learning techniques. They harnessed the capabilities of Bidirectional Long Short-Term Memory (BI-LSTM) networks to classify the intensity of concern regarding CC into three fundamental polarities: positive, negative, and neutral over 59,999 tweets. The emotional analysis was specifically applied to filtered positive tweets across three distinct case studies, each focusing on significant global events: Earth Day, Delhi Air Pollution, and the Australian Bushfire. Sham and Mohamed (2022) conducted experiments testing various hybrid combinations of lexicons and classic machine-learning classifiers. They evaluated SentiWordNet, TextBlob, VADER, SentiStrength, Hu and Liu, MPQA, and WKWSCI over a dataset consisting of climate-related tweets. Among these combinations, the hybrid approach using TextBlob in conjunction with Logistic Regression emerged as the most effective, achieving an F1-score of 75.3% providing evidence of the efficiency of hybrid approaches.

3.1.2 Topic Modeling

The second most used technique is TM, however, its usage as a standalone method was only observed in two studies both working on CC tangible manifestations. In their work, Zander et al. (2023a) employed the Gibbs sampling algorithm for the Dirichlet multinomial mixture (GSDMM) to analyze Twitter responses to heat waves, with the CC topic being discussed in 17% of the analyzed tweets. The authors contended that there was no misinformation disseminated regarding CC according to tweet source identification.

Moghadas et al. (2023) addressed the German flood disaster, which was directly attributed to CC by scientists. They used information collected from both tweets and online surveys to detect climate disaster capacities. The CC topic was present in all disaster phases (pre-, real-time, and post-event), highlighting its impact. Addressing disaster capacities adaptation and prevention involves the adoption of CC adaptation measures. Topic models were mostly combined with SA, text classification, and cluster analysis.

3.1.3 Sentiment Analysis and Topic Modeling Combined

An important percentage of papers examined coupled SA with TM, this technique was recommended by Ibrohim et al. (2023) or what is called topic-based SA where the SA is plotted against the topic to identify what people think or feel toward a specific topic and how it varies from one theme to another. By integrating these two approaches, we can understand not only the emotional tone of the text but also the context in which sentiments are expressed, enabling targeted sentiment assessment and informed decision-making. Moreover, this combination enhances content personalization efforts by delivering more relevant and tailored information to users based on both their interests and sentiments. Table 7.5 summarizes the topic models used in the analyzed works.

For SA, the lexicon-based approach was used in 86% of the papers, the other two works opted for deep learning techniques namely EmoRoBERTa and BiLSTM. For topic modeling, we observed the use of statistical models, ML, deep learning, and hybrid approaches, five techniques were used namely LDA (ten papers), Dirichlet Multinomial Regression (DMR) (1), Gibbs Sampling Algorithm for the Dirichlet Multinomial Mixture (GSDMM) (1), and Text Semantic Similarity (1) and BERTopic (1).

The usage of the VADER lexicon was identified in four studies, with three of them employing it in conjunction with Latent Dirichlet Allocation (LDA), and one

Table 7.5 Summary of topic models

	Algorithm/library	Paper
Classic TM	LDA	Mulyani et al. (2024), Camarillo et al. (2021), Zeng (2022), Shen and Wang (2023), Pani et al. (2023), Shen and Li (2023), Bennett et al. (2021), Moghadas et al. (2023), Dahal et al. (2019), Qiao and Williams (2022), Navarro and Tapiador (2023), Yang et al. (2021), Mouronte-López and Subirán (2022), Bui et al. (2023), Wu et al. (2023)
	Dirichlet Multinomial Regression (DMR)	Lee et al. (2023)
	Gibbs sampling algorithm for the Dirichlet multinomial mixture (GSDMM)	Zander et al. (2023a, 2023b)
	Text Semantic Similarity	Zhang et al. (2022)
Deep learning TM	BERTopic	Gjorshoska et al. (2023)

study employing BERTopic (Gjorshoska et al., 2023). Among these, two studies addressed the issue of CC mitigation by reducing greenhouse gas (GHG) emissions through the adoption of electric vehicles (EV) (Pani et al., 2023) and waste management (Gjorshoska et al., 2023).

Pani et al. (2023) used the VADER lexicon and LDA to analyze 7,102,362 tweets where the Changepoint analysis was employed to detect significant shifts in public sentiment. Findings revealed a prevailing positivity toward EV usage and emphasized its environmental benefits in GHG reduction. Conversely, Gjorshoska et al. (2023) observed a predominance of negative sentiment surrounding waste management, with discussions mainly focusing on inadequate waste disposal practices and the widespread use of plastic bags. These findings were derived through VADER SA and BERTopic applied to data from Twitter and news reports. The same negativity pattern was observed by Dahal et al. (2019) who conducted a comparative study on regional discussions on CC on Twitter. Their findings unveiled a predominantly negative sentiment within these discussions, intricately intertwined with "political situations" and "extreme weather conditions" topics. Furthermore, the research revealed that Americans exhibited relatively lower levels of interest in CC policies when contrasted with their counterparts in other nations. Mouronte-López and Subirán (2022) also found a prevalence of negative sentiment surrounding CC topics. They used VADER augmented with the Textblob lexicon to identify various topics, including Net zero, pandemic, Climate emergency, CC evidence, Government action, Policies and finances, CC awareness, CC activism, Sustainability, and Biodiversity. Most of these topics exhibited predominantly negative polarity, except for CC activism. Notably, the CC evidence topic displayed the highest level of negativity among all identified topics expressing CC denial.

TextBlob lexicon was also employed alongside the Gibbs sampling algorithm for the Dirichlet multinomial mixture (GSDMM) in Zander et al. (2023b). They investigated trends in 82,927 Twitter discussions about the Australian Bushfires during the black summer, as a tangible manifestation of CC. Surprisingly, despite the severity of the incidents, a general trend of positive sentiment was observed. However, negativity was predominantly found in topics related to government actions, protection, and prevention measures. This negativity toward government entities was observed in the Indonesian context regarding photovoltaic (PV) adoption in a study by Mulyani et al. (2024) through the analysis of 5792 YouTube and TikTok comments by BERT for SA and LDA for TM. Additionally, LDA was coupled with BiLSTM, another SA deep learning model, in Wu et al. (2023), focusing on Chinese public opinion on Sina Weibo with 169,592 posts. Surprisingly, the majority exhibited positive attitudes, emphasizing carbon reduction, neutrality, green development, and new energy solutions. Slight negativity was attributed to extreme weather events and abnormal climate conditions. These findings corroborate the conclusions drawn by Shen and Wang (2023), who observed that topics such as natural disasters, deforestation, and glacier melting tend to evoke predominantly negative sentiments. Conversely, clean energy and education generate more positive sentiments among the Chinese populace. Meanwhile, individuals within the G20 nations tend to express positive sentiment toward human intervention and global cooperation, while exhibiting negativity toward ideological stances. These findings emerged from the application of different techniques across different platforms. Twitter data were analyzed using an RNN model while sentiments on Sina Weibo were assessed using SnowNLP with a Chinese lexicon both alongside LDA for content analysis. The latter combination was employed by Zeng (2022) on 346,921 Weibo posts who similarly observed an overall optimistic sentiment among the Chinese population regarding CC. Negative opinions, as identified, tend to be predominantly associated with climate-related disasters and exhibit heightened negativity during the winter and fall seasons. The results further illuminate that Chinese individuals tend to perceive CC through a "top-down perspective," considering it a matter of national concern rather than an issue at the individual level.

The Linguistic Inquiry and Word Count (LIWC)-2022 lexicon was used by Lee et al. (2023) to analyze 61,370 Reddit threads with Dirichlet Multinomial Regression (DMR). Their analysis uncovered positive sentiments associated with careers related to wildlife, fires, elections, and green energy, among other topics. Conversely, negativity was observed in discussions concerning natural disasters, carbon emissions, UN CC, global warming, and academia.

For emotion analysis deep learning and lexicon-based models were employed by Bui et al. (2023) and Qiao and Williams (2022) respectively with LDA. EmoRoBERTa achieved high precision in extracting nuanced CC emotions during Hurricane Harvey on Twitter highlighting fear emotions arising from CC stressors such as care for family and friends and landfall danger (Bui et al., 2023). The NRC lexicon was applied in a more general context of global warming with LDA by Qiao and Williams (2022). Fear prevails across all topics, with trust following closely. Trust is highest when discussing global warming reality and necessary actions, but lowest for Hoax topics. Anticipation is minimal in Hoax discussions but peaks when addressing global warming impacts. Sadness is most prominent in conversations about global warming and Covid-19. Joy is highest in Reality discussions but lowest in Hoax. Notably, Hoax discussions are overwhelmingly negative, marked by high fear, disgust, and surprise, and low trust, joy, and anticipation.

Figure 7.2 presents a visual summary of the aforementioned topics in a word cloud, highlighting the most discussed subjects. The importance and frequency of each word are represented by its size in the cloud. We observe the significance of energy topics and sustainability. Additionally, these topics are plotted against the sentiments found in the papers, where a corresponding sentiment score of 1 is assigned to each topic, whether positive or negative. Some topics are discussed in the literature as exhibiting both polarities because they are found in different contexts (different countries, spans of time, divisions into sub-topics, etc.). For example, topics like bushfires and green housing are given a score of 1 for both sentiments. Climate change topics, as analyzed by the set of references, mostly exhibit a negative sentiment, which is presented by 60%.

Table 7.6 illustrates that the most prevalent sentiment polarity classification is used as a three-category polarity identification of negative, neutral, and positive sentiments, as adopted by 18 works, nine of which belong to SA-only studies and nine works with TM. Following these 13 studies used the binary classification of negative and positive with ten independent SA.

However, some studies opt for a different labeling approach, such as pro, anti, or neutral, primarily because in this specific work, the authors investigated the attitudes toward two topics vaccines and CC. This approach is more aligned with opinion analysis, tailored to the specific research goals of the authors. Additionally, Bergstedt et al. (2018) introduced a nuanced polarity classification, categorizing sentiments into negative, positive, low negative, and low positive. Furthermore, six studies employed a multi-polarity scenario, as exemplified by Ibrohim et al. (2023), blending emotion and SA to extract both basic emotions and corresponding sentiment polarities. The identification of emotions in these studies is often based on Plutchik or Ekman's basic emotions, with reliance on NRC lexicons for SA.

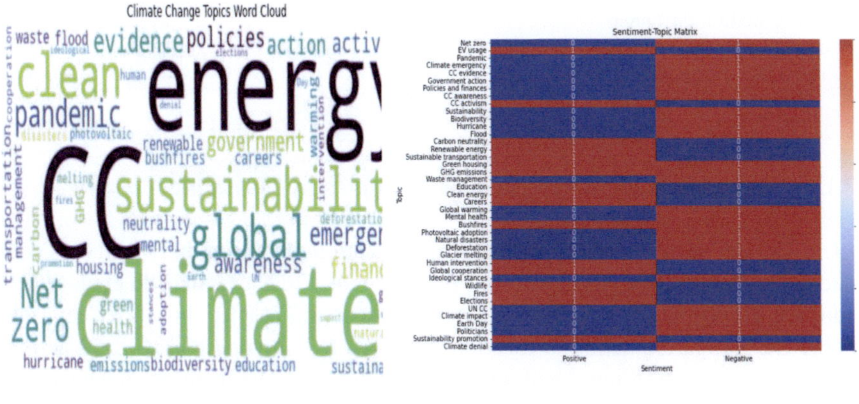

Fig. 7.2 Discussed topics and the corresponding sentiments

Table 7.6 Sentiment polarities summary

Polarity	SA as independent method	SA+ other methods
Negative/neutral/positive	Ballestar et al. (2022), Yao et al. (2022), Sham and Mohamed (2022), Abdar et al. (2020), Xiang et al. (2021), Klingenberger et al. (2022), Thukral et al. (2021), Kim et al. (2021), Meenar et al. (2023)	Mulyani et al. (2024), Shen and Wang (2023), Pani et al. (2023), Shen and Li (2023), Zhang et al. (2022), Gjorshoska et al. (2023), (Wu et al. (2023), Dahal et al. (2019), Zander et al. (2023b), Navarro and Tapiador (2023), Effrosynidis et al. (2022)
Negative/positive	Cheng et al. (2023), Smirnov and Hsieh (2022), Chen (2020), Barrios-O'Neill (2021), Noviello et al. (2023), Arce-García et al. (2023), Jeong et al. (2023), Loureiro and Alló (2020), Kirelli and Arslankaya (2020)	Zeng (2022), Lee et al. (2023), Mouronte-López and Subirán (2022), Zhang et al. (2021)
Pro-/anti-/neutral	Schonfeld et al. (2021)	
Negative/positive/low negative/low positive	Bergstedt et al. (2018)	
Believers/deniers	Lydiri et al. (2022)	
Anger/anticipation/disgust/fear/joy/sadness/surprise/trust	Chen (2020), Thukral et al. (2021), Arce-García et al. (2023), Bergstedt et al. (2018), Loureiro and Alló (2020)	
Anger/disgust/fear/joy/sadness/surprise/trust	Smirnov and Hsieh (2022)	
Pride, embarrassment, grief, relief, remorse, nervousness, joy, desire, confusion, disappointment, realization, amusement, disapproval, annoyance, disgust, excitement, love, gratitude, surprise, admiration, anger, sadness, optimism, approval, fear, curiosity, caring, neutral		Bui et al. (2023)

3.1.4 Topic Modeling with Text Classification and Clustering

LDA TM has been integrated with text classification/prediction in two studies conducted by Camarillo et al. (2021) and Bennett et al. (2021). Camarillo et al. (2021) adopted a two-phase approach: initially using Bi-LSTM to classify tweets into either CC mitigation action or inaction, followed by a second phase where action-related tweets were filtered and analyzed using LDA. The most prominent topics identified were "government actions," "environmental behaviors," "sustainable

Table 7.7 TM summary

	Technique	Paper
Classic TM	LDA	Mulyani et al. (2024), Camarillo et al. (2021), Zeng (2022), Shen and Wang (2023), Pani et al. (2023), Shen and Li (2023), Bennett et al. (2021), Moghadas et al. (2023), Dahal et al. (2019), Qiao and Williams (2022), Navarro and Tapiador (2023), Yang et al. (2021), Mouronte-López and Subirán (2022), Bui et al. (2023), Wu et al. (2023)
	Dirichlet Multinomial Regression (DMR)	Lee et al. (2023)
	Gibbs sampling algorithm for the Dirichlet multinomial mixture (GSDMM)	Zander et al. (2023a, 2023b)
	Text Semantic Similarity	Zhang et al. (2022)
Deep learning TM	BERTopic	Gjorshoska et al. (2023)

production," and "awareness." In the study by Bennett et al. (2021), 17 CC topics were identified, including climate impact, Earth Day, politicians, sustainability promotion, clean energy, and climate denial. These topics were regionally mapped, and their analysis was enhanced by random forest regression to predict consolidated opinions on CC. In another study by Yang et al. (2021), LDA was employed along with hierarchical clustering to extract 12 topics related to public concerns about air pollution and categorize them into egoistic, altruistic, biospheric concerns, and adaptation strategies. Table 7.7 summarizes topic models found in the analyzed works.

3.1.5 Text Classification

While supervised text classification remains one of the most used methods by researchers due to its superior performance, it's intriguing to note that only a handful of studies have independently employed it, alongside a few incorporating SA and TM. For instance, some studies have employed deep learning techniques. Ji et al. (2024), for example, leveraged the BERT model to categorize Sina Weibo posts related to CC into eight predefined topics. The labels are detailed in Table 7.8, along with the classification of information veracity into 'misinformation' and 'non-misinformation'. Notably, the model achieved an accuracy of 79.2% and a precision of 79% for topic classification and information veracity, respectively.

Classic ML algorithms were applied by Rizzoli (2023) to categorize Italian tweets into risk perception categories, specifically SVM and random forest. The results indicated an average accuracy of 86% and 70% respectively. In a different

Table 7.8 Labels used in text classification works

Paper	Labels
Ji et al. (2024)	(a) (1) politics and governance; (2) environmental and health impacts; (3) culture; (4) action advocacy; (5) beliefs and perception; (6) economy; (7) science and technology; and (8) other
	(b) Misinformation; nonmisinformation
Karimiziarani et al. (2023)	Caution; damage; evacuation;injury; help; sympathy
Rizzoli (2023)	Consciousness; justification; distance; denial
Styve et al. (2022)	Flood-related/not flood-related
Camarillo et al. (2021)	Action/inaction

approach, Karimiziarani et al. (2023) utilized an unspecified text classifier alongside lexicon-based SA to categorize tweets concerning Hurricane Harvey. Other studies have opted to compare different approaches. For instance, Styve et al. (2022) compared deep learning (CNN), classic ML (random forest, logistic regression), and transfer learning (Universal Language Model Fine-tuning, ULMFiT) for classifying flood-related tweets. Interestingly, the models trained using transfer learning via ULMFiT demonstrated the most promising results, achieving 95% accuracy, closely.

3.1.6 Cluster Analysis

The next method observed was the clustering analysis or text clustering, also known as text segmentation, which is an NLP technique used to group similar documents or pieces of text into clusters or categories based on their content (Hotho et al., 2003). Six works with four different methods as summarized in Table 7.9 were observed, two works used it independently (Han & Sun, 2024; Kaushal et al., 2022), two combined with SA, one with NER (Chen et al., 2023), and one with TM (Yang et al., 2021). *k*-means was most used in three works A BERT word embedding was used by Han and Sun (2024) followed by a fuzzy clustering method namely FCM to detect themes related to low carbon awareness. The combination of deep learning encoding and cluster analysis was also adopted by Kaushal et al. (2022) where USE sentence encoder and a *k*-means model were implemented for Reddits' discussions on CC extracting ten themes including and not limited to climate science, global warming, energy and carbon.

3.1.7 Named Entity Recognition (NER)

Named Entity Recognition (NER) is a method in NLP where entities such as person names, organization names, locations, and dates are identified and categorized within the text (Lample et al., 2016; Ratinov & Roth, 2009). NER systems employ

Table 7.9 Clustering techniques summary

Approach	Algorithm	Paper
Partitioning method	*k*-means	Chen et al. (2023), Zhang et al. (2021), Kaushal et al. (2022)
Fuzzy clustering	FCM	Han and Sun (2024)
Hierarchical method	Hierarchical clustering	Yang et al. (2021)
Non-specified	Co-occurrence clustering	Li et al. (2023)

ML algorithms or rule-based approaches to automatically detect and classify entities based on their linguistic features and context. In the analyzed studies, this technique has been used to evaluate attitudes and sentiments toward specific entities related to the CC debate using Standford NER tagger (Navarro & Tapiador, 2023) and to extract specific locations using Conditional Random Fields (CRF) techniques (Chen et al., 2023).

3.2 *Topical Dimension*

In this section RQ4 will be answered. From the analysis of the selected papers, a broad spectrum of climate-change-related topics and corresponding sub-topics were detected. We classified the papers as follows: CC as a general topic (18 papers), CC mitigation (17 papers), CC tangible manifestations (ten papers), CC activism and advocacy (six papers), and CC and health (five papers) as synthesized in Table 7.10.

3.2.1 CC as a General Topic

Various perspectives on the issues were mapped as they were addressed across different platforms, including Twitter, Sina Weibo, Reddit, and Twitch, along with the diverse keywords used for data collection. Consequently, disparities in the obtained results are apparent; however, some common findings have emerged. In numerous studies, researchers have sought to discern the prevailing public sentiment toward CC across various regional contexts. While many analyses have taken a broad approach, others have focused on specific regions. The research focused on China, for instance, has highlighted an overall optimistic outlook in Chinese discourse regarding CC, with less emphasis on its political dimensions (Zeng, 2022; Ji et al., 2024; Shen & Wang, 2023). Conversely, investigations concerning the US public have highlighted a keen interest in political aspects, alongside polarization and climate denial (Bennett et al., 2021), with other works focusing on particular states like Alaska (Bergstedt et al., 2018). Additionally, there have been inquiries into the attitudes of the Turkish population (Kirelli & Arslankaya, 2020).

Table 7.10 Detected themes

Theme	Sub-theme	Paper
CC as a general topic		Lydiri et al. (2022), Sham and Mohamed (2022), Lee et al. (2023), Shen and Wang (2023), Ji et al. (2024), Thukral et al. (2021), Kaushal et al. (2022), Bennett et al. (2021), Bergstedt et al. (2018), Kirelli and Arslankaya (2020), Wu et al. (2023), Effrosynidis et al. (2022), Dahal et al. (2019), Qiao and Williams (2022), Navarro and Tapiador (2023), Mouronte-López and Subirán (2022), Zeng (2022), Qiao and Jiang (2021)
CC mitigation	CC mitigation in general	Camarillo et al. (2021)
	Carbon neutrality	Li et al. (2023), Yao et al. (2022), Pani et al. (2023), Han and Sun (2024)
	Renewable energy	Mulyani et al. (2024), Abdar et al. (2020), Zhang et al. (2022), Jeong et al. (2023), Kim et al. (2021)
	Carbon taxation	Zhang et al. (2021)
	Sustainable transportation	Pani et al. (2023)
	Green housing	Shen and Li (2023)
	Corporate climate agenda	Klingenberger et al. (2022)
	GHG emissions	Zhang et al. (2022), Gjorshoska et al. (2023)
	Waste management	Gjorshoska et al. (2023)
	Energy policies	Loureiro and Alló (2020)
	CC policy	Xiang et al. (2021)
CC tangible manifestations	Floods	Styve et al. (2022), Moghadas et al. (2023)
	Waterlogging	Chen et al. (2023)
	Heat waves	Cheng et al. (2023), Zander et al. (2023a), Meenar et al. (2023)
	Natural hazards	Yigitcanlar et al. (2022)
	Extreme weather events	Noviello et al. (2023)
	Hurricanes	Karimiziarani et al. (2023)
	Bushfires	Zander et al. (2023b)
CC activism and advocacy	Greta Thunberg	El Barachi et al. (2021), Ballestar et al. (2022), Arce-García et al. (2023)
	Influencers marketing	Ballestar et al. (2022)
	NGOs	Barrios-O'Neill (2021), Ter-Mkrtchyan and Taylor (2023)
CC and health	COVID-19	Smirnov and Hsieh (2022)
	Vaccines	Schonfeld et al. (2021)
	Mental health	Bui et al. (2023), Chen (2020), Rizzoli (2023)
CC adaptation		Yang et al. (2021)

Despite its global scope, discussions on CC are marked by a mixture of sentiments. Advocates for addressing CC are prominent (Lydiri et al., 2022), with an overall prevalence of positive sentiment (Zeng, 2022; Sham & Mohamed, 2022), emphasizing a global approach and human capacity to intervene effectively (Shen & Wang, 2023). Conversely, negative attitudes toward CC often revolve around politics and extreme weather events (Shen & Wang, 2023).

3.2.2 CC Mitigation

Mitigation discussions emerged prominently in the analyzed papers as one of the most popular strategies for addressing CC. Mitigation involves the actions and efforts aimed at reducing greenhouse gas (GHG) emissions and bolstering carbon sinks (Introduction to Mitigation | UNFCCC). Eleven subtopics were extracted, with carbon neutrality and renewable energy emerging as the primary areas of research focus. Attitudes toward carbon neutrality ranged from optimistic findings (Li et al., 2023) in the Chinese context to pessimistic ones (Yao et al., 2022) among English and French users, with diverse sub-dimensions related to reducing carbon emissions such as economy, technology, sports, entertainment, and transportation. There was a consensus regarding the positivity of research on renewable energy worldwide (Mulyani et al., 2024; Abdar et al., 2020; Kim et al., 2021). However, the temporal aspect influenced the public's overall opinion, with differing views on short-term and long-term adoption of renewable energy. Works related to policies tackled both general climate policies (Xiang et al., 2021) and specific dimensions of energy (Loureiro & Alló, 2020).

Despite being the primary driver of CC and the focal point of CC mitigation efforts, it is surprising that only two works directly addressing GHG emissions were discovered, despite being included in keyword searches. This scarcity may stem from the strict quality selection criteria that may have led to the exclusion of relevant works. Twitter conversations predominantly centered around escalating emissions, the related climate crisis, and emission policies (Zhang et al., 2022).

3.2.3 CC Tangible Manifestations

Another important theme extracted is CC's tangible manifestations ranging from natural disasters to extreme weather events. Through SM, these manifestations serve as tangible touchpoints, highlighting the relationship between environmental realities and public opinion. As individuals grapple with the effects of CC their experiences are increasingly shared, debated, and reshaped within the digital sphere. Seven distinct subtopics were identified, with five associated with natural hazards and two with extreme weather events. With socio-technical advancements, such as those outlined by Moghadas et al. (2023), the monitoring of natural crises has become more refined, supporting responsible entities in implementing effective intervention strategies for climate resilience (Karimiziarani et al., 2023). This includes the development of models capable of accurately discerning content related to natural disasters (Styve et al., 2022) and assessing resilience capacities across preventive, anticipative, absorptive, adaptive, and transformative measures (Moghadas et al., 2023). Monitoring public opinion before, during, and after the disaster (Karimiziarani et al., 2023; Zander et al., 2023b). The analysis of extreme weather events-related content also provides valuable information into diverse perceptions of CC that was perceived negatively during these events attributing it as the major cause for such anomalies (Meenar et al., 2023).

3.2.4 CC Activism and Advocacy

Five studies focused on CC activism and advocacy. Within this category, three studies analyzed content related to the renowned climate activist Greta Thunberg (El Barachi et al., 2021; Ballestar et al., 2022; Arce-García et al., 2023). These works primarily employed sentiment and emotion analysis as the main method for content analysis, revealing predominant sentiments surrounding the climate activists. What's intriguing about this theme is its expansion beyond public opinion to include the analysis of organizational content disseminated through SM platforms, such as environmental NGOs. The findings underscore the consistent prioritization of CC issues on the SM agendas of environmental advocates (NGOs) (Ter-Mkrtchyan & Taylor, 2023), with a focus on major threats and the overexploitation of resources (Barrios-O'Neill, 2021).

3.2.5 CC and Health

In certain studies within this category, CC was juxtaposed with another contentious public issue, vaccines (Schonfeld et al., 2021). Findings revealed that pro-vaccine individuals tended to believe in anthropogenic CC, whereas this correlation did not apply in the opposite scenario. From public health to mental well-being, other studies (Bui et al., 2023; Chen, 2020) focused on users' emotional responses toward CC. This included experiencing the tangible effects of CC (Bui et al., 2023) or exposure to CC-related content (Chen, 2020).

3.3 Platforms and Languages

After conducting the analysis, the findings reveal a diverse array of platforms for which discourse analysis models have been proposed. The selection of the platform depends on the research objectives and the target population. For example, if the goal is to obtain broader and global insights into the issue, Twitter tends to be the optimal choice. Conversely, if the objective is to study a specific country, researchers often utilize platforms popular nationally among that population, such as Sina

Weibo for the Chinese population and Dcard for Taiwanese users. Among these platforms, Twitter emerges as the most extensively studied, with 42 papers dedicated to analyzing CC discourse on this platform. The platforms studied and their corresponding papers are summarized in Table 7.11.

X (previously Twitter), with its real-time and public nature, has become a primary platform for discussing a wide spectrum of issues (Wei et al., 2021). Literature explorations emphasize the importance of Twitter and its usefulness as a de facto data source in several decision-making scenarios although it was not originally designed to provide such services. We can say that research scholars have created methods to make use of Twitter content in studying several perspectives of CC issues. These features besides standardized SM benefits such as low-cost and two-way communications and freedom of speech set the perfect environment to study public opinion about a global issue. In addition, data acquisition from Twitter is quite easier than other SM platforms thanks to the Twitter Application Programming Interface (API). While the data collection process has become more challenging and costly in recent times, it remains relatively accessible to acquire data. Sina Weibo, a microblogging platform popular in China, has also received attention in the literature. While fewer in number compared to Twitter studies, research has highlighted the unique characteristics of CC discourse on this platform with works conducting comparative analysis between the discourse on these platforms and others (Shen & Wang, 2023). Reddit, known for its diverse communities and discussion forums, has been the focus of three studies. Other platforms, such as YouTube, Twitch, TikTok,

Table 7.11 Detected platforms

Platform	Paper
Twitter: 42 paper	El Barachi et al. (2021), Lydiri et al. (2022), Styve et al. (2022), El Barachi et al. (2021), Ballestar et al. (2022), Ter-Mkrtchyan and Taylor (2023), Zhang et al. (2021), Qiao and Jiang (2021), Camarillo et al. (2021), Yao et al. (2022), Sham and Mohamed (2022), Smirnov and Hsieh (2022), Schonfeld et al. (2021), Pani et al. (2023), Yigitcanlar et al. (2022), Bui et al. (2023), Abdar et al. (2020), Barrios-O'Neill (2021), Noviello et al. (2023), Xiang et al. (2021), Klingenberger et al. (2022), Thukral et al. (2021), Bennett et al. (2021), Arce-García et al. (2023), Zhang et al. (2022), Bergstedt et al. (2018), Gjorshoska et al. (2023), Kim et al. (2021), Zander et al. (2023a), Loureiro and Alló (2020), Kirelli and Arslankaya (2020), Effrosynidis et al. (2022), Rizzoli (2023), Meenar et al. (2023), Moghadas et al. (2023), Dahal et al. (2019), Qiao and Williams (2022), Karimiziarani et al. (2023), Zander et al. (2023b), Mouronte-López and Subirán (2022), Shen and Wang (2023)
Sina Weibo	Li et al. (2023), Chen et al. (2023), Zeng (2022), Cheng et al. (2023), Shen and Wang (2023), Ji et al. (2024), Shen and Li (2023)
Reddit	Lee et al. (2023), Kaushal et al. (2022), Jeong et al. (2023)
Youtube	Mulyani et al. (2024), Chen (2020), Klingenberger et al. (2022)
Twitch	Navarro and Tapiador (2023)
Tiktok	Mulyani et al. (2024)
LinkedIn	Klingenberger et al. (2022)
Dcard	Yang et al. (2021)

Table 7.12 Languages used

Language	# works
English	43
Chinese	7
Mandarin	1
French	1
Indonesian translated to English	1
Turkish	1
Italian	1
German	1
Spanish	1
Macedonian	1

LinkedIn, and Dcard, have also been explored, albeit to a lesser extent. Studies on these platforms, such as Mulyani et al. (2024), Chen (2020), Klingenberger et al. (2022), Navarro and Tapiador (2023), and Yang et al. (2021), highlight the diverse ways in which CC is discussed and perceived across different SM ecosystems.

Overall, while Twitter remains the most researched platform for CC discourse analysis, exploring other platforms offers a diverse perspective on the complexity of online discussions surrounding CC. This is because platform users' demographics vary; for example, TikTok and Reddit users tend to belong to Generation Z, while Facebook, Twitter, and LinkedIn are predominantly used by Millennials and Generation X (Chaffey, 2024). Consequently, each platform's nature may be somewhat biased toward specific demographic categories. Further research exploring the aforementioned platforms in greater depth can enhance our understanding of how people perceive CC. The choice of platform also influences language preferences in studies, with a significant proportion of papers using Twitter. This correlation is closely tied to the prevalence of the English language, which can be attributed to various factors such as the platform's user demographics, the widespread adoption of English globally, and the abundance of techniques and open-source libraries available in this language. Additionally, many pre-trained models and dictionaries are primarily trained on English corpora and data, further contributing to its dominance in SM research. Chinese language is the next most encountered, appearing in seven papers focused on Sina Weibo. Additionally, eight other languages were identified, as detailed in Table 7.12.

4 Research Gaps and Recommendations for Future Exploration

Through the analysis of existing studies, we found that there are still many limitations in current works that could provide a pathway to future research. The identified gaps are as follows:

Lack of optimal combinations: Lack of integration of different combination of methods despite proving greater performance in terms of interpretability or accuracy. The focus was on comparing the methods rather than strategically incorporating them and leveraging the advantages of each. This can be attributed to the complexity of optimally designing and implementing such models, except for the combination of TM and SA, integrating NER, text classification, and cluster analysis in a two-stage or a hybrid approach has the potential to provide a holistic view of the problem.

Limited scope: The analysis overlooked user demographics and external factors that shape CC opinions and attitudes. Additionally, there was insufficient consideration for the temporal dimension, except in studies focusing on specific events. Tracking opinions over time is crucial for gaining deeper insights into changes in attitudes and behaviors. The temporal aspect was only present in eight works (Chen et al., 2023; Wu et al., 2023; Effrosynidis et al., 2022; Cheng et al., 2023; Pani et al., 2023; Shen & Li, 2023; Tyagi et al., 2021) where most of them focusing on specific events.

Lack of ML-based network analysis: CC is a popular issue, scholars note users seek affirming information, shaping group polarization (Valenzuela et al., 2021). Debate exists on online platforms' role. Isolated information ecosystems strengthen polarization, notably in CC discussions (Wilson et al., 2020). Network analysis is vital for understanding information dissemination dynamics and interaction patterns, crucial in tackling polarization specifically in this issue. There was a limited incorporation of network analysis; only three studies (Zhang et al., 2022; Tyagi et al., 2021; Yao et al., 2022), proposed its inclusion. However, numerous ML (ML) and deep learning (DL) methods can enhance natural language processing (NLP) models with a network perspective, including graph neural networks and node classification techniques.

Language bias: The prevalence of the English language is apparent, attributed to several factors outlined in the results analysis. Only 25% of the analyzed works incorporated other languages in content analysis, 5% of which using a combination of English and another language, while only 20% focused on a specific language. Certain languages present numerous challenges, including difficulties in semantic analysis due to their rich and complex morphology, orthographic ambiguity, and orthographic noise, as observed in Chinese and Arabic. While Chinese appeared in seven works, no studies were identified on the Arabic language. Therefore, transcending language-centric models and introducing a language-independent model is groundbreaking and holds significant potential if implemented optimally.

Lack of annotated data related to different themes in CC: Through literature analysis, a notable gap has been identified concerning annotated data in the context of CC discourse. Specifically, while there has been some attention given to supervised ML-based SA, we observed a deficiency in labeled data specifically tailored to CC-related themes. Out of the corpus of literature examined, only a small fraction, comprising five works, chose to focus on text classification related to CC. This limited number of studies indicates a substantial gap in the availability of annotated data for training ML models to effectively analyze and understand CC discourse. Without access to adequately labeled data, it becomes difficult to train and evaluate ML algorithms effectively, hindering progress in understanding public perceptions, attitudes, and behaviors toward CC. For instance, thematic areas within CC discourse may include public perception of CC mitigation strategies, CC denial and skepticism, an annotated corpus of pro-environmental behavior or non-supportive attitudes, etc. Addressing this gap in the literature requires concerted efforts to collect and annotate large-scale datasets. Such initiatives could involve collaboration between researchers, policymakers, and domain experts to ensure the creation of high-quality annotated datasets that accurately capture the nuances and complexities of CC communication.

Lack of Local focus on Québec province: The analysis of the existing body of literature shows a significant research gap regarding the Quebec context and the French language. None of the reviewed works specifically address the unique socio-political and linguistic dimensions of climate change discussions in Quebec. This omission is critical given Quebec's distinct climate change mitigation and adaptation efforts over the last century. The province has implemented a series of policies and strategies aimed at reducing greenhouse gas (GHG) emissions and enhancing climate resilience. The province's financial commitment to climate action is substantial. According to the Government of Quebec's 2022 budget, the province allocated CAD 6.7 billion to its 2021–2026 Green Economy Plan. From 1990 to 2019, Quebec reduced its emissions by 8.7%, despite economic growth and population increases during this period.[2]

Given the substantial efforts, future research should prioritize this region to better understand the sentiment and topics discussed within this community. Analyzing climate change discourse in Quebec, particularly in the French language, can provide insightful recommendations into the specific concerns, priorities, and attitudes of its residents. This focus can also shed light on the effectiveness of Quebec's policies and initiatives.

The diverse methodologies presented can effectively harness province-specific data. As the sourced data will predominantly be in French, the province's primary language, it's imperative to prioritize the usage of NLP libraries tailored for French text processing. Essential libraries for this purpose include NLTK, SpaCy, Flair, and Camembert, among others.

To enhance the body of knowledge we recommend the following directions:

Analyze Quebec climate change discourse: Investigate the characteristics of climate change discourse in Quebec. This can involve analyzing the content of social media platforms with specific geolocation, and local news outlets like UnpointCinq[3] (The media of climate action in Quebec) to capture the local sentiment and topics accurately using sentiment modeling.

Assessing policy impact: Analyze the diffused content from governmental entities on their official social media accounts and the corresponding public responses by means of text classification, sentiment analysis, and NER to assess the effective-

[2] Press release (gouv.qc.ca).

[3] (unpointcinq.ca).

ness of public initiatives in shaping public attitudes toward climate change. This is facilitated thanks to the activeness of such entities on social media, citing accounts such as Environnement Québec and Ministère des Ressources naturelles et des Forêts.

Monitor public opinion for crisis management: It is crucial to note the recent wildfires spreading across the region as a manifestation of climate change. Further research can build on the findings discussed in Sect. 3.2, specifically regarding the tangible manifestations of climate change. This research should aim to track public opinion during such disasters and implement proactive measures accordingly based on this data. Examples include targeted public education campaigns, pre-emptive evacuations, and resource allocation to areas identified as high-risk by the public. Engaging with communities through transparent and responsive communication can also foster trust and cooperation, which are essential during crisis situations.

Previous research emphasizes the importance of such studies in tailoring effective climate policy strategies (Bennett et al., 2021) and implementing timely interventions and urgent adjustments (El Barachi et al., 2021). Applying this research to the Quebec context is promising for supporting the decision-making process.

5 Conclusion

This survey offers an extensive examination of the diverse ML and NLP methods employed in analyzing CC-related discourse across SM platforms. Through comprehensive literature review and analysis, we have identified six distinct approaches, each exhibiting varying frequencies of application. By synthesizing existing research, we have discerned the most effective and popular methods, as well as those requiring further investigation and potential enhancement. Notably, SA and TM emerged as the most employed methods, with classic approaches such as lexicon-based and LDA receiving extensive attention. This prevalence likely stems from the social aspect of CC discourse and the demand for interpretable results. Researchers prioritize the exploration of diverse dimensions and perspectives over the development of predictive AI models, emphasizing interpretability over performance accuracy. While deep learning approaches demonstrate exceptional performance, they are less prevalent in the survey. Stance detection, featured in only one study, holds promise as a future research avenue, given the contentious nature of CC discourse and the wide spectrum of opinions evident, particularly on SM platforms where climate and science skepticism are detected. NER also presents opportunities for feature extraction, particularly in studies employing a geospatial approach, as it facilitates the extraction of location names from text, overcoming challenges associated with extracting geolocations from SM data. Furthermore, the findings reveal a predominant focus on the English language due to the availability of open-source libraries and dictionaries, with Chinese emerging as another prominent language.

French, however, is relatively underrepresented, appearing in only one study. Combined with the lack of research on Quebec, a province-focused study analyzing Quebecers' climate change discourse would be a valuable area of investigation, filling both regional and linguistic gaps in the literature.

Climate change is primarily addressed as a global issue in the literature, with limited exploration of adaptation and mitigation. These aspects receive attention in only a few studies. Surprisingly, GHG emissions, the direct cause of CC and a primary focus of mitigation efforts, are only discussed in two studies. The results reveal a predominantly negative pattern in climate change-related topics. Negativity is generally associated with political and ideological stances, while positivity is linked to adaptation strategies and renewable energy solutions. Sentiment differences are also evident between regions. Asian discourse, particularly from China, tends to be optimistic. In contrast, Western discourse, mainly from the USA, is characterized by negativity, often rooted in political issues. Platform-wise, Twitter dominates as the primary source of data, raising concerns about potential biases in research outcomes. While Twitter data acquisition is relatively straightforward, a trade-off exists between result generalizability and data availability, posing a significant challenge in conducting research on a controversial topic like CC. In conclusion, this analysis highlights the need to address existing gaps in this research domain and emphasizes the importance of balanced methodological approaches to yield robust findings that serve as decision-making-support system tools for climate policymakers and different stakeholders.

References

Abdar, M., Basiri, M., Yin, J., Habibnezhad, M., Chi, G., Nemati, S., & Asadi, S. (2020). Energy choices in Alaska: Mining people's perception and attitudes from geotagged tweets. *Renewable & Sustainable Energy Reviews, 124*, 109781. https://doi.org/10.1016/j.rser.2020.109781

Agrawal, C., Pandey, A., & Goyal, S. (2021). A survey on role of machine learning and NLP in fake news detection on social media. In *2021 IEEE 4th International Conference on Computing, Power and Communication Technologies (GUCON)* (pp. 1–7). https://doi.org/10.1109/GUCON50781.2021.9573875

Alturayeif, N., Luqman, H., & Ahmed, M. (2023). A systematic review of machine learning techniques for stance detection and its applications. *Neural Computing and Applications, 35*(7), 5113–5144. https://doi.org/10.1007/s00521-023-08285-7

Arce-García, S., Díaz-Campo, J., & Cambronero-Saiz, B. (2023). Online hate speech and emotions on Twitter: A case study of Greta Thunberg at the UN Climate Change Conference COP25 in 2019. *Social Network Analysis and Mining, 13*(1), 48. https://doi.org/10.1007/s13278-023-01052-5

Asgari-Chenaghlu, M., Feizi-Derakhshi, M.-R., Farzinvash, L., Balafar, M.-A., & Motamed, C. (2021). Topic detection and tracking techniques on Twitter: A systematic review. *Complexity, 2021*, e8833084. https://doi.org/10.1155/2021/8833084

Balaji, T. K., Annavarapu, C. S. R., & Bablani, A. (2021). Machine learning algorithms for social media analysis: A survey. *Computer Science Review, 40*, 100395. https://doi.org/10.1016/j.cosrev.2021.100395

Ballestar, M., Martín-Llaguno, M., & Sainz, J. (2022). An artificial intelligence analysis of climate-change influencers' marketing on Twitter. *Psychology & Marketing, 39*(12), 2273–2283. https://doi.org/10.1002/mar.21735

Barrios-O'Neill, D. (2021). Focus and social contagion of environmental organization advocacy on Twitter. *Conservation Biology, 35*(1), 307–315. https://doi.org/10.1111/cobi.13564

Bennett, J., Rachunok, B., Flage, R., & Nateghi, R. (2021). Mapping climate discourse to climate opinion: An approach for augmenting surveys with social media to enhance understandings of climate opinion in the United States. *PLoS One, 16*(1), e0245319. https://doi.org/10.1371/journal.pone.0245319

Bergstedt, H., Ristea, A., Resch, B., & Bartsch, A. (2018). Public perception of climate change in Alaska: A case study of opinion-mining using Twitter. *GI_Forum, 6*(1), 47–64. https://doi.org/10.1553/GISCIENCE2018_01_S47

Blei, D. M., Ng, A. Y., & Jordan, M. I. (2003). Latent Dirichlet allocation. *Journal of Machine Learning Research, 3*, 993–1022.

Bui, T., Hannah, A., Madria, S., Nabaweesi, R., Levin, E., Wilson, M., & Nguyen, L. (2023). Emotional health and climate-change-related stressor extraction from social media: A case study using Hurricane Harvey. *Mathematics, 11*(24), 4910. https://doi.org/10.3390/math11244910

Camarillo, M., Ferguson, E., Ljevar, V., & Spence, A. (2021). Big changes start with small talk: Twitter and climate change in times of coronavirus pandemic. *Frontiers in Psychology, 12*, 661395. https://doi.org/10.3389/fpsyg.2021.661395

Chaffey, D. (2024). Global social media statistics research summary 2024 [Jan 2024]. *Smart Insights*. https://www.smartinsights.com/social-media-marketing/social-media-strategy/new-global-social-media-research/

Chen, N. (2020). Exploring the cognitive and emotional impact of online climate change videos on viewers. *Sustainability, 12*(22), 9571. https://doi.org/10.3390/su12229571

Chen, X., Zou, L., & Zhao, B. (2019). Detecting climate change deniers on Twitter using a deep neural network. In *Proceedings of the 2019 11th International Conference on Machine Learning and Computing* (pp. 204–210). https://doi.org/10.1145/3318299.3318382

Chen, Y., Hu, M., Chen, X., Wang, F., Liu, B., & Huo, Z. (2023). An approach of using social media data to detect the real time spatio-temporal variations of urban waterlogging. *Journal of Hydrology, 625*, 130128. https://doi.org/10.1016/j.jhydrol.2023.130128

Cheng, Y., Yu, Z., Xu, C., Manoli, G., Ren, X., Zhang, J., Liu, Y., Yin, R., Zhao, B., & Vejre, H. (2023). Climatic and economic background determine the disparities in urbanites' expressed happiness during the summer heat. *Environmental Science and Technology, 57*(30), 10951–10961. https://doi.org/10.1021/acs.est.3c01765

Corner, A., Whitmarsh, L., & Xenias, D. (2012). Uncertainty, scepticism and attitudes towards climate change: Biased assimilation and attitude polarisation. *Climatic Change, 114*(3), 463–478. https://doi.org/10.1007/s10584-012-0424-6

Dahal, B., Kumar, S. A. P., & Li, Z. (2019). Topic modeling and sentiment analysis of global climate change tweets. *Social Network Analysis and Mining, 9*(1), 24. https://doi.org/10.1007/s13278-019-0568-8

Devi, G., & Somasundaram, K. (2020). Literature review on sentiment analysis in social media: Open challenges toward applications. *International Journal of Advanced Science and Technology, 29*, 1462–1471.

Du, C., Sun, H., Wang, J., Qi, Q., & Liao, J. (2020). Adversarial and domain-aware BERT for cross-domain sentiment analysis. In D. Jurafsky, J. Chai, N. Schluter, & J. Tetreault (Eds.), *Proceedings of the 58th annual meeting of the Association for Computational Linguistics* (pp. 4019–4028). Association for Computational Linguistics. https://doi.org/10.18653/v1/2020.acl-main.370

El Barachi, M., AlKhatib, M., Mathew, S., & Oroumchian, F. (2021). A novel sentiment analysis framework for monitoring the evolving public opinion in real-time: Case study on climate change. *Journal of Cleaner Production, 312*, 127820. https://doi.org/10.1016/j.jclepro.2021.127820

Effrosynidis, D., Karasakalidis, A. I., Sylaios, G., & Arampatzis, A. (2022). The climate change Twitter dataset. *Expert Systems with Applications, 204*, WOS:000819313900007. https://doi.org/10.1016/j.eswa.2022.117541

Falkenberg, M., Galeazzi, A., Torricelli, M., Di Marco, N., Larosa, F., Sas, M., Mekacher, A., Pearce, W., Zollo, F., Quattrociocchi, W., & Baronchelli, A. (2022). Growing polarization around climate change on social media. *Nature Climate Change, 12*(12), 12. https://doi.org/10.1038/s41558-022-01527-x

Ghani, N. A., Hamid, S., Targio Hashem, I. A., & Ahmed, E. (2019). Social media big data analytics: A survey. *Computers in Human Behavior, 101*, 417–428. https://doi.org/10.1016/j.chb.2018.08.039

Gjorshoska, I., Dedinec, A., Prodanova, J., Dedinec, A., & Kocarev, L. (2023). Public perception of waste regulations implementation. Natural language processing vs real GHG emission reduction modeling. *Ecological Informatics, 76*, 102130. https://doi.org/10.1016/j.ecoinf.2023.102130

Glaz, A. L., Haralambous, Y., Kim-Dufor, D.-H., Lenca, P., Billot, R., Ryan, T. C., Marsh, J., DeVylder, J., Walter, M., Berrouiguet, S., & Lemey, C. (2021). Machine learning and natural language processing in mental health: Systematic review. *Journal of Medical Internet Research, 23*(5), e15708. https://doi.org/10.2196/15708

Han, C., & Sun, X. (2024). Enhancing low carbon awareness in social media discourse: A fuzzy clustering approach. *IEEE Access, 12*, 3200–3207. https://doi.org/10.1109/ACCESS.2023.3348123

Hotho, A., Staab, S., & Stumme, G. (2003). Explaining text clustering results using semantic structures. In I. N. Lavrač, D. Gamberger, L. Todorovski, & H. Blockeel (Eds.), *Knowledge discovery in databases: PKDD 2003* (pp. 217–228). Springer. https://doi.org/10.1007/978-3-540-39804-2_21

Howe, P., Mildenberger, M., Marlon, J., & Leiserowitz, A. (2015). Geographic variation in opinions on climate change at state and local scales in the USA. *Nature Climate Change, 5*, 596. https://doi.org/10.1038/nclimate2583

Hu, M., & Liu, B. (2004). Mining and summarizing customer reviews. In *Dans Proceedings of the tenth ACM SIGKDD international conference on knowledge discovery and data mining* (pp. 168–177). Association for Computing Machinery. https://doi.org/10.1145/1014052.1014073

Hwang, H., An, S., Lee, E., Han, S., & Lee, C. (2021). Cross-societal analysis of climate change awareness and its relation to SDG 13: A knowledge synthesis from text mining. *Sustainability, 13*(10), 10. https://doi.org/10.3390/su13105596

Ibrohim, M. O., Bosco, C., & Basile, V. (2023). Sentiment analysis for the natural environment: A systematic review. *ACM Computing Surveys, 56*(4), 1–37. https://doi.org/10.1145/3604605

Jelodar, H., Wang, Y., Yuan, C., Feng, X., Jiang, X., Li, Y., & Zhao, L. (2019). Latent Dirichlet allocation (LDA) and topic modeling: Models, applications, a survey. *Multimedia Tools and Applications, 78*(11), 15169–15211. https://doi.org/10.1007/s11042-018-6894-4

Jeong, D., Hwang, S., Kim, J., Yu, H., & Park, E. (2023). Public perspective on renewable and other energy resources: Evidence from social media big data and sentiment analysis. *Energy Strategy Reviews, 50*, 101243. https://doi.org/10.1016/j.esr.2023.101243

Ji, J., Hu, T., Chen, Z., & Zhu, M. (2024). Exploring the climate change discourse on Chinese social media and the role of social bots. *Asian Journal of Communication, 34*(1), 109–128. https://doi.org/10.1080/01292986.2023.2269423

Karimiziarani, M., Shao, W. Y., Mirzaei, M., & Moradkhani, H. (2023). Toward reduction of detrimental effects of hurricanes using a social media data analytic approach: How climate change is perceived? *Climate Risk Management, 39*, WOS:000924768800001. https://doi.org/10.1016/j.crm.2023.100480

Kasztelnik, K. (2020). Data analytics and social media as the innovative business decision model with natural language processing. *Journal of Business and Accounting, 13*, 136–153.

Kaushal, A., Acharjee, A., & Mandal, A. (2022). Machine learning based attribution mapping of climate related discussions on social media. *Scientific Reports, 12*(1), 19033. https://doi.org/10.1038/s41598-022-22034-1

Khanbhai, M., Anyadi, P., Symons, J., Flott, K., Darzi, A., & Mayer, E. (2021). Applying natural language processing and machine learning techniques to patient experience feedback: A systematic review. *BMJ Health & Care Informatics, 28*(1), e100262. https://doi.org/10.1136/bmjhci-2020-100262

Khatibi, F. S., Dedekorkut-Howes, A., Howes, M., & Torabi, E. (2021). Can public awareness, knowledge and engagement improve climate change adaptation policies? *Discover Sustainability, 2*(1), 18. https://doi.org/10.1007/s43621-021-00024-z

Kim, S., Ganesan, K., Dickens, P., & Panda, S. (2021). Public sentiment toward solar energy-opinion mining of Twitter using a transformer-based language model. *Sustainability, 13*(5), 2673. https://doi.org/10.3390/su13052673

Kirelli, Y., & Arslankaya, S. (2020). Sentiment analysis of shared tweets on global warming on Twitter with data mining methods: A case study on Turkish language. *Computational Intelligence and Neuroscience, 2020*, 1904172. https://doi.org/10.1155/2020/1904172

Kitchenham. (2004). *Barbara Ann Procedures for Performing Systematic Reviews.*, Keele University, Department of Computer Science, Keele University, Kelee, UK

Klingenberger, L., Shahi, S., Au, C.-D., Frere, E., & Zureck, A. (2022). Inclusive measurement of public perception of corporate low-carbon ambitions: Analysis of strategic positioning for sustainable development using natural language processing. *International Journal of Sustainable Development and Planning, 17*(1), 259–265. https://doi.org/10.18280/ijsdp.170126

Lample, G., Ballesteros, M., Subramanian, S., Kawakami, K., & Dyer, C. (2016). Neural architectures for named entity recognition. In K. Knight, A. Nenkova, & O. Rambow (Eds.), *Proceedings of the 2016 Conference of the North American Chapter of the Association for Computational Linguistics: Human Language Technologies* (pp. 260–270). Association for Computational Linguistics. https://doi.org/10.18653/v1/N16-1030

Laureate, C. D. P., Buntine, W., & Linger, H. (2023). A systematic review of the use of topic models for short text social media analysis. *Artificial Intelligence Review, 56*(12), 14223–14255. https://doi.org/10.1007/s10462-023-10471-x

Lee, G., & Kwak, Y. H. (2012). An Open Government Maturity Model for social media-based public engagement. *Government Information Quarterly, 29*(4), 492–503. https://doi.org/10.1016/j.giq.2012.06.001

Lee, K. N., Lee, H., Kim, J. H., Kim, Y., & Lee, S. H. (2023). Comparing social media and news articles on climate change: Different viewpoints revealed. *KSII Transactions on Internet and Information Systems, 17*(11), 2966–2986. https://doi.org/10.3837/tiis.2023.11.004

Li, R., Wang, Q., Zeng, L., & Chen, H. (2023). A study on public perceptions of carbon neutrality in China: Has the idea of ESG been encompassed? *Frontiers in Environmental Science, 10.* https://doi.org/10.3389/fenvs.2022.949959

Liu, B. (2010). Sentiment analysis and subjectivity. In N. Indurkhya & F. Damerau (Eds.), *Handbook of natural language processing* (2nd ed., p. 40). Chapman and Hall/CRC. https://www.taylorfrancis.com/chapters/mono/10.1201/9781420085938-36/sentiment-analysis-subjectivity-bing-liunitin-indurkhya-fred-damerau?context=ubx&refId=db504611-27e9-4228-9ab3-665ca6a8c470

Loureiro, M., & Alló, M. (2020). Sensing climate change and energy issues: Sentiment and emotion analysis with social media in the UK and Spain. *Energy Policy, 143*, 111490. https://doi.org/10.1016/j.enpol.2020.111490

Lydiri, M., El Mourabit, Y., El Habouz, Y., & Fakir, M. (2022). A performant deep learning model for sentiment analysis of climate change. *Social Network Analysis and Mining, 13*(1), 8. https://doi.org/10.1007/s13278-022-01014-3

Mavrodieva, A. V., Rachman, O. K., Harahap, V. B., & Shaw, R. (2019). Role of social media as a soft power tool in raising public awareness and engagement in addressing climate change. *Climate, 7*(10), 10. https://doi.org/10.3390/cli7100122

Meenar, M., Rahman, M., Russack, J., Bauer, S., & Kapri, K. (2023). "The urban poor and vulnerable are hit hardest by the heat": A heat equity lens to understand community perceptions of climate change, urban heat islands, and green infrastructure. *Land, 12*(12), 2174. https://doi.org/10.3390/land12122174

Moghadas, M., Fekete, A., Rajabifard, A., & Kötter, T. (2023). The wisdom of crowds for improved disaster resilience: A near-real-time analysis of crowdsourced social media data on the 2021 flood in Germany. *GeoJournal, 88*(4), 4215–4241. https://doi.org/10.1007/s10708-023-10858-x

Mouronte-López, M. L., & Subirán, M. (2022). What do Twitter users think about climate change? Characterization of Twitter interactions considering geographical, gender, and account typologies perspectives. *Weather Climate and Society, 14*(4), 1039–1064., WOS:000905283100002. https://doi.org/10.1175/WCAS-D-21-0163.1

Mullah, N. S., & Zainon, W. M. N. W. (2021). Advances in machine learning algorithms for hate speech detection in social media: A review. *IEEE Access, 9*, 88364–88376. https://doi.org/10.1109/ACCESS.2021.3089515

Mulyani, Y. P., Saifurrahman, A., Arini, H. M., Rizqiawan, A., Hartono, B., Utomo, D. S., Spanellis, A., Beltran, M., Banjar Nahor, K. M., Paramita, D., & Harefa, W. D. (2024). Analyzing public discourse on photovoltaic (PV) adoption in Indonesia: A topic-based sentiment analysis of news articles and social media. *Journal of Cleaner Production, 434*, 140233. https://doi.org/10.1016/j.jclepro.2023.140233

Nagarhalli, T. P., Vaze, V., & Rana, N. K. (2021). Impact of machine learning in natural language processing: A review. In *2021 Third International Conference on Intelligent Communication Technologies and Virtual Mobile Networks (ICICV)* (pp. 1529–1534). https://doi.org/10.1109/ICICV50876.2021.9388380

Navarro, A., & Tapiador, F. J. (2023). Twitch as a privileged locus to analyze young people's attitudes in the climate change debate: A quantitative analysis. *Humanities & Social Sciences Communications, 10*(1), WOS:001105828200008. https://doi.org/10.1057/s41599-023-02377-4

Noviello, A., Menghani, S., Choudhri, S., Lee, I., Mohanraj, B., & Noviello, A. (2023). Guiding environmental messaging by quantifying the effect of extreme weather events on public discourse surrounding anthropogenic climate change. *Weather Climate and Society, 15*(1), 17–30. https://doi.org/10.1175/WCAS-D-22-0053.1

Pani, A., Balla, S. N., & Sahu, P. K. (2023). Decoding consumer-centric transition to electric mobility based on sentiment, semantic and statistical analysis. *Research in Transportation Business and Management, 51*, 101069. https://doi.org/10.1016/j.rtbm.2023.101069

Qiao, F., & Jiang, K. (2021). Attitudes towards global warming on Twitter: A hedonometer-appraisal analysis. *Journal of Global Information Management (JGIM), 30*(7), 1–20. https://doi.org/10.4018/JGIM.296708

Qiao, F., & Williams, J. (2022). Topic modelling and sentiment analysis of global warming tweets: Evidence from big data analysis. *Journal of Organizational and End User Computing, 34*(3), 1–18. https://doi.org/10.4018/JOEUC.294901

Ratinov, L., & Roth, D. (2009). Design challenges and misconceptions in named entity recognition. In S. Stevenson & X. Carreras (Eds.), *Proceedings of the Thirteenth Conference on Computational Natural Language Learning (CoNLL-2009)* (pp. 147–155). Association for Computational Linguistics. https://aclanthology.org/W09-1119

Rizzoli, V. (2023). The risk co-de model: Detecting psychosocial processes of risk perception in natural language through machine learning. *Journal of Computational Social Science*. https://doi.org/10.1007/s42001-023-00235-6

Schonfeld, J., Qian, E., Sinn, J., Cheng, J., Anand, M., & Bauch, C. (2021). Debates about vaccines and climate change on social media networks: A study in contrasts. *Humanities & Social Sciences Communications, 8*(1), 322. https://doi.org/10.1057/s41599-021-00977-6

Shah, B., & Shah, M. (2021). A survey on machine learning and deep learning based approaches for sarcasm identification in social media. In I. K. Kotecha, V. Piuri, H. N. Shah, & R. Patel (Eds.), *Data science and intelligent applications* (pp. 247–259). Springer. https://doi.org/10.1007/978-981-15-4474-3_29

Sham, N., & Mohamed, A. (2022). Climate change sentiment analysis using lexicon, machine learning and hybrid approaches. *Sustainability, 14*(8), 4723. https://doi.org/10.3390/su14084723

Shen, C., & Li, P. (2023). Green housing on social media in China: A text mining analysis. *Building and Environment, 237*, 110338. https://doi.org/10.1016/j.buildenv.2023.110338

Shen, C., & Wang, Y. (2023). Concerned or Apathetic? Exploring online public opinions on climate change from 2008 to 2019: A comparative study between China and other G20 countries. *Journal of Environmental Management, 332*, 117376. https://doi.org/10.1016/j.jenvman.2023.117376

Smirnov, O., & Hsieh, P.-H. (2022). COVID-19, climate change, and the finite pool of worry in 2019 to 2021 Twitter discussions. *Proceedings of the National Academy of Sciences of the United States of America, 119*(43), e2210988119. https://doi.org/10.1073/pnas.2210988119

Stede, M., & Patz, R. (2021). The climate change debate and natural language processing. In A. Field, S. Prabhumoye, M. Sap, Z. Jin, J. Zhao, & C. Brockett (Eds.), *Proceedings of the 1st Workshop on NLP for Positive Impact* (pp. 8–18). Association for Computational Linguistics. https://doi.org/10.18653/v1/2021.nlp4posimpact-1.2

Styve, L., Navarra, C., Petersen, J. M., Neset, T.-S., & Vrotsou, K. (2022). A visual analytics pipeline for the identification and exploration of extreme weather events from social media data. *Climate, 10*(11), 174. https://doi.org/10.3390/cli10110174

Tan, K. L., Lee, C. P., & Lim, K. M. (2023). A survey of sentiment analysis: Approaches, datasets, and future research. *Applied Sciences, 13*(7), 4550. https://doi.org/10.3390/app13074550

Tedmori, S., & Awajan, A. (2019). Sentiment analysis main tasks and applications: A survey. *Journal of Information Processing Systems, 15*, 500–519. https://doi.org/10.3745/JIPS.04.0120

Ter-Mkrtchyan, A., & Taylor, M. (2023). An empirical mapping of environmental protection and conservation nonprofit discourse on social media. *Nonprofit and Voluntary Sector Quarterly*. https://doi.org/10.1177/08997640231202459

Thukral, T., Varshney, A., & Gaur, V. (2021). Intensity quantification of public opinion and emotion analysis on climate change. *International Journal of Advanced Technology and Engineering Exploration, 8*(83), 1351–1366. https://doi.org/10.19101/IJATEE.2021.874417

Toetzke, M., Probst, B., & Feuerriegel, S. (2023). Leveraging large language models to monitor climate technology innovation. *Environmental Research Letters, 18*(9), 091004. https://doi.org/10.1088/1748-9326/acf233

Tyagi, A., Uyheng, J., & Carley, K. M. (2021). Heated conversations in a warming world: Affective polarization in online climate change discourse follows real-world climate anomalies. *Social Network Analysis and Mining, 11*(1), 87. https://doi.org/10.1007/s13278-021-00792-6

UNFCCC. (n.d.). *Introduction to mitigation*. https://unfccc.int/topics/introduction-to-mitigation

Upadhyaya, A., Fisichella, M., & Nejdl, W. (2023). Towards sentiment and temporal aided stance detection of climate change tweets. *Information Processing & Management, 60*(4), 103325. https://doi.org/10.1016/j.ipm.2023.103325

Usai, A., Pironti, M., Mital, M., Mejri, A., & Chiraz. (2018). Knowledge discovery out of text data: A systematic review via text mining. *Journal of Knowledge Management, 22*(7), 1471–1488. https://doi.org/10.1108/JKM-11-2017-0517

Valenzuela, S., Bachmann, I., & Bargsted, M. (2021). The personal is the political? What do WhatsApp users share and how it matters for news knowledge, polarization and participation in Chile. *Digital Journalism, 9*(2), 155–175. https://doi.org/10.1080/21670811.2019.1693904

Wankhade, M., Rao, A. C. S., & Kulkarni, C. (2022). A survey on sentiment analysis methods, applications, and challenges. *Artificial Intelligence Review, 55*(7), 5731–5780. https://doi.org/10.1007/s10462-022-10144-1

Wei, Y., Gong, P., Zhang, J., & Wang, L. (2021). Exploring public opinions on climate change policy in « Big Data Era »—A case study of the European Union Emission Trading System (EU-ETS) based on Twitter. *Energy Policy, 158*, 112559. https://doi.org/10.1016/j.enpol.2021.112559

Wilson, A. E., Parker, V. A., & Feinberg, M. (2020). Polarization in the contemporary political and media landscape. *Current Opinion in Behavioral Sciences, 34*, 223–228. https://doi.org/10.1016/j.cobeha.2020.07.005

Wu, M., Long, R., Chen, F., Chen, H., Bai, Y., Cheng, K., & Huang, H. (2023). Spatio-temporal difference analysis in climate change topics and sentiment orientation: Based on LDA and BiLSTM model. *Resources, Conservation and Recycling, 188*, 106697. https://doi.org/10.1016/j.resconrec.2022.106697

Xiang, N., Wang, L., Zhong, S., Zheng, C., Wang, B., & Qu, Q. (2021). How does the world view China's carbon policy? A sentiment analysis on Twitter data. *Energies, 14*(22), 7782. https://doi.org/10.3390/en14227782

Yang, C. L., Huang, C. Y., & Hsiao, Y. H. (2021). Using social media mining and PLS-SEM to examine the causal relationship between public environmental concerns and adaptation strategies. *International Journal of Environmental Research and Public Health, 18*(10), WOS:000654915000001. https://doi.org/10.3390/ijerph18105270

Yao, Q., Li, R., & Song, L. (2022). Carbon neutrality vs. neutralite carbone: A comparative study on French and English users' perceptions and social capital on Twitter. *Frontiers in Environmental Science, 10*. https://doi.org/10.3389/fenvs.2022.969039

Yigitcanlar, T., Regona, M., Kankanamge, N., Mehmood, R., D'Costa, J., Lindsay, S., Nelson, S., & Brhane, A. (2022). Detecting natural hazard-related disaster impacts with social media analytics: The case of Australian states and territories. *Sustainability, 14*(2), 2. https://doi.org/10.3390/su14020810

Zander, K., Garnett, S., Ogie, R., Alazab, M., & Nguyen, D. (2023a). Trends in bushfire related tweets during the Australian "Black Summer" of 2019/20. *Forest Ecology and Management, 545*, 121274. https://doi.org/10.1016/j.foreco.2023.121274

Zander, K., Rieskamp, J., Mirbabaie, M., Alazab, M., & Nguyen, D. (2023b). Responses to heat waves: What can Twitter data tell us? *Natural Hazards, 116*, 3547. https://doi.org/10.1007/s11069-023-05824-2

Zeng, L. (2022). Chinese public perception of climate change on social media: An investigation based on data mining and text analysis. *Journal of Environmental and Public Health, 2022*, e6294436. https://doi.org/10.1155/2022/6294436

Zhang, Y., Abbas, M., & Iqbal, W. (2021). Analyzing sentiments and attitudes toward carbon taxation in Europe, USA, South Africa, Canada and Australia. *Sustainable Production and Consumption, 28*, 241–253. https://doi.org/10.1016/j.spc.2021.04.010

Zhang, Y., Abbas, M., & Iqbal, W. (2022). Perceptions of GHG emissions and renewable energy sources in Europe, Australia and the USA. *Environmental Science and Pollution Research, 29*(4), 5971–5987. https://doi.org/10.1007/s11356-021-15935-7

Hana Ghiloufi is a doctoral student in Science, Technology and Society at UQÀM. Hana holds a bachelor's and master's degree in business analysis. Her research focuses on the application of data science techniques to address sustainable development challenges, particularly environmental issues. She is also a student member of CÉODD and the scientific committee, and works as CIRODD's content creation and social network management assistant.

Nicolas Merveille holds a doctorate in social anthropology and ethnology from the École des Hautes Études en Sciences Sociales (Paris), and is a professor in the Strategy, Social and Environmental Responsibility Department at ESG-UQAM. He holds a master's degree in environmental engineering and management from the École nationale supérieure des Mines de Paris, and is a member of CIRODD's scientific committee.

Sehl Mellouli is a full professor in the Department of Organizational Information Systems in the Faculty of Administrative Sciences. He was director of the Department of Organizational Information Systems from 2015 to 2020. In 2018, he was appointed President of the Digital Government Society, an international association of researchers on digital government, for a

2-year term. As a professor, Mr. Mellouli has received several recognitions for the excellence of his teaching and research. He is the author of over 100 articles published in journals and has given nearly 80 papers at internationally renowned conferences. He is associate editor of the journal Digital Government Research and Practice and a member of the editorial board of the journal *Government Information Quarterly*. In 2020, he was appointed Associate Vice-President, Academic and Student Affairs.

Chapter 8
Performance Indicators for Sustainable Remanufacturing Closed-Loop Supply Chains

Camilo Mejía-Moncayo, Amin Chaabane, Jean-Pierre Kenne, and Lucas A. Hof

Abstract Québec is transitioning to a circular economy (CE) by promoting the implementation of CE strategies, such as remanufacturing. However, the adoption of remanufacturing practices to achieve sustainable implementation in an enterprise is demanding and highly challenging. It requires balancing economic, environmental, and social dimensions, and guaranteeing products' remanufacturability and system circularity along closed-loop supply chains (CLSC). Key performance indicators (KPI) emerge as decision-support tools for decision-makers to control and enhance system performance. Nevertheless, the multidimensional nature of sustainable remanufacturing makes it challenging to determine suitable KPIs to employ. Therefore, this study performs a systematic literature review to identify the main KPIs in sustainable remanufacturing and its scope along its CLSC. A total of 100 documents from the Scopus database were analyzed to reveal the most frequently used 42 key performance indicators (KPI), categorized as 25 economic, 14 environmental, and 3 social-related indicators. The KPIs were distributed among the different CLSC actors, providing insights on selection of the most useful KPIs to consider for each CLSC actor.

Keywords Remanufacturing · Sustainability · Closed-loop supply chains (CLSC) · Key performance indicators (KPI)

1 Introduction

The Quebec government has outlined a comprehensive strategy to promote circular economy (CE) as part of its broader agenda to foster sustainable economic development and resource efficiency (Gouvernement du Québec, 2024). Quebec aims to

---

C. Mejía-Moncayo · A. Chaabane · J.-P. Kenne · L. A. Hof (✉)
École de technologie supérieure, Montréal, QC, Canada
e-mail: camilo.mejia-moncayo.1@ens.etsmtl.ca; amin.chaabanae@etsmtl.ca; jean-pierre.kenne@etsmtl.ca; lucas.hof@etsmtl.ca

M. Cheriet et al. (eds.), *Accelerating the Socio-Ecological Transition*,
https://doi.org/10.1007/978-3-031-82896-6_8

incentivize businesses to adopt CE practices as reuse, recycling, repair, refurbishing, and remanufacturing. This strategy aligns with their commitment to reducing waste, conserving resources, and mitigating environmental impacts (Gouvernement du Québec, 2023). Despite efforts of Quebec's Ministry of the Environment, the Fight against Climate Change, Wildlife and Parks, according to their recently published report on the province's circularity gap, the circularity index[1] of Québec is currently 3.5% (Circle-Economy & Recyc-Québec, 2021). Hence, this relatively low index represents a huge opportunity for implementation of CE actions to enhance the province's circularity index.

In fact, according to van Buren et al. (2016), CE strategies include: Refuse, Reduce, Reuse, Repair, Refurbishment, Remanufacturing, Repurpose, Recycle, and Recover energy strategies.

Remanufacturing stands out among the others because it is able to restore a used product to its original condition, reducing the need for resources and energy compared to the manufacturing of a new product (Sundin, 2019). Through remanufacturing, it becomes feasible to prolong the expected end of life of a used item, thus starting a new full-service cycle (Ingarao, 2017; Mejía-Moncayo et al., 2023). This is enabled by implementing the following seven (7) operations on the used product or core: (1) inspection (of its quality condition), (2) cleaning, (3) disassembly, (4) repairing, (5) refurbishing, (6) reassembly, and (7) a final quality test (Karvonen et al., 2017).

Through these operations, the product's life cycle can be expanded beyond the initially projected end-of-life (EoL), offering a new full-service life cycle during each remanufacturing cycle as illustrated in Fig. 8.1 (Russell & Nasr, 2023). Each cycle provides the product renewed functionalities using less materials and energy compared to the manufacturing of a new product. In addition, remanufacturing enables the integration of other CE strategies as reuse, repair, refurbishment, and recycle contributing to enhance circularity.

Indeed, the remanufacturing cycle involves the flow of new, used, and discarded products between the key actors in the remanufacturing's closed-loop supply chains (CLSC) as demonstrated in Fig. 8.2. The extraction of raw materials is at the origin of the CLSC (see Fig. 8.2). In a second step, processing takes place by suppliers who convert the raw materials into components, parts, or materials for the manufacturing of new products. These produced goods are then further distributed to users. Once the product EoL arrives, users could dispose of the used products or cores in a landfill, or a reverse logistics (RL) system could collect and recover them, which is a preferred action from a sustainability perspective. In the next step, quality inspection of the product or core condition is performed. Considering the deterioration, contamination, and risk involved in the treatment of the used product, the used

[1] "The Circularity Metric, also known as Socioeconomic cycling, is the proportion of secondary materials out of an economy's total material consumption (raw or primary material consumption plus secondary material consumption). The Circularity Metric represents the amount of technical and biological materials that remain within the technical cycle" (https://www.circularity-gap.world/methodology).

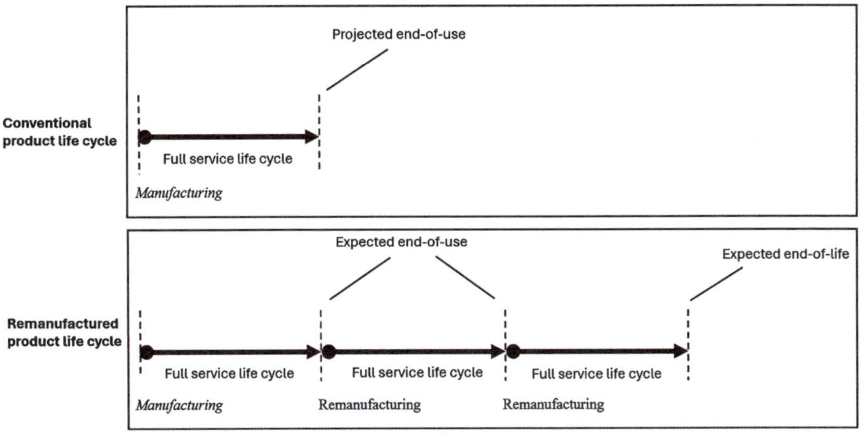

Fig. 8.1 Comparison between conventional and remanufactured product life cycle. (Adapted from Russell and Nasr (2023))

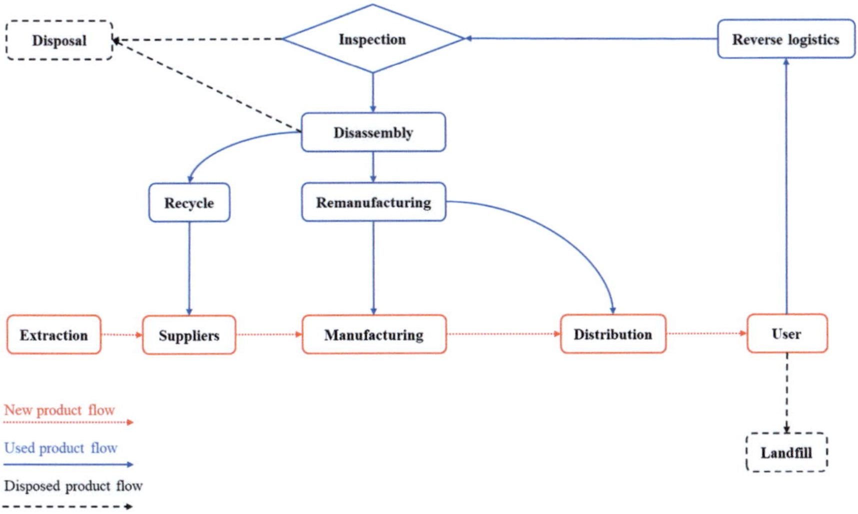

Fig. 8.2 Schematic overview of the remanufacturing's closed-loop supply chains (CLSC)

product can be disposed or disassembled. If the used product is disassembled, it can be further recycled or remanufactured. Even recycled materials can be returned to suppliers for manufacturing new parts, while components or parts for remanufacturing can be returned to manufacturing or distributed directly once remanufactured.

In fact, Fig. 8.2 provides insights into the complexity of the adoption of remanufacturing strategies. Indeed, there are different flows of materials and information among the CLSC actors requiring complex monitoring and controlling their operations and interactions.

In addition, there is a growing demand to address CE strategies like remanufacturing considering the potential economic, environmental, and societal impacts (Fatimah & Aman, 2018). Despite the increasing interest, the implementation of remanufacturing and other CE strategies continues to pose significant challenges. Economic considerations often take precedence over competing priorities, complicating the adoption of these sustainable practices (Lieder et al., 2017; Gusmerotti et al., 2019). For instance, social impacts are often neglected in CE implementations, with an emphasis placed on environmental and economic criteria (Kaya et al., 2022; Tsalis et al., 2022).

Key performance indicators (KPIs) are essential tools that guide decision-making in companies by measuring and controlling various areas of performance. During the transition to CE strategies like remanufacturing, KPIs play a critical role in making informed decisions at different stages of the process. The current literature presents a significant number of KPIs to assess sustainability in manufacturing, Mengistu and Panizzolo (2023), identified 1041 KPIs (290 economic, 410 environmental, and 341 social) to measure industrial sustainability, and something similar is evident in the EC, e.g., Saidani et al. (2019), synthesized and classified 55 KPIs to assess circularity at different levels and for different CE strategies. However, these indicators are usually presented separately for each subdomain, e.g., sustainability and CE. Selecting the most suitable KPIs for a particular application or scenario can be difficult, because of the vast number of options available. This is especially apparent in remanufacturing, due to the need to address circularity, remanufacturability, and supply chain challenges while integrating other CE strategies, such as reusing, recycling, repairing, and refurbishing.

Considering these challenges and issues, this study investigates the KPIs to evaluate sustainability in remanufacturing closed-loop supply chains (CLSC). A systematic literature review (SLR) was conducted to answer the following research questions:

RQ1: What are the main KPIs reported in the literature related to sustainable remanufacturing?
RQ2: What is the scope of the KPIs identified in **RQ1** along CLSCs?

The first question refers to identifying the main KPIs in the three dimensions of sustainability in remanufacturing. This approach aligns well with the need to address the fight against climate change in a sustainable manner. The second question allows identifying the KPIs' scope to understand its current use along the remanufacturing CLSC. Indeed, the performed SLR also contributes to the academic discussion by identifying research gaps and opportunities in sustainable remanufacturing.

2 Methodology

This study implements a SLR to identify the main KPIs in sustainable remanufacturing strategies and operations. The SLR process was formulated to answer two research questions using the methodology described by Tranfield et al. (2003).

Figure 8.3 describes the main steps of the methodology implemented in this study, including the search process, documents acceptance or rejection, and documents analysis as described in the following.

The process starts with a keyword search process performed using the Scopus database as a source of information. Scopus provides peer-reviewed academic document information from various areas of knowledge, which nourished the SLR. In order to capture the largest number of documents related to the study, the search process was carried out including journal articles, conference papers, and book chapters written in English, in a range since 2000 up to early 2024, in all areas of Scopus database, using the following combination of keywords that considers the different synonyms of sustainability and KPIs related to remanufacturing: TITLE-ABS-KEY (("remanufacturing") AND ("sustainab*" OR "TBL" OR "triple bottom line") AND ("metric*" OR "key performance indicator" OR "indicator" OR "criteria" OR "kpi").

The search process resulted in 179 documents, which were accepted or rejected considering their language and relevance to the research questions. In the first classification step, six documents were rejected due to their language: three in Chinese, two in German, and one in Portuguese. Then, considering the title and abstract of the documents, 34 documents were rejected. In fact, these documents were not directly related to remanufacturing, sustainability, circularity, remanufacturability, supply chain (for remanufacturing), or related to industries whose products are not remanufacturable, such as food, medicines, and clothing. Subsequently, in the reading and analysis process, 39 non-relevant documents were rejected because they did not introduce or use any KPIs, resulting in 100 documents used for further analysis in the present work. The KPIs considered in this study were extracted from these sources. Figure 8.4 illustrates the number of documents at each acceptance or rejection stages.

The analysis of the documents retained is central to this study. Each document underwent a comprehensive assessment, during which the KPIs identified or used in each study were extracted and compiled into a Microsoft Excel database. The database includes information such as the article's authors, title, source or journal name, year of publication, and country, as well as the main theme of the document, the

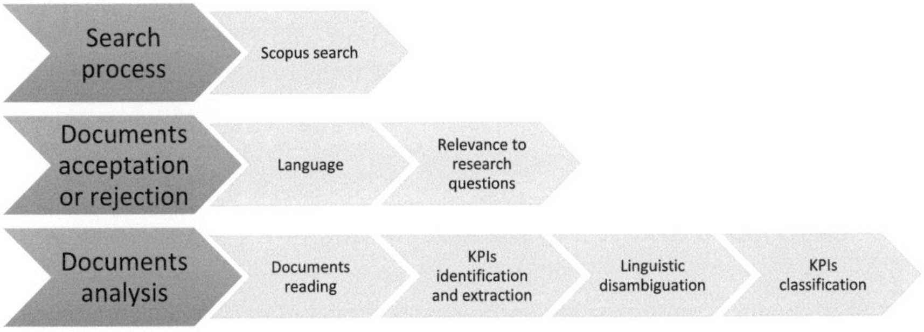

Fig. 8.3 Systematic Literature Review methodology steps implemented in this study

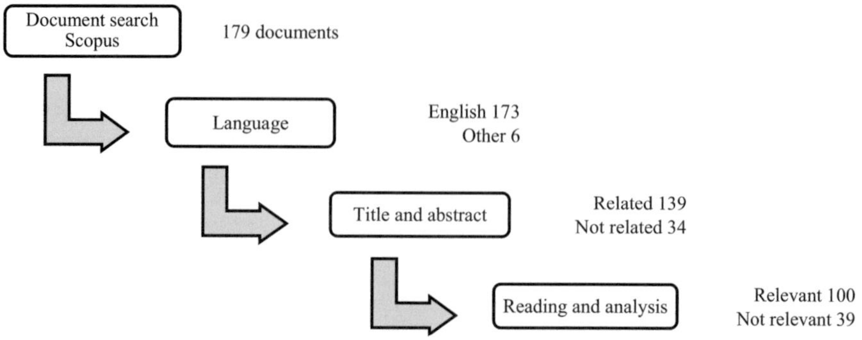

Fig. 8.4 Number of documents by stage of acceptance or rejection

performance indicator, its classification as an economic, environmental, or social indicator, and its primary use or field of application.

Once the database was compiled, a linguistic disambiguation process was carried out due to the different ways in which authors may refer to the same indicator. This process was performed by decomposing each KPI on the *core* of the indicator, its *category*, and the *object or scope* of its application field. For example, fixed cost of a warehouse facility—cost (*core*), fixed (*category*), warehouse facility (*object or scope*); energy consumption by cleaning—energy (*core*), consumption (*category*), cleaning (*object or scope*); and employee health and safety—health and safety (*core*), employee (*category, object, or scope*).

After filtering, we were able to group and categorize the KPIs to determine the most commonly used ones. Similarly, each study was categorized based on its primary objectives, including performance assessment, performance indicators, performance optimization, and sustainability assessment. The following section will present the results and discussion of this analysis process.

3 Results and Discussion

The findings from the review process of identifying KPIs for sustainable remanufacturing are presented. The presentation commences with a descriptive analysis of the documents, followed by an examination of the primary identified indicators and their respective objectives or purposes.

3.1 Descriptive Analysis of the Documents

The SLR conducted in this study encompasses a total of 100 documents sourced from journals, conference papers, and book chapters published between 2000 and 2024. Figure 8.5 illustrates the yearly distribution of published documents during this timeframe. It is evident that the volume of studies has been steadily growing

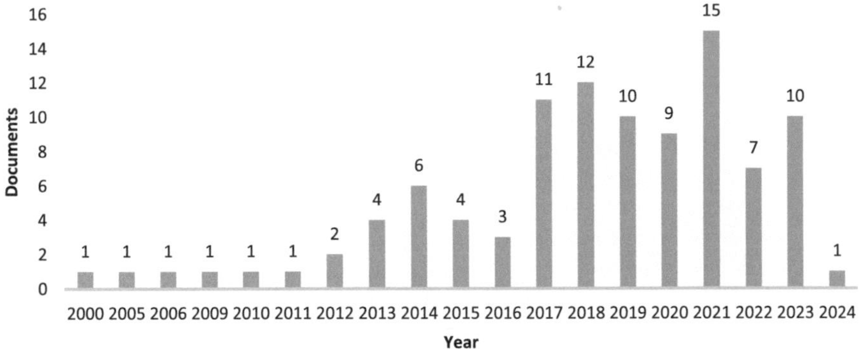

Fig. 8.5 Number of documents by year

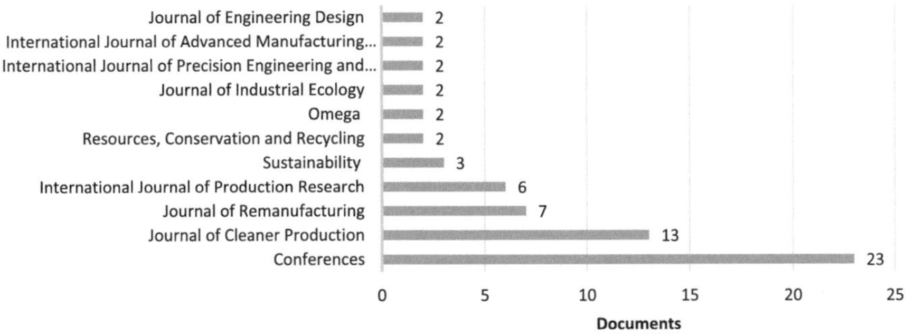

Fig. 8.6 Main sources of documents

since 2012, indicating the increasing relevance of this topic in academic and practitioner discussions.

Figure 8.6 summarizes the main sources of the reviewed documents. It can be noted that conference papers are the main source of documents for this study, specifically including conferences that focus on industrial engineering, manufacturing, sustainability, production management, materials, and information and communication technologies. The IIE annual conference and expo is the main contributor with four documents. Remaining reviewed conference studies are distributed among CIRP, IFIP, IEEE, and others.

The list of sources continues with the Journal of Cleaner Production, the Journal of Remanufacturing, the International Journal of Production Research, Sustainability, Resources-Conservation and Recycling, Omega, the International Journal of Precision Engineering and Manufacturing—Green Technology, the Journal of Industrial Ecology, the International Journal of Advanced Manufacturing Technology, and the Journal of Engineering Design, and other literature sources contribute with only one document and are not represented in this overview diagram.

The distribution of documents by country is depicted in Fig. 8.7, exposing the interest and efforts of some governments in promoting the CE, as China's Circular Economy Promotion Law, India's Resource Efficiency Policy, and the European Remanufacturing Network funded under Horizon 2020 among some examples.

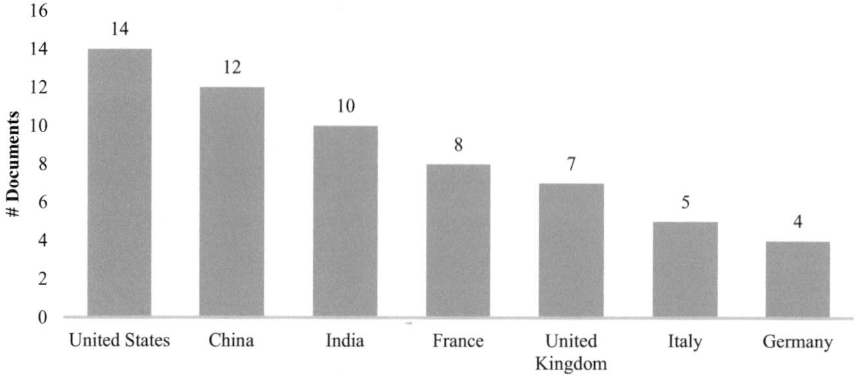

Fig. 8.7 Number of documents by country

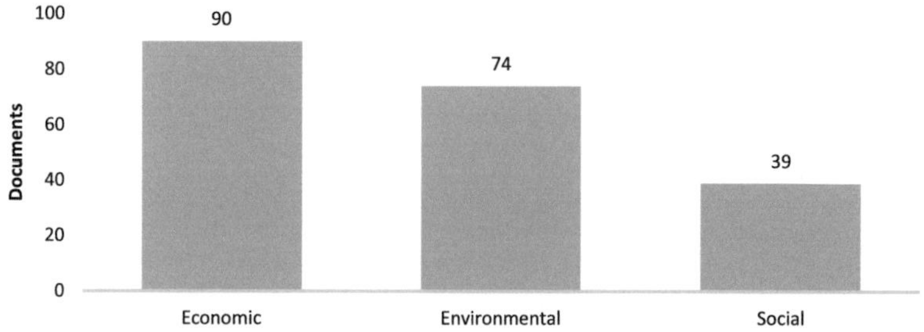

Fig. 8.8 Number of documents by the sustainability dimension

Figure 8.8 exposes the documents distribution by sustainability dimensions, where the Economic dimension takes a first place, followed by environmental and social dimensions. This sequence also shows that the economic dimension is, still, the main driver of remanufacturing.

3.2 Indicator Analysis

In this research work, a SLR was carried out to identify a total of 1153 KPIs specifically related to sustainable remanufacturing. These KPIs were further classified into 703 *economic* indicators, 334 *environmental* indicators, and 116 *social* indicators. The data for this analysis was extracted from 1569 records obtained from 100 different studies. A summarized presentation of the most frequently referenced KPIs in the analyzed literature, along with their associated document count, can be found in Table 8.1.

The high amount and frequency of economic KPIs in Table 8.1 unequivocally demonstrates that the economic dimension holds a prominent position in sustainable remanufacturing, outshining the environmental and social dimensions in terms

Table 8.1 Sustainable key performance indicators identified in the SLR in three or more documents

Sustainability dimension							
Economic			Environmental			Social	
Performance indicator		Documents	Performance indicator		Documents	Performance indicator	Documents
Cost	Transport	14	Energy	Consumption	20	Health and safety	9
	Fixed	10		Saved	3	Job creation	3
	Recovery	10		Embodied	3	Employment stability	3
	Remanufacturing	8		Renewable	3		
	Labor	7	Environmental impact		12		
	Disassembly	7	GHG emissions		12		
	Total	7	Recycling materials		6		
	Recycling	6	Remanufactured parts		5		
	Disposal	6	Acidification potential		5		
	Energy	5	Disposal materials		5		
	Distribution	5	Global warming potential		5		
	Investment	5	Waste		4		
	Production	5	Eutrophication potential		4		
	Operational	5	Reused parts		3		
	Materials	4					
	Cleaning	4					
Disassembly time		13					
Core quality condition		5					
Products Price	New	6					
	Remanufactured	4					
Capacity	Warehouse	5					
	Production	4					
Revenue		5					
Distance	Collection Centre	6					
	Transport	4					

of importance. The different cost categories represent the most frequently used KPIs, including CLSC operations or processes, such as transport cost (Ali et al., 2020; Alkhayyal, 2018; Grosse Erdmann et al., 2023), recovery cost (Taleizadeh et al., 2019; Yu & Solvang, 2017), distribution cost (Li et al., 2018), production cost (Zhao & Zhou, 2023), and operational cost (Deveci et al., 2021). Other cost categories are represented by CE strategies, such as remanufacturing cost (Mohamed Noor et al., 2018), and recycling cost (Jeng & Lin, 2017), and strategic decision-related costs, such as facilities fixed costs (Mota et al., 2018). Finally, remanufacturing operations constitute another category, for example, considering disassembly costs (Ren et al., 2021), cleaning costs (Ansari et al., 2020), and disposal costs (Yu & Solvang, 2017), as well as other relevant costs as labor, total, energy, and investment.

Disassembly time emerges as a second frequently deployed KPI after the cost categories. In fact, it is commonly used to evaluate the complexity level of the disassembly process (Mandolini et al., 2018). The core or used product quality condition also represents a critical factor toward remanufacturing (Kazancoglu & Ozkan-Ozen, 2020). The price of new and remanufactured products is also considered to evaluate remanufacturability (Zwolinski et al., 2006). Other studies consider production and warehousing capacities (Das & Mehta, 2015), distance to collection centers (Alkhayyal, 2018), and transport (Yanikara et al., 2014), representing important decisions in CLSC design and planning, similar to revenue (Aydin et al., 2017).

Energy consumption is the driving factor behind environmental KPIs (Prajapati et al., 2021; Vimal et al., 2021), followed by energy savings (Goepp et al., 2014), embodied energy (Justham et al., 2013), and renewable energy (Bhatia et al., 2019). Some studies merge different environmental KPIs as "environmental impact" (Goodall et al., 2014; Hummen & Wege, 2021; Tchertchian et al., 2013); however, it represents a barrier to perform a comparison or even a source of confusion. Greenhouse gas (GHG) emissions are used in many studies (Alamerew et al., 2020; Golinska & Kuebler, 2014; Inoue et al., 2020; Miyajima et al., 2019). Recycling materials (Boorsma et al., 2022), remanufactured (Ali et al., 2020), and reused parts (Goepp et al., 2014) are also considered as KPIs. Life cycle assessment (LCA) KPIs, such as acidification potential (Jannone Da Silva et al., 2012), disposal of materials (Govindan et al., 2019), global warming potential (Favi et al., 2021), waste (Tsiliyannis, 2014), and eutrophication potential (Spreafico, 2022) were identified in the literature as other used KPIs in the environment category, and these KPIs represent a reference for different purposes along remanufacturing's CLSC.

The social dimension is concentrated on strategic KPIs, such as health and safety for employees and users (Deveci et al., 2021; Sethanan et al., 2019; Ullah et al., 2016), while job creation (Fatimah & Aman, 2018; Taleizadeh et al., 2019) and employment stability (Li et al., 2018; Zarbakhshnia et al., 2018) are considered in the CLSC actors. Unfortunately, social KPIs are less considered in comparison with the other dimensions.

3.3 Descriptive Analysis of the Documents

The analyzed studies can be classified into four groups depending on the main objective or purpose of each study, including performance assessment, performance indicators, performance optimization, and sustainability assessment. The studies that incorporate a performance assessment approach use sustainable and CE KPIs to evaluate various aspects, such as remanufacturability (Ali et al., 2021; Aydin et al., 2017; De Barba et al., 2013; Goodall et al., 2014; Justham et al., 2013; Tian et al., 2017; Zhang et al., 2021b), reusability and recoverability (Aydin et al., 2017), and repairability (Alkouh et al., 2023). Typically, RL is analyzed considering logistic performance (Sagnak, 2020), network design (Alkhayyal, 2018; Grosse Erdmann et al., 2023; Yanikara et al., 2014), third-party RL provider selection (Govindan et al., 2019; Li et al., 2018; Mishra et al., 2023), product recovery process selection (Jindal & Singh Sangwan, 2016; Zhang et al., 2021a), and circularity (Benini et al., 2022; Tsiliyannis, 2014).

The reviewed studies focusing on remanufacturing performance assessment consider supply chain risks (Ansari et al., 2020; Zhao & Zhou, 2023), value created (Vogtlander et al., 2017) and retained (Russell & Nasr, 2023), economic performance (Arredondo-Soto et al., 2018), environmental performance (Golinska-Dawson & Pawlewski, 2015; Pan & Liu, 2009), economic and environmental impact (van Loon & Van Wassenhove, 2018), facility planning (Bhatia et al., 2019), and remanufacturing planning (Jeng & Lin, 2017). Disassembly analysis includes its complexity or effort (Das et al., 2000; Mouflih et al., 2023; Shrivastava et al., 2005), disassembly planning (Marconi et al., 2019; Zhang et al., 2022), and sequencing (Mandolini et al., 2018; Priyono et al., 2016).

Product design is also considered in many studies considering eco-design (Favi et al., 2021; Jannone Da Silva et al., 2012), modular design (Miyajima et al., 2019; Tchertchian et al., 2013), complexity (Mesa et al., 2018), circular design (Boorsma et al., 2022), axiomatic design (Chakraborty et al., 2017), material selection (Yang et al., 2017), product profiles (Zwolinski et al., 2006), economic analysis (Wang & Tseng, 2010), and LCA (Spreafico, 2022).

Some studies also introduce performance indicators focused on circularity (Alamerew et al., 2020; Bobba et al., 2023; Boyer et al., 2021; Mishra et al., 2022), longevity and circularity (Figge et al., 2018; Linder et al., 2017), circularity adoption (Yadav et al., 2020), environmental sustainability (Haupt & Hellweg, 2019), remanufacturing sustainability (Fatimah & Aman, 2018; Sethanan et al., 2019), remanufacturing supply chain (Ansari et al., 2022), remanufacturability (Hummen & Wege, 2021), CLSC (Kurt et al., 2021), RL (Prajapati et al., 2021), disassemblability (Erdmann et al., 2023; Vanegas et al., 2018), and economic product design (Mohamed Noor et al., 2018), and upgradability (Aziz et al., 2017).

The studies focused on performance optimization using optimization models for improving processes (Shakourloo, 2017), remanufacturing operations (Ullah et al., 2016), facility design (Mejía-Moncayo et al., 2021, 2024), and facility location (Deveci et al., 2021), CLSC design (Das & Rao Posinasetti, 2015; Mota et al., 2018;

Taleizadeh et al., 2019), RL optimization (Ali et al., 2020; Yu & Solvang, 2017), disassembly line balancing (Kazancoglu & Ozkan-Ozen, 2020; Ren et al., 2021), disassembly sequencing (Lu et al., 2020), disassembly planning (Ren et al., 2021; Xia et al., 2014), and remanufacturability (Jiang et al., 2011).

The reviewed studies focusing on sustainability assessment typically use economic, environmental or social sustainability dimensions to evaluate remanufacturing companies' sustainability (Golinska & Kuebler, 2014; Golinska-Dawson & Pawlewski, 2015), remanufacturing sustainability (Chen et al., 2023; Fatimah et al., 2013; Jensen et al., 2019; Zhang et al., 2021c, 2021d), environmental and economic impacts of CLSC (Das, 2020; Das & Mehta, 2015), retailers sales effort on CLSC (Yang et al., 2019), the impact of green supply chain practices on company performance (Sarwar et al., 2021), supplier selection in RL (Zarbakhshnia et al., 2018), and LCA for remanufacturability (Schau et al., 2012).

Finally, sustainability assessment of product designs is an important topic to promote remanufacturing. Different strategies have been implemented to integrate sustainability in product design, such as modularity in which the use of standardized components and suppliers' selection have a key role in sustainability performance (Inoue et al., 2020). Modularity in electric vehicles is as well an emerging discussion topic as an enabler of sustainability (Lampón, 2023). Material and EoL strategy selection also represent important decisions on product design (Jayakrishna & Vinodh, 2017), while data models represent an important tool to support product design (Goepp et al., 2014).

3.4 KPIs Distribution Along CLSC

The distribution of sustainable KPIs in the remanufacturing's CLSC is schematically outlined in Fig. 8.9, where economic indicators are presented in blue, environmental indicators in green, and social indicators in red. A first group is presented in the lower part of Fig. 8.9, which corresponds to all the indicators that are common to all CLSC actors. This group presents a first reference for a decision-maker on the indicators they need to consider when evaluating its performance, regardless of which of the CLSC actors it is. From these, it is possible to establish that the location of the facilities is an essential factor to consider for any of the actors. This is because transportation costs, transportation distance, fixed costs, investment costs, energy consumption, environmental impact, and GHG emissions depend on where the facilities are located, their proximity to their suppliers or customers, and government incentives and/or regulations. This also highlights the need to promote local or regional supply chains, to reduce the impacts of global transportation. The health and safety of the different CLSC actors and job creation are social factors that must be considered.

A second group of indicators is presented, linked to each actor or operation of the CLSC. The price of new products and revenue play a critical role in manufacturing. Moreover, the price of new products serves as a reference for remanufactured

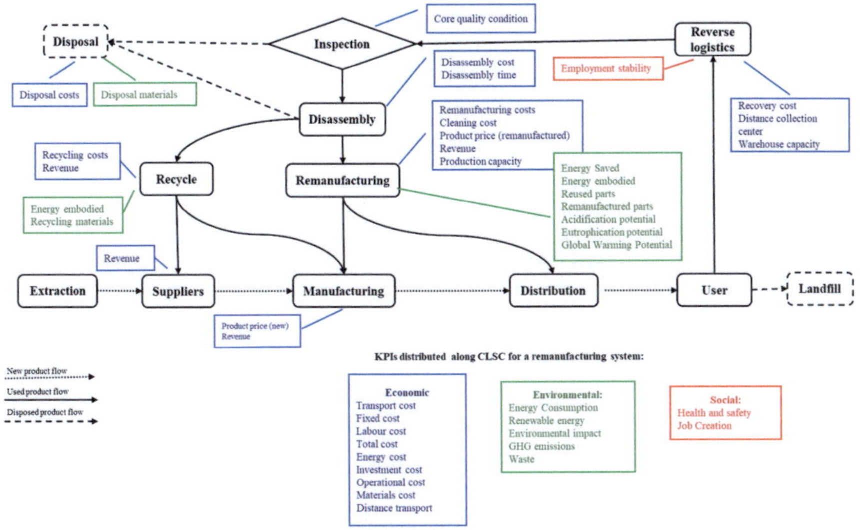

Fig. 8.9 Schematic representation of the distribution of sustainable KPIs along remanufacturing's CLSCs, where economic KPIs are identified in blue, environmental KPIs in green, and social KPIs in red

products. In reverse logistics, it is necessary to consider collection costs, distance to the collection center, storage capacity, and employment stability.

The quality conditions of the core or used product are evaluated in the inspection operation. The definition of quality conditions can be done by considering product deterioration, technical standards, regulations, maintenance data, and the experience and expertise of the organization. In the disposal operation, disposal costs and materials must be considered. The disposal process and its cost are determined depending on the types of materials and their conditions.

The disassembly operation depends on the time of the operation. This has a direct relationship with the complexity of the operation and affects its cost. The largest number of indicators is concentrated in remanufacturing. Among the economic indicators, it is necessary to consider remanufacturing costs, cleaning costs, the price of remanufactured products, income, and production capacity. In environmental terms, the energy saved and embodied in the product allows to quantify the total energy saved by remanufacturing. Also are considered the amount of reused and remanufactured parts, potential acidification, potential eutrophication, and the global warming potential. Each of these indicators contributes to evaluating the viability of remanufacturing.

In the recycling operation, costs and revenue are economic criteria to consider. These have a direct relationship with recycled materials and with the energy embodied in each of them. This energy is a criterion to be considered in terms of energy savings for subsequent transformation processes. Finally, revenue stands out among suppliers.

For those interested in adopting remanufacturing, this study summarizes the factors to consider for a sustainable operation. Figure 9 integrates the set of the most frequent KPIs in the analyzed literature. However, it is possible that, in particular cases, it may be necessary to incorporate other indicators depending on the characteristics of the products and CLSC.

4 Research Implications

The literature on decision-making tools is rich and includes a wide variety of KPIs to evaluate sustainability in remanufacturing, evidence of that is the1153 KPIs identified in this study. This research identifies the KPIs most used for this purpose across the CLSC. Adopting this approach enables a comprehensive understanding of sustainability and helps to understand which KPIs must be considered by each actor within the CLSC.

The three dimensions of sustainability were used to categorize the KPIs identified in the literature. This process shows the existence of a dominant position of economic criteria, followed by environmental and social criteria. This imbalance between dimensions of sustainability raises questions about the priorities of different stakeholders along the closed supply chain, and even highlights the need to analyze the selection of stakeholders that must be considered in sustainable remanufacturing.

This study will help decision-makers evaluate the impact of remanufacturing from a sustainable perspective and implement the right decisions that are economically, environmentally, and socially acceptable. Indeed, the study outcomes are particularly useful in scenarios such as sustainability performance assessments, or process, operation, and product design optimization.

5 Conclusions

To evaluate the sustainability of remanufacturing practices, this study presents a comprehensive literature review that specifically focused on identifying and analyzing the key performance indicators (KPIs) that are commonly employed in this field. The review process of one hundred systematically selected published documents resulted in the identification of a total of 42 KPIs. In terms of frequency of use, the KPIs are distributed among economic, environmental, and social factors, in descending order. The studies that were consulted have identified four primary areas of focus within the various actors of the remanufacturing's CLSC. These include remanufacturing operations or processes, product design, and the evaluation of remanufacturability and circularity.

The conducted SLR presents a scenario regarding sustainable remanufacturing where the main concerns in terms of the identified KPIs are the costs of remanufacturing (remanufactured products, remanufacturing processes, CLSC), resources and energy consumed or saved, the environmental impacts of the remanufacturing, and the safety and health of stakeholders, as well as stable job opportunities. It is crucial to note that this study has certain limitations in its scope, as it exclusively draws on documents from the Scopus database, thereby potentially excluding other valuable perspectives.

Finally, the presented SLR also identified research opportunities. These include developing methodological tools to guide the use of KPIs in practice, understanding the reasons for the low use of social KPIs, and creating models that integrate different KPIs for sustainable control of remanufacturing operations or CLSC.

References

Alamerew, Y. A., Kambanou, M. L., Sakao, T., & Brissaud, D. (2020). A multi-criteria evaluation method of product-level circularity strategies. *Sustainability (Switzerland), 12*(12), 5129. https://doi.org/10.3390/su12125129

Ali, S. S., Paksoy, T., Torğul, B., & Kaur, R. (2020). Reverse logistics optimization of an industrial air conditioner manufacturing company for designing sustainable supply chain: A fuzzy hybrid multi-criteria decision-making approach. *Wireless Networks, 26*(8), 5759–5782. https://doi.org/10.1007/s11276-019-02246-6

Ali, A., Enyoghasi, C., & Badurdeen, F. (2021). Quantitative assessment of product remanufacturability. In K. K., P. K. Ghate A. (Ed.), *IISE Annual Conference and Expo 2021* (pp. 477–482). Institute of Industrial and Systems Engineers, IISE. https://www.proquest.com/scholarly-journals/quantitative-assessment-product/docview/2560889073/se-2

Alkhayyal, B. A. (2018, March 6). Carbon emissions policies impact on reverse supply chain network. In *8th Annual International Conference on Industrial Engineering and Operations Management*.

Alkouh, A., Keddar, K. A., & Alatefi, S. (2023). Revolutionizing repairability of industrial electronics in oil and gas sector: A mathematical model for the index of repairability (IOR) as a novel technique. *Electronics (Switzerland), 12*(11), 2461. https://doi.org/10.3390/electronics12112461

Ansari, Z. N., Kant, R., & Shankar, R. (2020). Evaluation and ranking of solutions to mitigate sustainable remanufacturing supply chain risks: A hybrid fuzzy SWARA-fuzzy COPRAS framework approach. *International Journal of Sustainable Engineering, 13*(6), 473–494. https://doi.org/10.1080/19397038.2020.1758973

Ansari, Z. N., Kant, R., & Shankar, R. (2022). Remanufacturing supply chain: An analysis of performance indicator areas. *International Journal of Productivity and Performance Management, 71*(1), 25–57. https://doi.org/10.1108/IJPPM-01-2020-0038

Arredondo-Soto, K. C., Sanchez-Leal, J., Reyes-Martinez, R. M., Salazar-Ruíz, E., & Maldonado-Macias, A. A. (2018). World class remanufacturing productions systems: An analysis of Mexican maquiladoras. *Advances in Intelligent Systems and Computing, 606*, 153–161. https://doi.org/10.1007/978-3-319-60474-9_14

Aydin, R., Brown, A., Ali, A., & Badurdeen, F. (2017). Assessment of end-of-life product lifecycle "ilities." In C. K. , C. E. Nembhard H.B. (Ed.), *67th Annual Conference and Expo of the Institute of Industrial Engineers* 2017 (pp. 1691–1696). Institute of Industrial Engineers. https://www.proquest.com/scholarly-journals/assessment-end-life-product-lifecycleilities/docview/1951119941/se-2?accountid=27231

Aziz, N. A., Wahab, D. A., & Ramli, R. (2017). Establishment of engineering metrics for upgradable design of brake caliper. *Smart Innovation, Systems and Technologies, 68*, 87–97. https://doi.org/10.1007/978-3-319-57078-5_9

Benini, L., Leroy, Y., Tolio, T., & Magnanini, M. C. (2022). Proposal of a strategic model to unlock the circular potential in industrial practice. *Procedia CIRP, 109*, 233–238. https://doi.org/10.1016/j.procir.2022.05.242

Bhatia, M. S., Dora, M., & Jakhar, S. K. (2019). Appropriate location for remanufacturing plant towards sustainable supply chain. *Annals of Operations Research.* https://doi.org/10.1007/s10479-019-03294-z

Bobba, S., Eynard, U., Maury, T., Ardente, F., Blengini, G. A., & Mathieux, F. (2023). Circular Input Rate: Novel indicator to assess circularity performances of materials in a sector—Application to rare earth elements in e-vehicles motors. *Resources, Conservation and Recycling, 197*, 107037. https://doi.org/10.1016/j.resconrec.2023.107037

Boorsma, N., Polat, E., Bakker, C., Peck, D., & Balkenende, R. (2022). Development of the circular product readiness method in circular design. *Sustainability (Switzerland), 14*(15), 9288. https://doi.org/10.3390/su14159288

Boyer, R. H. W., Mellquist, A. C., Williander, M., Fallahi, S., Nyström, T., Linder, M., Algurén, P., Vanacore, E., Hunka, A. D., Rex, E., & Whalen, K. A. (2021). Three-dimensional product circularity. *Journal of Industrial Ecology, 25*(4), 824–833. https://doi.org/10.1111/jiec.13109

Chakraborty, K., Mondal, S., & Mukherjee, K. (2017). Analysis of product design characteristics for remanufacturing using Fuzzy AHP and Axiomatic Design. *Journal of Engineering Design, 28*(5), 338–368. https://doi.org/10.1080/09544828.2017.1316014

Chen, Q., Lai, X., Chen, J., Yao, Y., Guo, Y., Zhai, M., Han, X., Lu, L., & Zheng, Y. (2023). Comparative environmental impacts of different hydrometallurgical recycling and remanufacturing technologies of lithium-ion batteries considering multi-recycling-approach and temporal-geographical scenarios in China. *Separation and Purification Technology, 324*, 124642. https://doi.org/10.1016/j.seppur.2023.124642

Circle-Economy, & Recyc-Québec. (2021). *The circularity gap report-Quebec*. https://www.recyc-quebec.gouv.qc.ca/sites/default/files/documents/rapport-indice-circularite-en.pdf

Das, K. (2020). Planning environmental and economic sustainability in closed-loop supply chains. *Operations and Supply Chain Management: An International Journal, 13*(1), 64–81. https://doi.org/10.31387/oscm0400253

Das, K., & Mehta, M. (2015). Integrating environmental and economic sustainability in closed loop supply chain. *IIE Annual Conference. Proceedings; Norcross, 2*(2), 272–281.

Das, K., & Rao Posinasetti, N. (2015). Addressing environmental concerns in closed loop supply chain design and planning. *International Journal of Production Economics, 163*, 34–47. https://doi.org/10.1016/j.ijpe.2015.02.012

Das, S. K., Yedlarajiah, P., & Narendra, R. (2000). An approach for estimating the end-of-life product disassembly effort and cost. *International Journal of Production Research, 38*(3), 657–673. https://doi.org/10.1080/002075400189356

De Barba, D. J., de Oliveira Gomes, J., Salis, J. I., & Bork, C. A. S. (2013). Remanufacturing versus manufacturing—Analysis of requirements and constraints for a study case: Control arm of a suspension system. In *Re-engineering manufacturing for sustainability* (pp. 669–673). Springer. https://doi.org/10.1007/978-981-4451-48-2_109

Deveci, M., Simic, V., & Torkayesh, A. E. (2021). Remanufacturing facility location for automotive Lithium-ion batteries: An integrated neutrosophic decision-making model. *Journal of Cleaner Production, 317*, 128438. https://doi.org/10.1016/j.jclepro.2021.128438

Erdmann, J. G., Koller, J., Brimaire, J., & Döpper, F. (2023). Assessment of the disassemblability of electric bicycle motors for remanufacturing. *Journal of Remanufacturing, 13*(2), 137–159. https://doi.org/10.1007/s13243-023-00124-1

Fatimah, Y. A., & Aman, M. (2018). Remanufacturing sustainability indicators: An Indonesian small and medium enterprise case study. *IOP Conference Series: Materials Science and Engineering, 403*(1), 012055. https://doi.org/10.1088/1757-899X/403/1/012055

Fatimah, Y. A., Biswas, W., Mazhar, I., & Islam, M. N. (2013). Sustainable manufacturing for Indonesian small- and medium-sized enterprises (SMEs): The case of remanufactured alternators. *Journal of Remanufacturing, 3*(1), 6. https://doi.org/10.1186/2210-4690-3-6

Favi, C., Marconi, M., Rossi, M., & Cappelletti, F. (2021). Product eco-design in the era of circular economy: Experiences in the design of espresso coffee machines. In *Lecture notes in mechanical engineering* (pp. 194–199). https://doi.org/10.1007/978-3-030-70566-4_31

Figge, F., Thorpe, A. S., Givry, P., Canning, L., & Franklin-Johnson, E. (2018). Longevity and circularity as indicators of eco-efficient resource use in the circular economy. *Ecological Economics, 150*, 297–306. https://doi.org/10.1016/J.ECOLECON.2018.04.030

Goepp, V., Zwolinski, P., & Caillaud, E. (2014). Design process and data models to support the design of sustainable remanufactured products. *Computers in Industry, 65*(3), 480–490. https://doi.org/10.1016/j.compind.2014.02.002

Golinska, P., & Kuebler, F. (2014). The method for assessment of the sustainability maturity in remanufacturing companies. *Procedia CIRP, 15*, 201–206. https://doi.org/10.1016/j.procir.2014.06.018

Golinska-Dawson, P., & Pawlewski, P. (2015). Simulation modeling approach to environmental issues in supply chain with remanufacturing. *Studies in Computational Intelligence, 598*, 363–372. https://doi.org/10.1007/978-3-319-16211-9_37

Goodall, P., Rosamond, E., & Harding, J. (2014). A review of the state of the art in tools and techniques used to evaluate remanufacturing feasibility. *Journal of Cleaner Production, 81*, 1–15. https://doi.org/10.1016/j.jclepro.2014.06.014

Gouvernement du Québec. (2023, June 20). *2030 Plan for a Green Economy*. https://www.quebec.ca/en/government/policies-orientations/plan-green-economy

Gouvernement du Québec. (2024). *Accélérer le développement de l'économie circulaire feuille de route* (pp. 1–59). Bureau de coordination du développement durable du ministère de l'Environnement, de la Lutte contre les changements climatiques, de la Faune et des Parcs (MELCCFP). https://www.quebec.ca/gouvernement/politiques-orientations/developpement-durable/strategie-gouvernementale/feuille-route-gouvernementale-economie-circulaire

Govindan, K., Kadziński, M., Ehling, R., & Miebs, G. (2019). Selection of a sustainable third-party reverse logistics provider based on the robustness analysis of an outranking graph kernel conducted with ELECTRE I and SMAA. *Omega (United Kingdom), 85*, 1–15. https://doi.org/10.1016/j.omega.2018.05.007

Grosse Erdmann, J., Koller, J., Amir, S., Mihelič, A., & Döpper, F. (2023). Simulation-based analysis of (reverse) supply chains in circular product-service-systems. In *Lecture notes in mechanical engineering* (pp. 111–118). Springer Science and Business Media Deutschland GmbH. https://doi.org/10.1007/978-3-031-28839-5_13

Gusmerotti, N. M., Testa, F., Corsini, F., Pretner, G., & Iraldo, F. (2019). Drivers and approaches to the circular economy in manufacturing firms. *Journal of Cleaner Production, 230*, 314–327. https://doi.org/10.1016/j.jclepro.2019.05.044

Haupt, M., & Hellweg, S. (2019). Measuring the environmental sustainability of a circular economy. *Environmental and Sustainability Indicators, 1–2*, 100005. https://doi.org/10.1016/j.indic.2019.100005

Hummen, T., & Wege, E. (2021). Remanufacturing of energy using products makes sense only when technology is mature: Introducing a circular economy indicator for remanufacturing based on a parameterized LCA and LCC assessment of a circulation pump. In *Sustainable production, life cycle engineering and management* (pp. 67–82). Springer Science and Business Media Deutschland GmbH. https://doi.org/10.1007/978-3-030-50519-6_6

Ingarao, G. (2017). Manufacturing strategies for efficiency in energy and resources use: The role of metal shaping processes. *Journal of Cleaner Production, 142*, 2872–2886. https://doi.org/10.1016/j.jclepro.2016.10.182

Inoue, M., Yamada, S., Miyajima, S., Ishii, K., Hasebe, R., Aoyama, K., Yamada, T., & Bracke, S. (2020). A modular design strategy considering sustainability and supplier selection. *Journal of Advanced Mechanical Design, Systems and Manufacturing, 14*(2). https://doi.org/10.1299/jamdsm.2020jamdsm0023

Jannone Da Silva, E., Ometto, A. R., Rozenfeld, H., Lopes Silva, D. A., Cristina, D., Pigosso, A., Ricco, V., & Reis, A. (2012). *Prototypical implementation of a remanufacturing oriented grinding machine*. www.bragecrim.weebly.com

Jayakrishna, K., & Vinodh, S. (2017). Application of grey relational analysis for material and end of life strategy selection with multiple criteria. *International Journal of Materials Engineering Innovation, 8*(3/4), 250. https://doi.org/10.1504/IJMATEI.2017.090241

Jeng, S. Y., & Lin, C. W. R. (2017). Fuzzy cradle to cradle remanufacturing planning for a recycled toner cartridge industry. *International Journal of Industrial and Systems Engineering, 25*(4), 423. https://doi.org/10.1504/IJISE.2017.083039

Jensen, J. P., Prendeville, S. M., Bocken, N. M. P., & Peck, D. (2019). Creating sustainable value through remanufacturing: Three industry cases. *Journal of Cleaner Production, 218*, 304–314. https://doi.org/10.1016/j.jclepro.2019.01.301

Jiang, Z., Zhang, H., & Sutherland, J. W. (2011). Development of multi-criteria decision making model for remanufacturing technology portfolio selection. *Journal of Cleaner Production, 19*(17–18), 1939–1945. https://doi.org/10.1016/j.jclepro.2011.07.010

Jindal, A., & Singh Sangwan, K. (2016). A fuzzy-based decision support framework for product recovery process selection in reverse logistics. *International Journal of Services and Operations Management, 25*(4), 413.

Justham, L. M., Rosamond, E. L., Goodall, P. A., Conway, P. P., & West, A. A. (2013). A proposed novel knowledge framework for remanufacturing viability in a modern supply chain. In *Advanced concurrent engineering* (pp. 861–870). https://doi.org/10.1007/978-1-4471-4426-7_73

Karvonen, I., Jansson, K., Behm, K., Vatanen, S., & Parker, D. (2017). Identifying recommendations to promote remanufacturing in Europe. *Journal of Remanufacturing, 7*(2–3), 159–179. https://doi.org/10.1007/s13243-017-0038-2

Kaya, S. K., Ayçin, E., & Pamucar, D. (2022). Evaluation of social factors within the circular economy concept for European countries. *Central European Journal of Operations Research, 31*, 73. https://doi.org/10.1007/s10100-022-00800-w

Kazancoglu, Y., & Ozkan-Ozen, Y. D. (2020). Sustainable disassembly line balancing model based on triple bottom line. *International Journal of Production Research, 58*(14), 4246–4266. https://doi.org/10.1080/00207543.2019.1651456

Kurt, A., Cortes-Cornax, M., Cung, V. D., Front, A., & Mangione, F. (2021). A classification tool for circular supply chain indicators. In *IFIP advances in information and communication technology, 634 IFIP* (pp. 644–653). https://doi.org/10.1007/978-3-030-85914-5_68

Lampón, J. F. (2023). Efficiency in design and production to achieve sustainable development challenges in the automobile industry: Modular electric vehicle platforms. *Sustainable Development, 31*(1), 26–38. https://doi.org/10.1002/sd.2370

Li, Y.-L., Ying, C.-S., Chin, K.-S., Yang, H.-T., & Xu, J. (2018). Third-party reverse logistics provider selection approach based on hybrid-information MCDM and cumulative prospect theory. *Journal of Cleaner Production, 195*, 573–584. https://doi.org/10.1016/j.jclepro.2018.05.213

Lieder, M., Asif, F. M. A., Rashid, A., Mihelič, A., & Kotnik, S. (2017). Towards circular economy implementation in manufacturing systems using a multi-method simulation approach to link design and business strategy. *International Journal of Advanced Manufacturing Technology, 93*(5–8), 1953–1970. https://doi.org/10.1007/s00170-017-0610-9

Linder, M., Sarasini, S., & van Loon, P. (2017). A metric for quantifying product-level circularity. *Journal of Industrial Ecology, 21*(3), 545–558. https://doi.org/10.1111/JIEC.12552

Lu, Q., Ren, Y., Jin, H., Meng, L., Li, L., Zhang, C., & Sutherland, J. W. (2020). A hybrid metaheuristic algorithm for a profit-oriented and energy-efficient disassembly sequencing problem. *Robotics and Computer-Integrated Manufacturing, 61*, 101828. https://doi.org/10.1016/j.rcim.2019.101828

Mandolini, M., Favi, C., Germani, M., & Marconi, M. (2018). Time-based disassembly method: How to assess the best disassembly sequence and time of target components in complex products. *International Journal of Advanced Manufacturing Technology, 95*(1–4), 409–430. https://doi.org/10.1007/S00170-017-1201-5/METRICS

Marconi, M., Germani, M., Mandolini, M., & Favi, C. (2019). Applying data mining technique to disassembly sequence planning: A method to assess effective disassembly time of industrial products. *International Journal of Production Research, 57*(2), 599–623. https://doi.org/10.1080/00207543.2018.1472404

Mejía-Moncayo, C., Kenné, J. P., & Hof, L. A. (2021). A hybrid architecture for a reconfigurable cellular remanufacturing system. In *IFIP advances in information and communication technology, 631 IFIP* (pp. 488–496). https://doi.org/10.1007/978-3-030-85902-2_52

Mejía-Moncayo, C., Kenné, J.-P., & Hof, L. A. (2023). On the development of a smart architecture for a sustainable manufacturing-remanufacturing system: A literature review approach. *Computers & Industrial Engineering, 180*, 109282. https://doi.org/10.1016/j.cie.2023.109282

Mejía-Moncayo, C., Kenné, J.-P., & Hof, L. A. (2024). A reconfigurable cellular remanufacturing architecture: A multi-objective design approach. *Journal of Remanufacturing, 14*, 185. https://doi.org/10.1007/s13243-024-00139-2

Mengistu, A. T., & Panizzolo, R. (2023). Analysis of indicators used for measuring industrial sustainability: A systematic review. *Environment, Development and Sustainability, 25*(3), 1979–2005. https://doi.org/10.1007/s10668-021-02053-0

Mesa, J. A., Illera, D., Esparragoza, I., Maury, H., & Gómez, H. (2018). Functional characterisation of mechanical joints to facilitate its selection during the design of open architecture products. *International Journal of Production Research, 56*(24), 7390–7404. https://doi.org/10.1080/00207543.2017.1412530

Mishra, A., Verma, P., & Tiwari, M. K. (2022). A circularity-based quality assessment tool to classify the core for recovery businesses. *International Journal of Production Research, 60*(19), 5835–5853. https://doi.org/10.1080/00207543.2021.1973135

Mishra, A. R., Rani, P., Saha, A., Pamucar, D., & Hezam, I. M. (2023). A q-rung orthopair fuzzy combined compromise solution approach for selecting sustainable third-party reverse logistics provider. *Management Decision, 61*(6), 1816–1853. https://doi.org/10.1108/MD-01-2022-0047

Miyajima, S., Yamada, S., Yamada, T., & Inoue, M. (2019). Proposal of a modular design method considering supply chain: Comprehensive evaluation by environmental load, cost, quality and lead time (Vol. 13, No. 1, pp. 119–132). https://jamt.utem.edu.my/jamt/article/view/5257

Mohamed Noor, A. Z., Md Fauadi, M. H. F., Jafar, F. A., Mohamad, N. R., Zulkifli, M. W. Z., Hasbulah, M. H., Othman, M. A., Mat Ali, M., Goh, J. B., & Morthui, R. (2018). Decision making support system using intelligence tools to select best alternative in design for remanufacturing (economy indicator). In *Lecture notes in mechanical engineering* (pp. 401–413). Pleiades Journals. https://doi.org/10.1007/978-981-10-8788-2_36.

Mota, B., Gomes, M. I., Carvalho, A., & Barbosa-Povoa, A. P. (2018). Sustainable supply chains: An integrated modeling approach under uncertainty. *Omega (United Kingdom), 77*, 32–57. https://doi.org/10.1016/j.omega.2017.05.006

Mouflih, C., Gaha, R., Bosch, M., & Durupt, A. (2023). Improving the sustainability of manufactured products: Literature review and challenges for digitalization of disassembly and dismantling processes. In *IFIP Advances in Information and Communication Technology, 667 IFIP* (pp. 630–640). https://doi.org/10.1007/978-3-031-25182-5_61

Pan, F., & Liu, L. (2009). The environmental performance research of automotive remanufacturing industry based on entropy weight fuzzy synthesis evaluation in China. In *2009 International Conference on Management and Service Science* (pp. 1–4). https://doi.org/10.1109/ICMSS.2009.5301204

Prajapati, H., Kant, R., & Shankar, R. (2021). Devising the performance indicators due to the adoption of reverse logistics enablers. *Journal of Remanufacturing, 11*(3), 195–225. https://doi.org/10.1007/s13243-020-00098-4

Priyono, A., Ijomah, W., & Bititci, U. (2016). Disassembly for remanufacturing: A systematic literature review, new model development and future research needs. *Journal of Industrial Engineering and Management, 9*(4), 899. https://doi.org/10.3926/jiem.2053

Ren, Y., Jin, H., Zhao, F., Qu, T., Meng, L., Zhang, C., Zhang, B., Wang, G., & Sutherland, J. W. (2021). A multiobjective disassembly planning for value recovery and energy conserva-

tion from end-of-life products. *IEEE Transactions on Automation Science and Engineering, 18*(2), 791–803. https://doi.org/10.1109/TASE.2020.2987391
Russell, J. D., & Nasr, N. Z. (2023). Value-retained vs. impacts avoided: The differentiated contributions of remanufacturing, refurbishment, repair, and reuse within a circular economy. *Journal of Remanufacturing, 13*(1), 25–51. https://doi.org/10.1007/s13243-022-00119-4
Sagnak, M. (2020). Assessment of logistics performance in sustainable supply chain: Case from emerging economy. In *Supply chain sustainability: Modeling and innovative research frameworks* (pp. 73–87). De Gruyter. https://doi.org/10.1515/9783110628593-004
Saidani, M., Yannou, B., Leroy, Y., Cluzel, F., & Kendall, A. (2019). A taxonomy of circular economy indicators. *Journal of Cleaner Production, 207*, 542–559. https://doi.org/10.1016/J.JCLEPRO.2018.10.014
Sarwar, A., Zafar, A., Hamza, M. A., & Qadir, A. (2021). The effect of green supply chain practices on firm sustainability performance: Evidence from Pakistan. *Uncertain Supply Chain Management, 9*(1), 31–38. https://doi.org/10.5267/j.uscm.2020.12.004
Schau, E. M., Traverso, M., & Finkbeiner, M. (2012). Life cycle approach to sustainability assessment: A case study of remanufactured alternators. *Journal of Remanufacturing, 2*(1), 5. https://doi.org/10.1186/2210-4690-2-5
Sethanan, K., Pitakaso, R., Kosacka-Olejnik, M., Werner-Lewandowska, K., & Wasyniak, A. (2019). Remanufacturing sustainability indicators: A study on diesel particulate filter. *Logforum, 15*(3), 413–423. https://doi.org/10.17270/J.LOG.2019.349
Shakourloo, A. (2017). A multi-objective stochastic goal programming model for more efficient remanufacturing process. *The International Journal of Advanced Manufacturing Technology, 91*(1–4), 1007–1021. https://doi.org/10.1007/s00170-016-9779-6
Shrivastava, P., Zhang, H. C., Li, J., & Whitely, A. (2005). Evaluating obsolete electronic products for disassembly, material recovery and environmental impact through a decision support system. In *Proceedings of the 2005 IEEE International Symposium on Electronics and the Environment, 2005* (pp. 221–225). https://doi.org/10.1109/ISEE.2005.1437029
Spreafico, C. (2022). An analysis of design strategies for circular economy through life cycle assessment. *Environmental Monitoring and Assessment, 194*(3), 180. https://doi.org/10.1007/s10661-022-09803-1
Sundin, E. (2019). The role of remanufacturing in a circular economy. In *Remanufacturing in the circular economy: operations, engineering and logistics* (pp. 31–60). Wiley. https://doi.org/10.1002/9781119664383.CH2
Taleizadeh, A. A., Haghighi, F., & Niaki, S. T. A. (2019). Modeling and solving a sustainable closed loop supply chain problem with pricing decisions and discounts on returned products. *Journal of Cleaner Production, 207*, 163–181. https://doi.org/10.1016/j.jclepro.2018.09.198
Tchertchian, N., Millet, D., & Pialot, O. (2013). Modifying module boundaries to design remanufacturable products: The modular grouping explorer tool. *Journal of Engineering Design, 24*(8), 546–574. https://doi.org/10.1080/09544828.2013.776671
Tian, G., Zhang, H., Feng, Y., Jia, H., Zhang, C., Jiang, Z., Li, Z., & Li, P. (2017). Operation patterns analysis of automotive components remanufacturing industry development in China. *Journal of Cleaner Production, 164*, 1363–1375. https://doi.org/10.1016/j.jclepro.2017.07.028
Tranfield, D., Denyer, D., & Smart, P. (2003). Towards a methodology for developing evidence-informed management knowledge by means of systematic review. *British Journal of Management, 14*(3), 207–222. https://doi.org/10.1111/1467-8551.00375
Tsalis, T., Stefanakis, A. I., & Nikolaou, I. (2022). A framework to evaluate the social life cycle impact of products under the circular economy thinking. *Sustainability (Switzerland), 14*(4), 2196. https://doi.org/10.3390/su14042196
Tsiliyannis, C. A. (2014). Cyclic manufacturing: Necessary and sufficient conditions and minimum rate policy for environmental enhancement under growth uncertainty. *Journal of Cleaner Production, 81*, 16–33. https://doi.org/10.1016/j.jclepro.2014.06.028
Ullah, S. M. S., Muhammad, I., & Ko, T. J. (2016). Optimal strategy to deal with decision making problems in machine tools remanufacturing. *International Journal of Precision Engineering and Manufacturing-Green Technology, 3*(1), 19–26. https://doi.org/10.1007/s40684-016-0003-9

van Buren, N., Demmers, M., van der Heijden, R., & Witlox, F. (2016). Towards a circular economy: The role of Dutch logistics industries and governments. *Sustainability (Switzerland), 8*(7), 1–17. https://doi.org/10.3390/su8070647

van Loon, P., & Van Wassenhove, L. N. (2018). Assessing the economic and environmental impact of remanufacturing: A decision support tool for OEM suppliers. *International Journal of Production Research, 56*(4), 1662–1674. https://doi.org/10.1080/00207543.2017.1367107

Vanegas, P., Peeters, J. R., Cattrysse, D., Tecchio, P., Ardente, F., Mathieux, F., Dewulf, W., & Duflou, J. R. (2018). Ease of disassembly of products to support circular economy strategies. *Resources, Conservation and Recycling, 135*, 323–334. https://doi.org/10.1016/J.RESCONREC.2017.06.022

Vimal, K. E. K., Kandasamy, J., & Duque, A. A. (2021). Integrating sustainability and remanufacturing strategies by remanufacturing quality function deployment (RQFD). *Environment, Development and Sustainability, 23*(9), 14090–14122. https://doi.org/10.1007/s10668-020-01211-0

Vogtlander, J. G., Scheepens, A. E., Bocken, N. M. P., & Peck, D. (2017). Combined analyses of costs, market value and eco-costs in circular business models: Eco-efficient value creation in remanufacturing. *Journal of Remanufacturing, 7*(1), 1–17. https://doi.org/10.1007/s13243-017-0031-9

Wang, W., & Tseng, M. M. (2010). Economic analysis of product end-of-life strategies to achieve design for sustainable manufacturing. In *5th International Conference on Responsive Manufacturing—Green Manufacturing (ICRM 2010)* (pp. 268–272). https://doi.org/10.1049/cp.2010.0444

Xia, K., Gao, L., Li, W., Wang, L., & Chao, K.-M. (2014, June 9). A Q-learning based selective disassembly planning service in the cloud based remanufacturing system for WEEE. In *Materials; Micro and nano technologies; Properties, applications and systems; Sustainable manufacturing* (Vol. 1). https://doi.org/10.1115/MSEC2014-4008

Yadav, G., Mangla, S. K., Bhattacharya, A., & Luthra, S. (2020). Exploring indicators of circular economy adoption framework through a hybrid decision support approach. *Journal of Cleaner Production, 277*, 124186. https://doi.org/10.1016/j.jclepro.2020.124186

Yang, S. S., Nasr, N., Ong, S. K., & Nee, A. Y. C. (2017). Designing automotive products for remanufacturing from material selection perspective. *Journal of Cleaner Production, 153*, 570–579. https://doi.org/10.1016/j.jclepro.2015.08.121

Yang, D., Zhang, L., Wu, Y., Guo, S., Zhang, H., & Xiao, L. (2019). A sustainability analysis on retailer's sales effort in a closed-loop supply chain. *Sustainability (Switzerland), 11*(1), 8. https://doi.org/10.3390/su11010008

Yanikara, F. S., Kuhl, M. E., & Thorn, B. K. (2014). Comparing reverse logistics network configurations using simulation. In *IIE Annual Conference. Proceedings; Norcross* (pp. 2864–2873).

Yu, H., & Solvang, W. D. (2017). A carbon-constrained stochastic optimization model with augmented multi-criteria scenario-based risk-averse solution for reverse logistics network design under uncertainty. *Journal of Cleaner Production, 164*, 1248–1267. https://doi.org/10.1016/j.jclepro.2017.07.066

Zarbakhshnia, N., Soleimani, H., & Ghaderi, H. (2018). Sustainable third-party reverse logistics provider evaluation and selection using fuzzy SWARA and developed fuzzy COPRAS in the presence of risk criteria. *Applied Soft Computing, 65*, 307–319. https://doi.org/10.1016/J.ASOC.2018.01.023

Zhang, X., Li, Z., Wang, Y., & Yan, W. (2021a). An integrated multicriteria decision-making approach for collection modes selection in remanufacturing reverse logistics. *Processes, 9*(4), 631. https://doi.org/10.3390/pr9040631

Zhang, X., Tang, Y., Zhang, H., Jiang, Z., & Cai, W. (2021b). Remanufacturability evaluation of end-of-life products considering technology, economy and environment: A review. *Science of the Total Environment, 764*, 142922. https://doi.org/10.1016/j.scitotenv.2020.142922

Zhang, X., Xu, L., Zhang, H., Jiang, Z., & Wang, Y. (2021c). Emergy based sustainability evaluation model for retired machineries integrating energy, environmental and social factors. *Energy, 235*, 121331. https://doi.org/10.1016/j.energy.2021.121331

Zhang, Z., Matsubae, K., & Nakajima, K. (2021d). Impact of remanufacturing on the reduction of metal losses through the life cycles of vehicle engines. *Resources, Conservation and Recycling, 170*, 105614. https://doi.org/10.1016/j.resconrec.2021.105614

Zhang, L., Wu, Y., Li, Z., Zheng, Y., Ren, Y., & Zhu, L. (2022). A systematic approach in remanufacturing for high efficiency and low cost: The selective parallel disassembly sequence planning. *Proceedings of the Institution of Mechanical Engineers, Part B: Journal of Engineering Manufacture, 236*(5), 572–585. https://doi.org/10.1177/09544054211041036

Zhao, Y., & Zhou, H. (2023). Remanufacturing vs. greening: Competitiveness and harmony of sustainable strategies of supply chain under uncertain yield. *Computers & Industrial Engineering, 179*, 109233. https://doi.org/10.1016/j.cie.2023.109233

Zwolinski, P., Lopez-Ontiveros, M. A., & Brissaud, D. (2006). Integrated design of remanufacturable products based on product profiles. *Journal of Cleaner Production, 14*(15–16), 1333–1345. https://doi.org/10.1016/J.JCLEPRO.2005.11.028

Camilo Mejía-Moncayo is currently pursuing a doctoral degree in the mechanical engineering department at the École de Technologie Supérieure (ÉTS). His research focuses on design and control of smart sustainable remanufacturing systems. He obtained both his master's and bachelor's degrees in mechanical engineering from Universidad Nacional de Colombia. His experience in industrial and academics lies in manufacturing processes and systems, Industry 4.0, optimization, and engineering education.

Amin Chaabane, P. Eng, Ph.D., is a full professor in the Department of Systems Engineering at École de Technologie Supérieure (ÉTS, Montreal). He obtained a Doctor of Philosophy (Ph.D.) in Industrial Engineering from École de technologie supérieure (ÉTS). Prof. Chaabane is an active member of the Sustainable Digital Green (SDG) Innovation Network, a Quebec research network in industry X.0. He is also a regular member of CIRRELT, an interuniversity Research Center on Enterprise Networks, Logistics, and Transportation, and a member of CIRODD, the first strategic research group in sustainable development operationalization in Québec. Furthermore, he is an active researcher of RRECQ, a Quebec Circular Economy Research Network. Professor Chaabane has made significant contributions to operations and supply chain management. These contributions have resulted in practical applications across diverse industries such as food, logistics, retail, energy, oil and gas, construction logistics, and forestry. As an expert in applied research, Professor Chaabane collaborates with governmental bodies and industrial partners to develop advanced data-oriented analytics methods that facilitate the design and management of efficient, resilient, intelligent, and sustainable supply chains.

Jean-Pierre Kenne, P. Eng, Ph.D., is a Professor in the Mechanical Engineering Department at École de Technologie Supérieure, University of Quebec since 2000. He received his M.S.A and Ph.D in Mechanical Engineering both from Ecole Polytechnique de Montreal in 1991 and 1998 respectively. He was project manager in automation and control at GEBO Canada and Logitrol Inc. for two years, before joining Ecole de Technologie Supérieure. Professor Kenne's teaching activities include control of dynamic systems, optimization and stochastic control, design and control of hydraulic systems. He has been working on the modeling of dynamic systems and the development of control policies for more than twenty years. He proposed different strategies for controlling and real time validation of such systems. Over the past three decades, Professor Kenné and his colleagues have developed expertise in integrating the basic concepts of optimal stochastic control theory with simulation-based optimisation models, experimental designs and response surface methodology.

Lucas A. Hof, P. Eng, Ph.D., is an Associate Professor in the Department of Mechanical Engineering, École de Technologie Supérieure (ÉTS), Canada. He obtained his PhD degree in Mechanical Engineering from Concordia University, Canada, and obtained both his Master's and Bachelor's degrees in Mechanical Engineering at Delft University of Technology, The Netherlands. Lucas A. Hof, ing., Ph.D., is an Associate Professor in the Mechanical Engineering Department at École de technologie supérieure, Canada. He obtained his PhD in Mechanical Engineering at Concordia University, Montréal, Canada, and both his Masters and Bachelor in Mechanical Engineering at Delft University of Technology, The Netherlands. His research evolves from the need of a transformation from linear to circular manufacturing practices and concentrates on developing smart remanufacturing technologies for ceramic, metallic and polymer composite materials. Prof. Hof has developed extensive experience on highprecision machining of hard-to-machine materials, advanced manufacturing systems and processes with a specific focus on circular production methods using industry 4.0 technologies. As well, he has co-developed a novel glass micromachining technology to industrial level allowing the lean production of ultracustomized glass parts. In addition, he has accumulated over ten years of industrial project management experience and produced over 120 journal and conference papers and two patents (one pending) and he is supervising or co-supervising more than 25 PhD and MASc graduate students. He has a diverse portfolio on industrial research collaborations, varying from small to medium and large sized - international and local - businesses. In addition, he has accumulated over ten years of industrial project management experience and produced over 120 journal and conference papers and two patents (one pending) and he is (co-)supervising more than 25 PhD and MASc graduate students. He has a diverse portfolio on industrial research collaborations, varying from small to medium and large sized - international and local - businesses.

Chapter 9
Advancing Sustainability Through Digital Maturity: An Open Approach for Evaluating Quebec Organizations' Environmental Responsibility

Guillaume Bourgeois, Géraldine Angulo, Hassana El-Zein, Vincent Courboulay, and Mohamed Cheriet

Abstract The exponential growth of digital technologies has brought about substantial benefits in operational efficiency and communication. However, it has also led to increased energy consumption and greenhouse gas emissions (GHG), highlighting a growing concern: the environmental impact of information systems (IS). This chapter introduces an innovative approach designed to assess the environmental impact of digital technologies within organizations, tailored specifically to the North American context with a focus on Quebec. Key considerations include the local energy mix, regulations, and cultural factors. It addresses the challenges, opportunities, and prerequisites for successful adaptation, ensuring relevance to regional specificities. The proposed platform aims to facilitate collective efforts toward environmental preservation by promoting responsible and sustainable digital practices aligned with the United Nations Sustainable Development Goals. Adapted from the WeNR platform developed by the Institute of Responsible Digital Technology at the University of La Rochelle in France, this chapter discusses the requirement, description, and adaptation of the platform in Quebec. It addresses the pressing challenges of socio-ecological transition in the digital age and fulfills the pressing need for organizations to measure and mitigate the digital carbon footprint of their IS. Utilizing a life cycle assessment (LCA) methodology, the platform considers all phases of electronic equipment and data lifecycle. It emphasizes accessibility with an intuitive user interface and an open database featuring regularly

G. Bourgeois (✉)
École de technologie supérieure (ETS), Montréal, QC, Canada

La Rochelle Université, La Rochelle, France
e-mail: guillaume.bourgeois.3@ens.etsmtl.ca

G. Angulo · H. El-Zein · M. Cheriet
École de technologie supérieure (ETS), Montréal, QC, Canada
e-mail: mohamed.cheriet@etsmtl.ca

V. Courboulay
La Rochelle Université, La Rochelle, France
e-mail: vcourbou@univ-lr.fr

M. Cheriet et al. (eds.), *Accelerating the Socio-Ecological Transition*,
https://doi.org/10.1007/978-3-031-82896-6_9

updated impact factors. Users can complete a concise questionnaire to receive a comprehensive report on their digital carbon footprint, maturity level in responsible digital practices, and recommendations for reducing their carbon footprint. In conclusion, this chapter not only raises awareness but also actively promotes the implementation of responsible digital usage. By offering an innovative and adaptable platform, it addresses the environmental impact of digital technology within organizations, supporting their transition toward a more sustainable and environmentally respectful digital future.

Keywords Life cycle assessment · Digital carbon footprint assessment · Sustainable digital practices

1 Introduction

The digital age has fundamentally reshaped societies, revolutionizing interactions, work methods, and our relationship with information. Information and communication technologies (ICT) are pivotal in driving this transformation, enabling new professional opportunities, facilitating remote collaboration, and optimizing business efficiency. They promise continuous connectivity, unlimited access to knowledge, and innovative solutions to address future challenges, fostering a more interconnected, inclusive, and prosperous future (Imamov & Semenikhina, 2021).

As of 2023, 65.7% of the global population had internet access, surpassing the number of ICT devices, which already exceeded the human population. Projections indicate that by 2025, there will be approximately 55.9 billion devices (Brouillard, 2023), highlighting the expanding utilization of ICT services.

However, alongside these advancements, a concerning reality emerges: the escalating environmental impact of IS that supports everyday activities. Each digital interaction—from streaming online videos to booking flights or engaging with social media—consumes substantial energy and resources, particularly for metal extraction and refining, and requires significant quantities of freshwater.

The advent of artificial intelligence (AI) systems further exacerbates these challenges (OECD, n.d.). The intensive computational demands for training sophisticated deep learning models consume considerable energy resources, leaving a pronounced environmental footprint.

In Quebec, the digital sector contributes approximately 5% to the region's carbon footprint[1] (Pinsard & Toussaint, 2020). While seemingly modest compared to other sectors, the rapid growth in digital consumption, including data volume and equipment, necessitates careful consideration of its environmental implications.

[1] The carbon footprint quantifies the total amount of GHG emitted by an individual, organization, product, or activity. These gases, including carbon dioxide (CO_2), contribute to climate change by trapping heat in the Earth's atmosphere. Selin, N.E., (2024, June 19). *Carbon Footprint*. Encyclopedia Britannica.

Despite the escalating environmental concerns associated with digital technology, it remains a peripheral issue in the priorities of many sectors. A 2023 survey by The Information and Communications Technology Council (ICTC) revealed that over 40% of Canadian organizations did not factor in the environmental impact of ICT.[2]

In Quebec, the prevalence of hydroelectricity—an energy source with minimal GHG emissions compared to fossil fuels—partly explains this complacency. However, hydroelectricity production is not devoid of environmental impacts (Lu et al., 2020). The construction of dams and hydroelectric plants can disrupt river ecosystems, lead to habitat loss, and alter watercourses. Furthermore, hydroelectric reservoirs can emit methane, a potent GHG, due to the decomposition of organic matter in stagnant water.

An additional constraint we face is the lack of specific studies on the environmental footprint of digital technology in Quebec. This gap necessitates reliance on hypothetical data derived from global energy measurements, which often overlooks comprehensive life cycle analyses and associated criteria, potentially leading to inaccurate assessments of digital responsibility requirements.

To address this issue, understanding the environmental impacts of digital technology at the local level is crucial, utilizing platforms tailored specifically to the Quebec context. It is within this framework that the WeNR-QC platform has been developed. This platform is designed to assess the digital footprint of organizations in Quebec, regardless of size, focusing on three main sources of pollution: users, networks, and data centers. Additionally, it evaluates the organization's maturity level in terms of responsible digital practices.

The WeNR-QC platform meets the urgent need for organizations to quantify and mitigate their digital carbon footprint. Built on the principles of LCA, it comprehensively evaluates all phases of electronic equipment and data usage. Accessibility is prioritized through an intuitive user interface and an open database regularly updated with impact factors. Users can complete a brief questionnaire to receive a detailed report on their digital carbon footprint along with recommendations for improvement.

2 Environmental Footprint of IS

The rapid advancement of digital technology has profoundly transformed our lifestyles and work practices. However, it has also introduced significant environmental challenges, particularly concerning the ecological impact of IS. The ecological footprint of IS refers to the environmental impact caused by the entire lifecycle of information technology (IT) systems, encompassing stages from manufacturing and usage to disposal.

[2]Clark, A. & Matthews, M. (2023). *ICTC Policy Brief: How to Advance Environmentally Sustainable ICT in Canada* (p. 14), Information and Communications Technology Council (ICTC). Ottawa, Canada.

As our dependence on digital technology expands, understanding and mitigating this footprint becomes imperative for environmental sustainability.

The French Agency for Ecological Transition (ADEME) estimates that the digital sector currently contributes to 3–4% of global GHG. This footprint spans all phases of the lifecycle,[3] including energy consumption during data center operations, manufacturing of electronic devices, and the overall life cycle of IT equipment.

2.1 *Stages of IS Lifecycle*

The IS lifecycle encompasses multiple phases that oversee the development, deployment, and management of technology solutions. A comprehensive understanding of these stages is crucial for efficient project management and sustainable IT practices. Specifically, we will delve into the three primary phases where the environmental impact of IS is most pronounced: Manufacturing, use, and end-of-life disposal.

2.1.1 Manufacturing

The production of electronic devices, such as servers, computers, and smartphones, involves significant quantities of raw materials, energy, and water, as depicted in Fig. 9.1. Extraction of minerals, components assembly, and transportation all contribute to the environmental footprint.

The manufacturing phase of digital equipment plays a pivotal role, accounting for a substantial portion of the sector's carbon footprint. This impact is highlighted by the "Ecological Backpack"[4] methodology, which quantifies the hidden

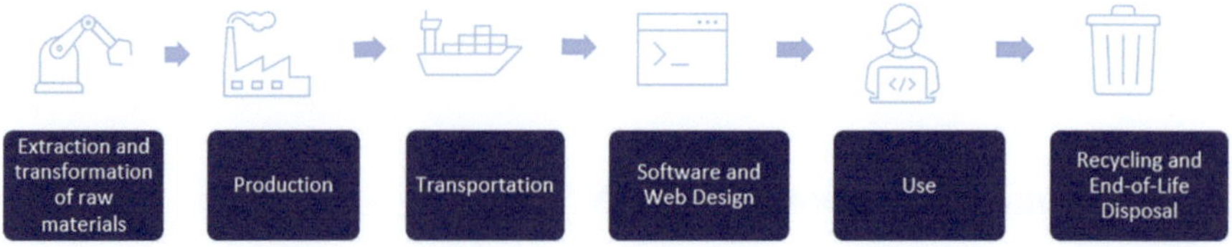

Fig. 9.1 Digital product life cycle

[3] The lifecycle of digital equipment comprises six distinct phases, each with its environmental implications. This document focuses on the production phase, involving raw material extraction, the usage phase, and the end-of-life phase.

[4] The concept of the ecological backpack represents the cumulative natural resources required to manufacture a product. It includes materials extracted, petroleum and water usage, and the biosphere impact per unit of the product. The calculation utilizes the MIPS (Measure of the amount of Indispensable materials per unit of service). *Ecological Backpack* (2024, April 20). The Digital Collage.

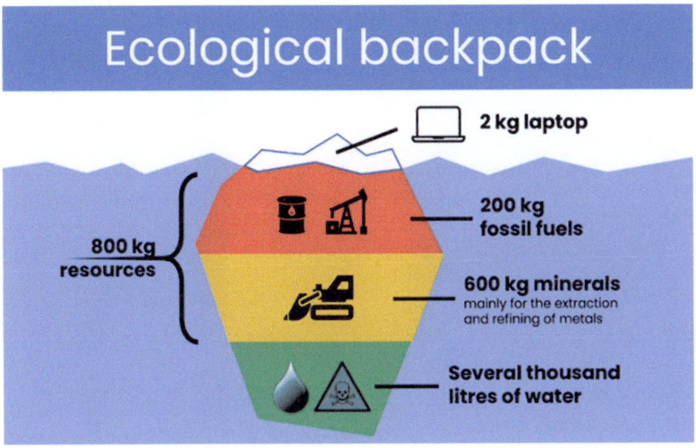

Fig. 9.2 Diagram of ecological backpack (see footnote 4)

environmental costs of production. For instance, Fig. 9.2 illustrates that manufacturing a 2 kg laptop requires approximately 800 kg of resources.

According to a 2020 ADEME report,[5] the estimated carbon dioxide equivalents (CO_2eq) emissions for various devices are as follows:

- Laptop (office use) = 169 kg CO_2eq.
- Desktop tower (office use) = 189 kg CO_2eq.
- High-performance desktop tower = 394 kg CO_2eq.
- 21.5-in. monitor = 236 kg CO_2eq.

As a reminder, the carbon footprint, expressed in CO_2eq, quantifies the impact of various GHG on climate warming.

2.1.2 Use

The global digital industry, comprising around 34 billion devices serving 4.1 billion users, consumes approximately 1300 TWh of electricity annually, accounting for 5.5% of the world's total electricity consumption.[6] This consumption is equivalent to the combined annual electricity production of France, Germany, and Belgium. As depicted in Fig. 9.3, end-user terminals contribute 38%, networks 11%, and data centers 14% to the overall carbon footprint, totaling 63%. These impacts encompass climate change, depletion of natural resources, and pollution of air, water, and soil.

[5] From *Modélisation et évaluation des impacts environnementaux de produits de consommation et biens d'équipement* (p. 25), ADEME, 2018.

[6] From The Environmental Footprint of the Digital World (p. 10), by Frédéric Bordage, 2019.

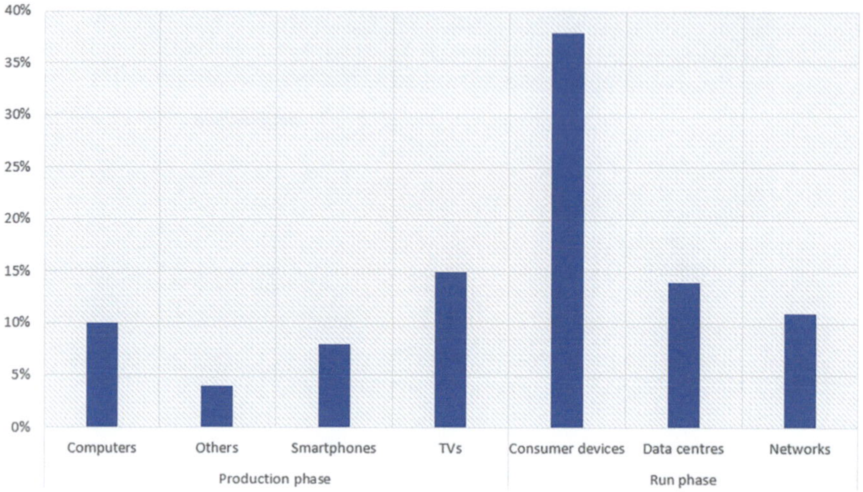

Fig. 9.3 Distribution of the global digital carbon footprint in 2019, by item for the production (40%) and usage (60%) phases. Taken from The Shift Project, 2021. Consulted on April 20, 2024

2.1.3 End-of-Life Disposal

When ICT hardware and devices reach the end of their life, they become electronic waste, known as WEEE (Waste Electrical and Electronic Equipment). WEEE poses a significant environmental risk due to the presence of toxic substances that can leach into soil and water, contaminating ecosystems and threatening wildlife and plant life. Additionally, improper disposal can pollute water resources, making them unfit for consumption and agricultural purposes.

Figure 9.4 illustrates the process where collected electronic waste enters the recycling system. While a small proportion of the materials can be separated and recycled, the majority is non-recyclable and often ends up in landfills or is incinerated. Recycling, therefore, provides only a partial solution, and there are inherent limitations to its effectiveness.

Since 2012, Quebec has implemented an electronic waste recycling system aimed at improving management practices. This system operates under an "extended producer responsibility" program, requiring manufacturers and retailers of electronic devices to ensure the recycling of end-of-life products (Québec Science, 2018).

Despite these efforts, significant challenges persist. A considerable amount of electronic waste still ends up in landfills or is illegally exported, bypassing safe recycling practices. Moreover, reliable data regarding the exact recovery rate of electronic waste in Quebec is limited.

In conclusion, addressing the environmental footprint of digital technology is crucial for sustainable development (Cecere & Pénard, 2020). Adopting eco-responsible practices and implementing innovative solutions can mitigate this impact and contribute to environmental preservation. Effective collaboration among

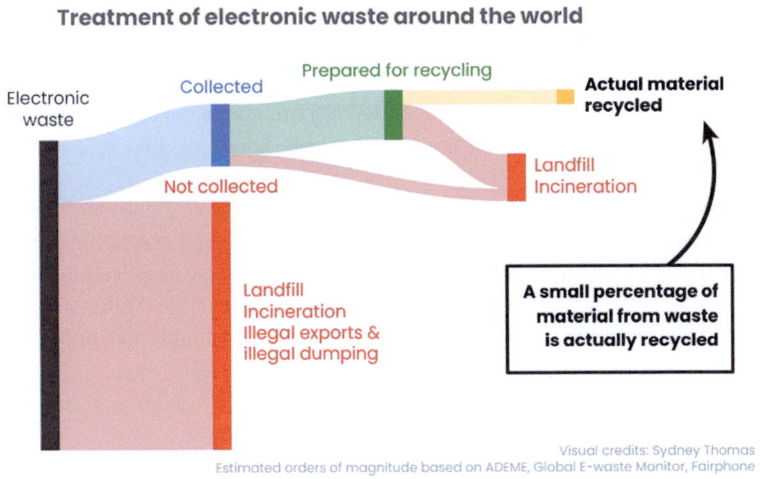

Fig. 9.4 Treatment of electronic waste around the world. (From The Digital Collage (2024, April 20))

digital stakeholders, governments, and citizens is essential to successfully implement these solutions.

3 Discussion of Existing Platforms and Their Limitations

Assessing the environmental footprint of digital technologies is crucial for improvement and accountability. This field has seen the development of several platforms aimed at measuring and reducing the ecological impact of IS. Below is an analysis of key platforms currently available and their respective limitations:

1. **CNRS Ecodiag**

   Description: Ecodiag, developed by France's Centre National de la Recherche Scientifique (CNRS), is a platform designed to assess the environmental footprint of IS.

   Limitations: While Ecodiag employs a robust quantitative approach, it lacks qualitative measures for evaluating the maturity of responsible digital practices. Moreover, irregular updates to its databases pose challenges, affecting the accuracy of results.
2. **The Boavizta platform**

   Description: Boavizta provides a digital carbon footprint assessment platform that leverages open-source and collaborative data to measure the environmental impact of information technologies.

   Limitations: Boavizta's platform is innovative in utilizing open-source data but lacks a structured methodology for assessing the qualitative maturity of organizations' digital practices. Its complexity often requires specialized technical expertise, which may limit accessibility for some organizations.

3. **NegaOctet**

   Description: NegaOctet aims to assist organizations in reducing their digital footprint through a methodology and database based on lifecycle assessment principles.

   Limitations: While NegaOctet excels as a quantitative database, it falls short in evaluating the qualitative maturity of responsible digital practices. Its methodology is occasionally opaque and relies on generalized assumptions that may not align with specific organizational or regional contexts. Additionally, limited access to current open-source data, predominantly available through fee-based models, can compromise result accuracy.

3.1 Limitations Common to Existing Platforms

Measuring Qualitative Maturity: Many existing platforms lack comprehensive assessments of the qualitative maturity of responsible digital practices. This omission hinders a thorough understanding of organizations' commitment to sustainability.

Lack of Structured Methodology: Some platforms lack structured and standardized methodologies, leading to inconsistencies in assessments and making comparisons between different organizations challenging.

Open Source and Open Data: The availability and regular updating of open-source data are often inadequate. This limitation can compromise the accuracy and reliability of environmental assessments.

3.2 How WeNR Addresses These Limitations

The WeNR platform addresses critical limitations observed in existing platforms through several key features. Firstly, WeNR incorporates an extensive qualitative assessment of responsible digital practices, providing a thorough understanding of organizational commitments to sustainability. Secondly, WeNR employs a structured and standardized methodology rooted in lifecycle assessment principles, ensuring consistency and facilitating comparisons across different organizations. Thirdly, the platform distinguishes itself with the use of regularly updated open-source data, enhancing the accuracy and reliability of environmental assessments. Emphasizing accessibility, WeNR features an intuitive user interface and an open database, catering to organizations of various sizes and sectors.

4 The WeNR Platform

Stemming from the efforts of the Institute for Sustainable IT (ISIT), the WeNR platform stands as a pivotal instrument in fostering an environmentally sustainable digital landscape. Its methodology integrates a holistic approach that embraces three pillars of sustainable development: societal impacts, ecological footprint, and the generation of economic and social value (Fisk, 2010).

This comprehensive approach is operationalized through an accessible online questionnaire, designed for organizations of all sizes and sectors. Covering diverse aspects of responsible digital practices, from infrastructure energy consumption to user awareness of sustainable digital behaviors, the WeNR platform provides a thorough assessment of organizational digital practices and identifies priority areas for improvement.

Table 9.1 illustrates the three versions of WeNR, tailored to meet the needs and capabilities of users. These versions facilitate understanding of the environmental and social impacts of digital technologies, allowing organizations to gauge their maturity in digital responsibility. Furthermore, WeNR supports ongoing progress monitoring and aids in the formulation of effective sustainability strategies over time.

The collective nature of WeNR is further demonstrated by its capacity to foster collaboration and the exchange of best practices among organizations. It facilitates performance benchmarking within sectors, provides guidances from ISIT network experts, and offers opportunities to join a community dedicated to responsible digital practices.

4.1 History and Development of WeNR in Europe

The WeNR initiative emerged from the Green IT collective's efforts, launched between 2014 and 2018 to assess the environmental footprint of the digital sector in large French companies. These initial evaluations played a crucial role in raising awareness about the ecological challenges associated with digital transformation.

Table 9.1 Different versions of WeNR

WeNR Light	This questionnaire assesses organizational maturity in Digital Sustainability (People-Planet-Profit)
WeNR Standard	WeNR comprises quantitative and qualitative questionnaires completed by participating organizations using a provided file template from ISIT. Accessible online, data is treated confidentially and results are delivered in a first-level report approximately 2–3 months after submission
WeNR Plus	Exclusive to ISIT member organizations, WeNR + enhances the standard model with comprehensive quantitative, qualitative, and comparative reports within the sector. Analysis platforms aid in identifying actionable steps to establish a sustainable IT strategy

In 2018, WeNR gained significant momentum with the publication of the WeGreenIT study, a collaborative effort with the World Wide Fund for Nature (WWF) (WeGreenIT Study, 2018). Presented to Mrs. Brune Poirson, Secretary of State for Ecological Transition in 2019, this study served as a pivotal moment, building upon earlier benchmarks to establish the foundational principles of the current WeNR. This laid the groundwork for a more environmentally responsible approach to digitalization in France.

The development of WeNR was made possible through collaboration and funding from key institutions in France, including La Rochelle University, the Generalist Engineering School—La Rochelle (EIGSI), the Nouvelle-Aquitaine Region, the Interministerial Directorate of Digital Affairs (DINUM), and ADEME.

WeNR's scientific partnerships extend internationally to institutions as Swiss ISIT, the Belgian ISIT, and the University of Louvain (UCLouvain). This diverse collaboration enriches WeNR's knowledge base, enhancing its credibility and effectiveness.

5 WeNR-QC: Adapting WeNR to the Quebec Context

The Quebec province, at the forefront of digital innovation with developments in AI,[7] quantum computing, and data center expansion, plays a pivotal role in both economic growth and environmental stewardship. ICT infrastructure is indispensable to Quebec's economy and society, necessitating careful management of its often-overlooked environmental impacts.

A 2020 report revealed that digital technology in Quebec accounted for:

- 4 million tons of CO_2e contributing to the greenhouse effect.
- 40,000 tons of electronic waste.
- 0.8 million tons of displaced earth.
- 2000 tons equivalent of antimony abiotic resource depletion (ADP).
- 12 TWh of final energy consumption (Pinsard & Toussaint, 2020).

While representing 5% of the province's total carbon footprint, the digital sector's rapid expansion in data volume and hardware demands thorough environmental assessment. Key concerns include resource extraction for component manufacturing and electronic waste management, which remain inadequately addressed due to limited regional measurement platforms.

WeNR emerges as a critical platform for Quebec, facilitating GHG reduction efforts and addressing data gaps. It enables measurement of digital footprints, informs policy and strategy development, fosters innovation in sustainable digital technology, and promotes the adoption of responsible digital practices.

[7] Benessaieh, K. (2022, March 9) Le Québec se classe au 7e au monde. *La Presse*.

5.1 Contextual Background of the WeNR-QC Adaptation

The momentum of innovation in Quebec converges significantly at the "Grand Rendez-vous de l'innovation québécoise", orchestrated by the Ministry of Economy, Innovation, and Energy (MEIE). This event marks a pivotal step in shaping the new Quebec Strategy for Research and Investment in Innovation, emphasizing sustainable development (Government of Quebec, 2023).

At the core of this event, the CIRODD (Centre interdisciplinaire de recherche en opérationnalisation du développement durable) and its partners, Ouranos and Écotech, organized a major thematic forum on climate change, with a specific focus on responsible digitalization (CIRODD, n.d.). Three workshops emerged from this forum, each offering unique perspectives: sustainable digitalization, adaptation and resilience, and green technologies. These workshops integrated principles of responsible digital practices into their discussions and proposals.

Under the auspices of the Quebec Innovation Council and the Ministry of Economy and Innovation, this event drew participation from 2600 individuals on November 18 and 1400 on November 19, 2021, demonstrating a strong commitment from diverse stakeholders to responsible and sustainable innovation. One significant outcome was the Launch of the Quebec Innovation Barometer—Beta Version, in June 2022 by the Quebec innovation council in collaboration with CIRODD, offering insightful indicators for the climate change pillar with a focus on responsible digitalization.

The Quebec Innovation Barometer is a platform developed to evaluate and track progress in critical areas of innovation (Innovation Barometer, 2022). In June 2022, the beta version of this barometer has been launched, highlighting key indicators specifically related to the climate change pillar. This barometer was designed to provide detailed and useful information on how climate change initiatives are being implemented and their impact on society.

Moreover, the barometer incorporates a dimension of responsible digitalization, recognizing the importance of technology and digital platforms in combating climate change while ensuring these platforms are used ethically and sustainably. Through this barometer, policymakers, researchers, and businesses can gain a better understanding of advancements and challenges in these crucial areas, thereby facilitating more informed and effective actions.

This annual gathering lays the foundation for responsible digitalization, which is where WeNR finds its origins. The adoption and adaptation of WeNR to the Quebec context stemmed from an international collaboration between CIRODD and La Rochelle University in May 2023, as part of Dr. Guillaume Bourgeois's doctoral thesis titled "Analysis and modeling of the environmental impact of IS." (Bourgeois et al., 2022; Bourgeois, 2023). This initiative highlighted the environmental impacts of digital technologies and led to the creation of WeNR-QC.[8]

[8]WeNR-QC aims to integrate the specificities of Quebec to provide a more precise and useful assessment of performance in terms of responsible digital practices. This adaptation highlights

CIRODD, as the key player in this initiative, played a crucial role in adapting the WeNR platform to Quebec's context, considering its cultural, economic, and environmental nuances. Expertise in sustainable development, circular economy, and responsible digital practices ensured the platform's adaptation was both successful and relevant.

This adaptation was primarily based on LCA, a methodology for evaluating environmental impacts throughout a product or service's lifecycle. To enhance accuracy in Quebec, local data on the energy mix was integrated to determine the GHG from energy production and transportation, accounting for regional transportation modes, distances, and associated emissions. This approach enabled a more precise evaluation of the IT-related environmental impacts in Quebec.

Additionally, software engineering played a pivotal role in this adaptation, encompassing development, deployment, and maintenance practices tailored to Quebec's specificities, including local best practices and regulations. AI and machine learning powered the analytical engine, enhancing decision-making capabilities.

Localization of technical terms to Quebec French, known as "québécisation"[9] was essential for clear communication within the WeNR-QC platform. Adapting terminology to Quebec's linguistic particularities ensured clarity and comprehension among users.

5.1.1 Energy Mix

In adapting the WeNR platform for québécisation, a pivotal focus was Quebec's unique energy mix, specifically its dominant reliance on hydroelectric power. Hydroelectric resources account for 94% of Quebec's electricity production, establishing the province as a global leader in renewable and clean energy (Levasseur et al., 2021). The remaining energy mix includes contributions from wind, solar, biomass, natural gas, and oil, distinguishing Quebec from other Canadian provinces and European nations, which often rely more heavily on fossil fuels or nuclear energy. As a result, Quebec maintains a notably low carbon footprint, with approximately 34 g of CO_2 emitted per kWh, in stark contrast to Canada's average of 140 g per kWh and Europe's average of 295 g per kWh (Hydro-Québec, n.d.).

In the process of adaptation, special attention was given to these energy dynamics, crucial for calculating the GHG impact of IS. This included refining emission

hydroelectricity, the Quebecois system for recycling digital equipment, adherence to local standards and regulations, as well as international comparability. From *Analyse et modélisation de l'impact environnemental du système d'information* (Bourgeois, 2023).

[9] The term "québéciser" refers to adapting to Quebec norms, whether in terms of laws, language, or other aspects, of an organization, product, or service. It can also involve modifying foreign words or expressions or other regional variants of French to conform to Quebec French usage. For example, "parking" can be québécisé to "stationnement". This verb derives from the noun "Québec" to which the suffix "-iser" is added (vitrinelinguistique, n.d.).

coefficients used to estimate both direct and indirect emissions associated with digital equipment's electricity consumption, alongside integrating data on Quebec's diverse energy sources. Furthermore, alignment with Quebec's governmental objectives and measures for energy transition was integral. These initiatives aim to diminish reliance on fossil fuels and amplify the proportion of renewable energy within Quebec's energy portfolio (Esmia Consultants, n.d.).

5.1.2 Transport

The WeNR-QC platform also addresses transportation, a significant contributor to the environmental footprint of digital technology. Transporting digital equipment, raw materials, and waste products produces varying levels of GHG depending on the modes and distances involved. Quebec's unique geography, climate, and transportation infrastructure distinguish it from Europe in several ways. Given Quebec's expansive territory, transportation routes often necessitate longer and more fuel-intensive journeys compared to Europe. Additionally, Quebec's less developed railway network limits opportunities for more environmentally friendly rail transport, which contrasts with Europe's more extensive rail infrastructure. Furthermore, Quebec's harsh weather conditions can impact the performance and durability of digital equipment during transport. In response, adjustments to GHG coefficients have been tailored to suit the Quebec context and ensure regulatory compliance.

5.2 *WeNR-QC Scope*

The primary goal of WeNR is to conduct a comprehensive assessment of IT systems' environmental impact in terms of GHG over a calendar year. This includes assigning a "responsible digital maturity" score to each participating organization and building a database to facilitate meaningful comparisons across entities. Our approach thus encompasses both quantifying GHG emissions and evaluating the digital responsibility of involved organizations.

The initial phase of our approach involves clearly delineating the motivations and expected benefits of this digital footprint assessment. Equally important is defining the study's scope, which encompasses the activities, processes, products, and services related to ICT that contribute to GHG emissions.

In terms of measurement, we use kilograms of CO_2 equivalent per year (kgCO_2e/year) as our reference unit. This choice allows for consistent comparison of emissions from various GHGs based on their global warming potential, aligning with recommendation L.1450 and ADEME's IS RCP for methodological coherence (ITU, n.d.; ADEME, n.d.).

Geographically, the WeNR-QC platform specifically targets the ICT sector within Quebec-based organizations. This ensures analyses tailored to local conditions while delivering pertinent insights to stakeholders.

Establishing the study's boundaries entails specifying which elements are included or excluded within the defined scope. Thus, WeNR-QC considers the entire lifecycle of digital services, from manufacturing through to end-of-life, accounting for GHG emissions relevant to each phase's geographical location. For instance, a significant portion of electronic equipment is manufactured in Asia before being transported to consumer markets. WeNR-QC integrates emissions associated with the transportation modes—be it by air, sea, rail, or road—for accurate assessment.

5.3 Operation of WeNR-QC: LCA of IS

Quantifying the carbon footprint of IT systems through WeNR-QC is a crucial step toward understanding and mitigating their environmental impact. This measurement encompasses all GHG generated by IT activities, providing a precise analysis of key emission sources. It facilitates targeted reduction efforts, fostering awareness among sector stakeholders about climate change challenges and promoting adoption of eco-responsible practices.

WeNR-QC comprehensively assesses the environmental impact of digital technology across several dimensions:

- **Data Centre Management**: Data centers, pivotal to IT infrastructure, are significant energy consumers. WeNR-QC evaluates their energy efficiency to identify opportunities for reducing their carbon footprint, such as adopting renewable energy sources or optimizing cooling systems (Masanet et al., 2020).
- **Digital Infrastructure**: Beyond data centers, WeNR-QC assesses the environmental impact of telecommunication networks, computer equipment, and connected devices (Lange et al., 2011).
- **Electronic Waste Management:** Digital equipment contributes to substantial electronic waste, containing potentially harmful materials (Rautela et al., 2021). WeNR-QC advocates best practices for recycling and reuse to minimize environmental impact.

Simultaneously, the platform identifies strategies to mitigate organizations' IT systems' environmental footprint, including:

- **Promoting Eco-Responsible Practices**: Encouraging companies to adopt internal policies that reduce energy consumption, encourage equipment reuse, and promote electronic waste recycling.
- **Advancing Green Technologies**: Supporting research and development of eco-efficient IT technologies like virtualization, cloud computing, and innovative cooling systems.
- **Raising Awareness**: Educating IS users about environmental impacts associated with their technology use and promoting responsible behaviors, such as optimizing device settings and reducing data consumption.

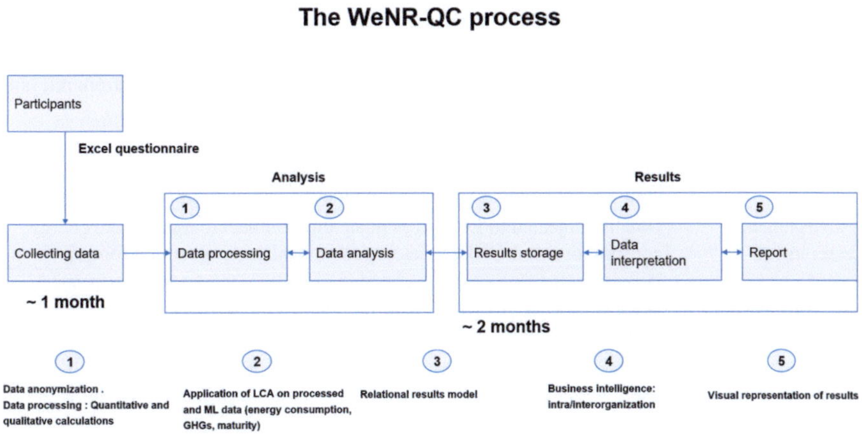

Fig. 9.5 The WeNR-QC process

5.4 WeNR-QC Methodology

5.4.1 WeNR-QC Process

As depicted in Fig. 9.5, the WeNR-QC methodology follows a structured process comprising multiple stages dedicated to gathering, processing, evaluating, and interpreting data to generate high-quality reports.

The WeNR-QC process seamlessly incorporates machine learning and business intelligence across all stages of data collection, processing, analysis, and interpretation. This integration ensures the generation of high-quality reports and meaningful insights tailored to the participating organizations. Refer to Table 9.2 for a detailed definition of each step in the process.

To delve deeper, our initial will be on data collection and analysis.

5.4.1.1 Data Collection and Analysis

As integral to the WeNR-QC application, gathering data on office equipment and data centers is essential for evaluating their environmental impact. Appendix 1 includes a table detailing the minimum required equipment for data collection and analysis. Figure 9.6 illustrates the various key parameters involved in this process.

- **Quantity and Operational Lifespan**: Gathering data on the number of units of each equipment type and their average operational lifespan is essential. This information allows for estimating equipment replacement or disposal frequency, influencing their overall carbon footprint.
- **GHG Emitter Settings:** Alongside quantity and lifespan, acquiring data on other GHG emitter parameters such as electricity consumption and Power Usage

Table 9.2 Steps of the WeNR-QC process

Data collection	Participants initiate data collection using a predefined Excel questionnaire. This step marks the outset of gathering pertinent information from relevant sources
Data processing	The collected data undergoes a thorough processing and analysis phase, lasting approximately 1 month. This includes anonymization for confidentiality and quantitative/qualitative calculations to extract insights
Environmental assessment	Data is subjected to LCA to evaluate environmental impacts like energy consumption and GHG, assessing the overall environmental footprint
Application of machine learning	Deep learning techniques, specifically recurrent neural networks (RNN), analyze process or system maturity, identifying nuanced trends that traditional methods may overlook
Creation of a relational model	Analysis outcomes are synthesized into a relational model, maintained for approximately 2 months. This model provides a structured depiction of relationships among studied variables
Data interpretation	Utilizing business intelligence, processed data and the relational model are interpreted to derive actionable insights for intra- and inter-organizational applications
Visual reporting	A comprehensive visual report is prepared to present analysis results clearly and concisely, outlining conclusions and recommendations derived from collected and processed data

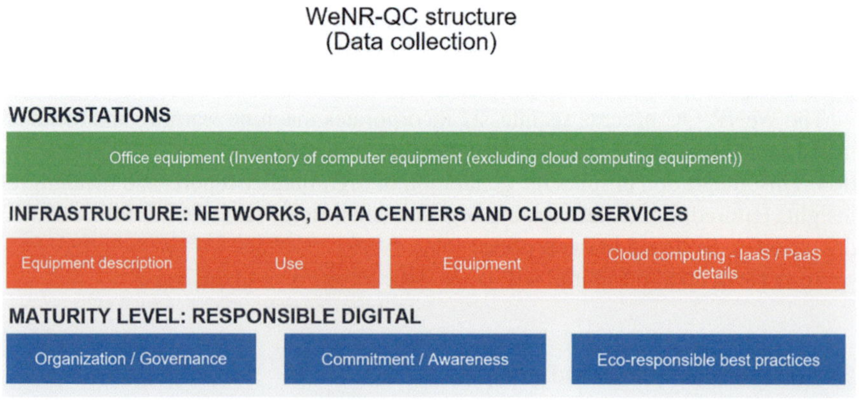

Fig. 9.6 WeNR-QC structure (data collection)

Effectiveness (PUE) for data centers is crucial. This data aids in assessing equipment's contribution to GHG emissions throughout its lifecycle.

- **GHG Calculation:** Utilizing collected data to calculate average annual GHG for each equipment category includes direct emissions during use, as well as those from manufacturing, distribution, end-of-life disposal, and other emission factors such as regional energy sources and online service usage.

Thorough interpretation of results also hinges on gathering relevant contextual data. This includes quantifying data usage and flow, essential for calculating subscriber numbers and ICT service-generated data traffic. Understanding the nature and volume of data flows, particularly in cloud computing contexts, is paramount.

Finally, a critical analysis of the data ensures the reliability of WeNR-QC results:

- **Verification of Data Sources and Consistency:** Ensuring data reliability by verifying sources, dates, and consistency, prioritizing information from reputable and updated organizations.
- **Oversight of Methods and Calculations:** Monitoring the methods and calculations employed, emphasizing data derived from actual measurements rather than models.
- **Addressing Data Uncertainty:** Recognizing data uncertainty and variability by presenting ranges or confidence intervals when necessary to strengthen the reliability of findings.

5.4.1.2 Calculation of Carbon Footprint with WeNR-QC

Determining the WeNR-QC footprint of technological goods is a meticulous process that involves a detailed analysis of each category, including computers, servers, networks, and mobile terminals. Evaluating parameters such as the number of goods in operation, their average lifespan, and energy consumption facilitates estimating the carbon footprint associated with each category, encompassing all phases from manufacturing through use to recycling.

Once individual footprints are established, the overall WeNR-QC footprint is calculated by aggregating this data. This calculation extends beyond summing up direct technological equipment footprints to include indirect emissions from digital services like cloud computing, which significantly impact the total environmental footprint.

Ultimately, WeNR-QC offers a comprehensive and precise assessment of the ICT's environmental impact by addressing both direct factors related to technological equipment and the indirect effects arising from digital service utilization. This holistic approach provides valuable insights into the environmental consequences of our growing dependence on digital technologies.

5.4.1.3 Interpretation of WeNR-QC Results

It is crucial to elucidate any disparities in the obtained results by identifying discrepancies in methodologies, scopes considered, and available data sources. Key references include ADEME's GHG assessment resource center and the ENERGY STAR program, which provide data on the use phase. Additionally, emissions from equipment were combined using information from the inter-organizational working group Boavizta and the online platform BeNUTS, alongside data supplied by manufacturers.

These explanations are vital for ensuring transparency and credibility in the analysis. They enable decision-makers to fully grasp the implications of the results and formulate appropriate policies and strategies to meet GHG reduction targets at provincial, national, and international levels.

5.4.1.4 Creation of WeNR-QC Dashboards

The WeNR-QC report provides a comprehensive overview of the study's objectives, scope, and methodology, meticulously justifying each step and clearly defining key concepts. It articulates underlying assumptions and emphasizes specific constraints within the ICT domain. Crucially, the report thoroughly references all data used, enhancing transparency and bolstering the analysis's reliability.

Regarding data access to rights and confidentiality, the report affirms its commitment to upholding current ethical and legal standards. This commitment includes safeguarding personal and sensitive data, as well as respecting intellectual property rights. The report underscores adherence to licensing requirements and restrictions on data reuse, underscoring the importance placed on maintaining the integrity and legality of the entire process.

5.4.2 WeNR-QC Methodology

The WeNR-QC methodology is grounded in robust ethical and legal principles, which are integrated into every phase of the analysis. Based on these principles, the report employs various calculation formulas that combine quantitative and qualitative approaches to ensure a comprehensive evaluation:

- The initial phase focuses on quantitative assessment, aiming for precise results despite inherent data complexity that may introduce some level of uncertainty. This transparency in methodology enhances the credibility of the conclusions drawn.
- The subsequent stage adopts a qualitative approach to evaluate the organization's maturity in responsible digital practices. This analysis identifies organizational strengths and weaknesses, offering insights for continuous improvement and effective integration of best practices.

In summary, the WeNR-QC report distinguishes itself through its unwavering commitment to ethical standards and legal compliance, aiming to deliver robust findings that advance responsible digital practices.

6 Experimental Results

This section presents the experimental results of a case study conducted with the WeNR-QC platform. It outlines the experimental protocol employed by a Quebec-based organization, emphasizing transparency and ensuring the reproducibility of results.

6.1 Experimental Protocol

6.1.1 Objective and Scope

The objective was to assess the environmental impact of digital technologies within a large organization (over 1000 users) by measuring both the digital carbon footprint and the maturity level of responsible digital practices. The study encompassed office equipment, data centers, network infrastructure, and user devices.

6.1.2 Collecting Data

Data was collected via a structured questionnaire. A comprehensive inventory of digital equipment was conducted, detailing each item's type, quantity, and specifications. Information on operational lifespan, usage patterns, and energy consumption was recorded. Contextual data, such as the organization's energy sources, methods of equipment transportation, and waste management practices, was also documented.

6.1.3 Methodology

The experimental protocol encompassed several key steps. Initially, LCA methodology was employed to quantify GHG emissions across each stage of the equipment life cycle, utilizing emission factors specific to Quebec. Subsequently, the carbon footprint of each equipment category was computed, encompassing both direct emissions from energy consumption and indirect emissions from manufacturing and disposal processes.

Collected data underwent rigorous processing using statistical analysis platforms to ensure accuracy and consistency. Anonymization techniques were applied to safeguard sensitive information. Machine learning algorithms, notably RNN, were leveraged to discern trends and patterns in the data, facilitating an assessment of the organization's maturity in responsible digital practices.

A relational model was constructed to depict interrelationships among variables such as equipment type, energy consumption, and GHG emissions. This model was maintained for a duration of 2 months to facilitate comprehensive analysis. Finally, the processed data was interpreted using business analysis platforms like Microsoft PowerBI to derive actionable insights. A visual report was then generated to present findings clearly and succinctly, emphasizing key areas for enhancement and offering strategic recommendations.

6.2 Results

The experimental findings revealed a significant digital carbon footprint within the organization, primarily attributed to emissions during the manufacturing phase of office equipment and data centers. Desktops, laptops, and monitors accounted for a substantial portion of these emissions, typically lasting between 3 and 5 years in operational lifespan. Data centers emerged as major energy consumers, operating with a Power Usage Effectiveness (PUE) that aligned with industry averages, yet offering potential for enhancement through improved cooling systems and the integration of renewable energy sources. Network devices like routers and switches also contributed to emissions, albeit to a lesser extent compared to office equipment and data centers. Smartphones and tablets, despite their smaller size, collectively increased the carbon footprint due to high usage rates and shorter lifespans.

Appendix 2 of the report presents a detailed example of the experimental results. Graphs in the appendix delineate the breakdown of carbon emissions by equipment type and life-cycle stage. A maturity assessment of responsible digital practices highlights organizational strengths and identifies areas for improvement. Strategic recommendations are proposed to mitigate the carbon footprint, including the adoption of energy-efficient technologies, extending equipment lifecycles, and enhancing waste management practices.

6.3 Conclusion

The experimental results underscore the necessity of ongoing monitoring and enhancement of responsible digital practices. Leveraging the WeNR-QC platform empowered the organization to grasp its digital environmental footprint and pinpoint actionable steps toward a sustainable future. The meticulous experimental protocol guarantees transparency and reproducibility, establishing a robust framework for other organizations striving to achieve digital sustainability.

7 Potential of WeNR for Quebec and the Rest of the Country

With its deep expertise in digital data, WeNR-QC has the potential to emerge as a pivotal player academia. By sharing its expertise, WeNR-QC can facilitate researchers' understanding and quantification of the environmental impact of digital technologies. This collaborative approach could pave the way for advancements and innovations in the responsible digital sector.

Furthermore, WeNR-QC could forge collaborations with Quebec's public sector, which has committed to implementing an indicator for responsible digital maturity index. Such collaboration could assist the government in achieving its objectives for responsible digital transformation, as outlined in the Government Sustainable Development Strategy (2023–2028). Initiatives like training programs, conferences, and publications would further facilitate knowledge transfer. With its leadership and expertise, Quebec can play a pivotal role in promoting digital sustainability at the federal level. Through platforms for exchange and collaboration, and by fostering innovation, Quebec, leveraging platforms like WeNR, can significantly advance the adoption and promotion of responsible digital practices across Canada.

Moreover, WeNR's application potential could be extended to other Canadian provinces facing similar environmental challenges, such as climate change and natural resource depletion. For instance, the Canadian federal government has committed to achieving carbon neutrality by 2050 (Canada, 2020), a goal that WeNR supports by providing organizations with practical means to reduce their GHG.

8 Conclusion and Recommendations

This document underscores the critical importance of adopting a proactive monitoring platform such as WeNR in Quebec to significantly reduce the ecological footprint of digital technology and advance the province's sustainable development objectives. By fully integrating this innovative platform into IT practices, Quebec cannot only decrease carbon emissions but also accelerate the shift toward an environmentally friendly digital transformation aligned with the region's ecological goals.

Moreover, substantial avenues for improvement must be explored. WeNR-QC advocates for widespread adoption of responsible digital practices by organizations, stressing the pivotal role of eco-design in IT strategies. By embedding these principles into decision-making processes now, Quebec organizations can actively contribute to forging a sustainable digital future that harmonizes technological innovation with environmental stewardship for societal benefit.

Looking ahead, it is crucial to continue developing and validating WeNR-QC, promoting its adoption among Quebec and Canadian organizations and governments, and forging partnerships to customize and disseminate the platform nationally. It is essential to strengthen the platform by integrating the principles of software engineering and AI to guarantee informed decisions.

In conclusion, this chapter not only raises awareness but also drives tangible actions toward responsible digital usage. By offering an adaptable and innovative solution, WeNR-QC addresses the environmental impact of digital technology within organizations, supporting their evolution toward a sustainable and environmentally conscious digital future.

Acknowledgment We would like to thank Camille Cardin-Goyer for the English review, rewriting, and correction, as well as Valentina Poch for her review. We also extend our gratitude to the entire CIRODD team.

Appendix 1

Lifecycle stage	Category	Equipment
Manufacturing	Workstation environment	Desktop computers
		Laptop computers
		Computer monitor
		Docking station
		Television
		Tablet
		Workstation peripherals
		Personal printer
	Impression	Multifunction printer
		Landline phone
	Telephony	Smartphone
		Switch/router
	Local network	Wifi terminal
		Switch/router
	Extended network	Security systems (firewall, WAF, gateways, probes, etc.)
		Internet terminal
		Physical server
		Empty bay
	Computer centers	Storage bay
		Switch/router
		FC switch (storage)
		Security systems (firewall, WAF, gateways, probes, etc.)
		Network equipment
		Virtual machine
		Object storage
		Block storage
	Cloud (IaaS and PaaS and SaaS)	Runtime (PaaS)
		Managed database (PaaS)
		Any underlying infrastructure to provide the service
Use		Electricity consumption—digital equipment
		Electricity consumption—other (e.g., technical room air conditioning)

Appendix 2

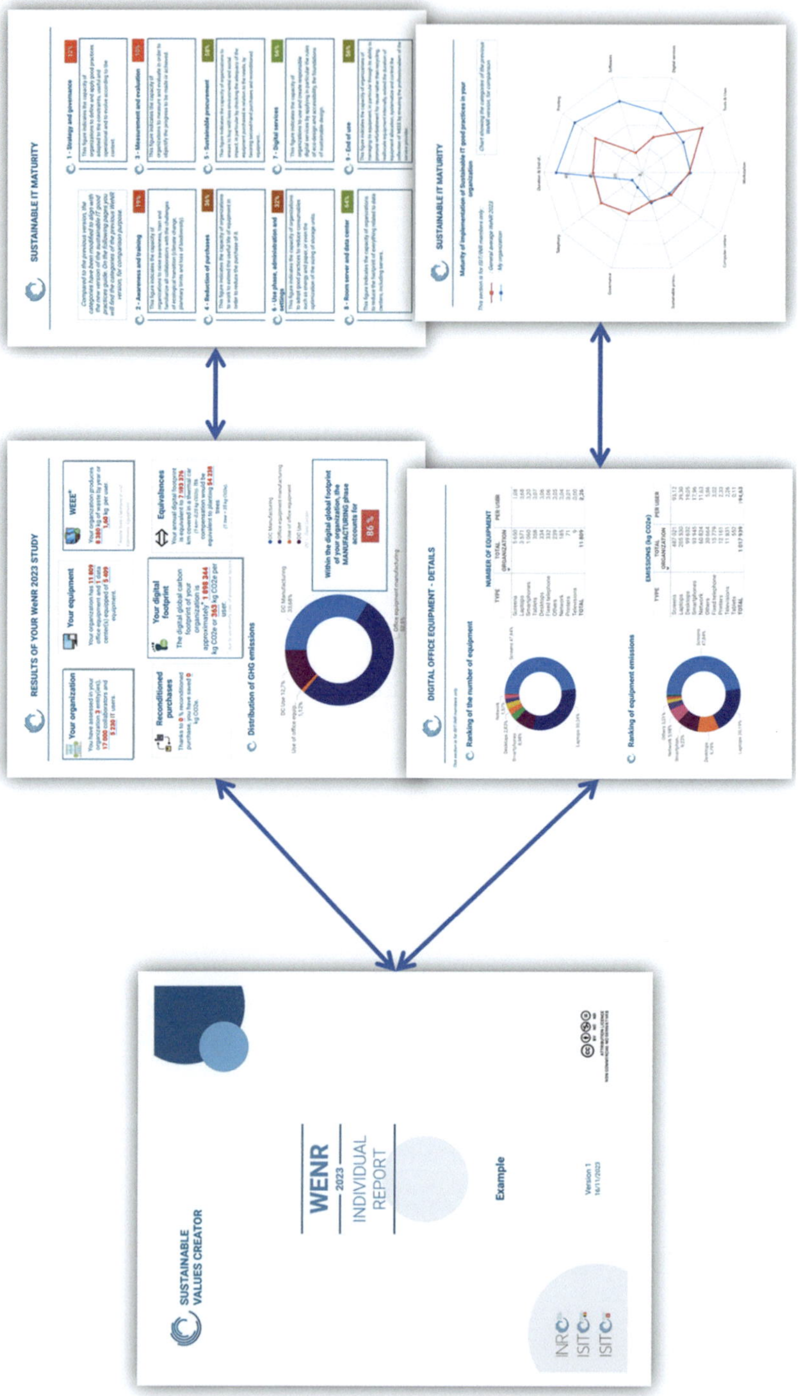
SUSTAINABLE IT MATURITY
RESULTS OF YOUR WeNR 2023 STUDY
DIGITAL OFFICE EQUIPMENT - DETAILS
SUSTAINABLE VALUES CREATOR
WENR
2023
INDIVIDUAL REPORT
Example
Version 1
16/11/2023

References

ADEME Library. (n.d.). *Methodological reference framework for environmental assessment of Information Systems (IS)*. Retrieved April 17, 2024, from https://librairie.ademe.fr/consommer-autrement/6649-referentiel-methodologique-d-evaluation-environnementale-des-systemes-d-information-si.html

Bourgeois, G., Benjamin, D., & Vincent, C. (2022). Review of the impact of IT on the environment and solution with a detailed assessment of the associated gray literature. *Sustainability, 14*(4), 2457. https://doi.org/10.3390/su14042457

Bourgeois, G. (2023). Analyse et modélisation de l'impact environnemental du système d'information [Phdthesis, Université de La Rochelle]. https://theses.hal.science/tel-04470070

Brouillard, P. (2023, February 6). *Environmental impacts of digital technology*. The Shift Project. https://theshiftproject.org/en/article/environmental-impacts-of-digital-technology-5-year-trends-and-5g-governance/

Canada. (2020, November 19). *Carbon neutrality by 2050*. https://www.canada.ca/fr/services/environnement/meteo/changementsclimatiques/plan-climatique/carboneutralite-2050.html

Cecere, G., & Pénard, T. (2020). Introduction to the special issue: "From the digital economy to the digitalization of the economy". *Revue d'économie Industrielle, 172*, 11–17. https://doi.org/10.4000/rei.9389

CIRODD. (n.d.). *CIRODD in brief. Centre interdisciplinaire de recherche en opérationnalisation du développement durable*. Retrieved April 17, 2024, from https://cirodd.org/mission/

Esmia Consultants. (n.d.). *Optimal trajectories for reducing Quebec's GHG emissions—Horizons 2030 and 2050 (updated 2021)*. Retrieved April 17, 2024, from https://esmia.ca/fr/projet/optimal-ghg-reduction-trajectories-for-quebec-horizon-2030-and-2050-update-2021/

Fisk, P. (2010). *People planet profit: How to embrace sustainability for innovation and business growth*. Kogan Page.

Government of Quebec. (2023). *Implementation plan*. Retrieved April 17, 2024, from https://www.quebec.ca/gouvernement/politiques-orientations/plan-economie-verte/plan-mise-en-oeuvre

Hydro-Québec. (n.d.). *GHG emission rates associated with electricity*. Retrieved April 17, 2024, from https://www.hydroquebec.com/developpement-durable/documentation-specialisee/taux-emission-ges.html

Imamov, M., & Semenikhina, N. (2021). The impact of the digital revolution on the global economy. *Linguistics and Culture Review, 5*(S4), 968–987. https://doi.org/10.21744/lingcure.v5nS4.1775

Innovation Barometer. (2022). *The innovation barometer*. Retrieved April 17, 2024, from https://lebarometre.ca/

ITU. (n.d.). *L.1450: Methodologies for assessing the environmental impact of the information and communication technology sector*. Retrieved April 17, 2024, from https://www.itu.int/rec/T-REC-L.1450-201809-I/fr

Lange, C., Kosiankowski, D., Weidmann, R., & Gladisch, A. (2011). Energy consumption of telecommunication networks and related improvement options. *IEEE Journal of Selected Topics in Quantum Electronics, 17*(2), 285–295. https://doi.org/10.1109/JSTQE.2010.2053522

Levasseur, A., Mercier-Blais, S., Prairie, Y. T., Tremblay, A., & Turpin, C. (2021). Improving the accuracy of electricity carbon footprint: Estimation of hydroelectric reservoir greenhouse gas emissions. *Renewable and Sustainable Energy Reviews, 136*, 110433. https://doi.org/10.1016/j.rser.2020.110433

Lu, S., Dai, W., Tang, Y., & Guo, M. (2020). A review of the impact of hydropower reservoirs on global climate change. *Science of the Total Environment, 711*, 134996. https://doi.org/10.1016/j.scitotenv.2019.134996

Masanet, E., Shehabi, A., Lei, N., Smith, S., & Koomey, J. (2020). Recalibrating global data center energy-use estimates. *Science, 367*(6481), 984–986. https://doi.org/10.1126/science.aba3758

OECD. (n.d.). *Measuring the environmental impacts of artificial intelligence compute and applications: The AI footprint*. Retrieved April 17, 2024, from https://www.oecd.org/publications/measuring-the-environmental-impacts-of-artificial-intelligence-compute-and-applications-7babf571-en.htm

Pinsard, M., & Toussaint, J. (2020). *DiagnosTIC : L'impact environnemental du numérique au Québec et au Canada*. https://doi.org/10.5281/zenodo.4284860

Québec Science. (2018, October 4). *Where do electronic waste go?* https://www.quebecscience.qc.ca/environnement/ou-vont-dechets-electroniques/

Québéciser. (n.d.). Retrieved April 17, 2024, from https://vitrinelinguistique.oqlf.gouv.qc.ca/fiche-gdt/fiche/26501200/quebeciser

Rautela, R., Arya, S., Vishwakarma, S., Lee, J., Kim, K.-H., & Kumar, S. (2021). E-waste management and its effects on the environment and human health. *Science of the Total Environment, 773*, 145623. https://doi.org/10.1016/j.scitotenv.2021.145623

WWF France. (2018). *WeGreenIT Study: What is the environmental impact of digital technology in companies?* Retrieved April 17, 2024, from https://www.wwf.fr/vous-informer/actualites/etude-wegreenit-quel-impact-environnemental-du-numerique-dans-les-entreprises

Guillaume Bourgeois holds a Ph.D. in computer science from the University of La Rochelle in France. He also has a Master's in computer science, specializing in computer systems for engineering, obtained from the University of Pau and the Adour region in France. He is currently pursuing a postdoctoral fellowship in computer science, jointly supervised by the University of La Rochelle in France and the École de technologie supérieure (ETS) in Canada. He is currently focusing his research on the environmental impact of information systems.

Géraldine Angulo is a pioneer of digital responsibility in Quebec, with over 18 years' experience in information technology. A graduate in computer science from the École de technologie supérieure (ÉTS), she is dedicated to promoting eco-responsible digital practices. Géraldine combines innovation and sustainability to build an ethical technological future. As an expert, she participates in various responsible digital projects for private and public organizations.

Hassana El-Zein holds a Ph.D. in Life Cycle Assessment (LCA) from Polytechnique Montréal. She has worked on various projects related to energy, sustainability, clean mobility, and innovation. Active in the LCA community since 2014, she has presented at and helped organize numerous national and international conferences and workshops.

Vincent Courboulay is an engineer and senior lecturer in computer science at La Rochelle University. For the past 10 years, he has specialized in responsible digital practices. In 2018, he co-founded the Institute for Responsible Digital Technology, where he serves as the scientific director. He is currently focused on the concept of responsible artificial intelligence.

Mohamed Cheriet earned his Master's and Ph.D. in computer science from the Université Pierre et Marie Curie (Paris VI) in 1985 and 1988, respectively. Since 1992, he has been a professor in the Department of Systems Engineering at the École de technologie supérieure (Université du Québec) in Montréal, where he was appointed full professor in 1998. He is founder and director of Synchromedia laboratory since 1998 and was former Canada Research Chair tier 1 in Smart Sustainable Eco-Cloud, 2013–2020. He is general director of the FRQNT strategic cluster, CIRODD (2019–2026): Interdisciplinary Research Centre on the Operationalization of Sustainability Development. His research focuses on future networks, cloud computing and green ICT.

Chapter 10
Navigating the Shift: Strategies Beyond "Build It and They Will Come" for Sustainable Mobility in Quebec

Jérôme Laviolette, Owen Waygood, and Anne-Sophie Gousse-Lessard

Abstract Passenger transport is an important contributor to unsustainable urban systems. To achieve the necessary socio-ecological transition will require overcoming the entrenched system of automobility. Composed of several mutually reinforcing components, this system has conferred psychosocial dimensions to car ownership and use that leads to important institutional, political, and individual resistance to change both car-centric transportation infrastructure and individual travel behaviour. For this reason, a growing consensus suggests that transitioning to a sustainable mobility system requires a more holistic approach that applies a synergistic integration of "hard" supply-side measures and "soft" demand-side solutions. This means increasing non-automobile accessibility and supporting such change with soft travel behaviour change solutions that target social-psychological barriers to change. While such approaches have demonstrated their effectiveness around the world, this second category of interventions remains underutilized, particularly in North America. Drawing from social psychology and a North American case study, this chapter proposes a theory-to-practice guide for practitioners to designing effective voluntary travel behaviour change interventions based on the Stage Model of Self-Regulated Behaviour Change (SSBC). A four-level integration framework for intervention design based on the SSBC is proposed. The framework proposes intervention approaches from using the model as a simple diagnostic tool to a complete integration to deliver a fully individualized and stage-tailored intervention. Stage-specific messages and strategies are described to shift people away from car use towards active, collective, and shared mobility options. The chapter concludes on suggestions for collaborative efforts between researchers and practitioners to design,

---

J. Laviolette (✉)
Université McGill, Montréal, QC, Canada
e-mail: jerome.laviolette@mcgill.ca

O. Waygood
Polytechnique Montréal, Montréal, QC, Canada
e-mail: owen.waygood@polymtl.ca

A.-S. Gousse-Lessard
Université du Québec à Montréal, Montréal, QC, Canada
e-mail: gousse-lessard.anne-sophie@uqam.ca

M. Cheriet et al. (eds.), *Accelerating the Socio-Ecological Transition*,
https://doi.org/10.1007/978-3-031-82896-6_10

evaluate, and enhance the effectiveness of these interventions, thus moving beyond infrastructure-only solutions to foster a successful transition to sustainable mobility in Québec.

Keywords Sustainable transport · Travel behaviour change · Soft intervention · Stage model of change · SSBC · Car dependency · Automobility · Travel demand management

1 Introduction

One imperative of the socio-ecological transition needed to avert the worst consequences of climate change is the transformation of our mobility systems. In the province of Québec, the transportation sector produces an outsized share of the province's carbon emission: 43% for the entire sector and 22% for the road passenger subsector (Gouvernement du Québec, 2022). Contrary to all the other sectors except agriculture, transport emissions have been increasing substantially in the last decades (+20.6% vs. −22.9% for all other sectors combined between 1990 and 2021).[1] Despite a strong push for transportation electrification (Gouvernement du Québec, 2023), it is increasingly clear that a successful socio-ecological transition requires a paradigm shift in mobility that moves us away from car dependency (Laviolette et al., 2020). The "system of automobility", or car dependency, is described as a transportation system that favours car travel and speed over alternative modes of transport. A system in which the car is considered a key satisfier of human needs for most of the population (Mattioli, 2016). The complexity of this system stems from several interlinked components that reinforce each other (e.g., Mattioli et al., 2020; Sheller & Urry, 2000), making any major shift in the system excessively challenging and politically difficult to accomplish. Furthermore, its well-documented consequences extend far beyond carbon emissions: put shortly, the automobility system poses a massive threat to public health, quality of life and well-being, equity and social justice (i.e., who bears the burden of car harms), climate change and the environment more broadly (e.g., Gärling & Steg, 2007; Glazener et al., 2021; Miner et al., 2024). Most of these will not be solved by simply switching to electric vehicles (Miner et al., 2024) and are too often overlooked when considering transportation solutions.

Most solutions proposed and discussed in the public sphere to reduce car dependency focus on improving accessibility by redesigning built environments, expanding public transit and active travel infrastructure, enhancing sustainable transport level-of-service, and offering shared mobility alternatives. However, there is a growing consensus that these "hard" infrastructure changes alone may not be enough (e.g., Banister, 2008; Handy, 2017; Javaid et al., 2020; Mattioli et al., 2020;

[1] Calculation by the authors based on the report by Gouvernement du Québec (2022).

Soza-Parra & Cats, 2024). This is because some of the institutional, political, and individual resistance to change is rooted in a car-centric mobility culture that reinforces the psychosocial dimensions of car travel, a core aspect of the automobility system (Mattioli et al., 2020; Sheller & Urry, 2000).

Based on this premise that "hard supply-side" solutions are not enough, a crucial question arises: how can strategically designed interventions effectively support behaviour change? To answer this question, this chapter serves two objectives. First, drawing insights from social psychology literature, we explore key elements of behaviour change theories, notably the Stage Model of Self-Regulated Behaviour Change (SSBC; Bamberg, 2013b), the role of habits and habit disruption, and the role of other psychological constructs such as norms, attitudes and beliefs. Second, we propose an integrated approach to guide practitioners in conducting a diagnostic of the issues, identify the potential for change within the targeted population and then provide a comprehensive framework for designing theory-driven, context-oriented, personalized behaviour change interventions to foster a successful transition to sustainable mobility in Québec.

1.1 Why We Need a Holistic and Interdisciplinary Approach for a Paradigm Shift in Mobility

Transportation and urban planning researchers have produced an extensive amount of evidence on what type of built environments, street design and provision of mobility options are associated with a reduced share of trips and distance travelled by car and lower car ownership needs (e.g., Ewing & Cervero, 2010; Javaid et al., 2020; Mishra et al., 2023). As Handy (2017) explains, "the built environment plays a role in determining the *choices available* to the individuals. […] But it doesn't mean that [it] determines the *choices made* by the individuals". It gives them *the possibility* to choose those options, which is an essential first step. These supply-side solutions (also called "hard" measures) are well known and widely discussed beyond academic circles. In already built-up cities, the policy challenge for those "hard" measures resides more in retrofitting. Indeed, many of those changes require shifting priorities and funding away from components of the system of automobility (Gössling, 2020). As such, a lot of the challenge of overcoming car dependency is political (Gallez & Dupuy, 2018) as is our local case in Québec (Laviolette et al., 2020). We argue that challenging and overcoming the system of automobility requires a more holistic approach that includes activating various levers to change the culture of mobility including encouraging, fostering, and supporting behaviour change away from the private car.

Many motivations exist for using cars (see Mattioli et al. (2020), Schwanen and Lucas (2011) and Javaid et al. (2020) for more comprehensive overviews). We highlight three reasons we believe are important here: habits, the normative aspect of automobile culture, and the affective and symbolic associations to cars. Habits such as travel mode choice (e.g., Gärling & Axhausen, 2003) arise from routine and once established are hard to change. Even with new infrastructure supporting new behaviours, car habits will resist change. New residents or young people might establish behaviours based on the new context, but those with established habits will need additional interventions.

A second reason is the presence of a strong automobility culture which imposes a normativity around cars where their ownership and use are perceived as the normal practice that every "normal" person does. Such social norms are strong explanatory factors of behaviour, though we often underestimate their influence (Nolan et al., 2008). The higher the rate of car ownership and the higher the modal share of cars, the stronger this barrier is likely to be.

A third reason is the perceived affective and symbolic benefits of the car (e.g., Sheller, 2004; Steg, 2005). Cars provide a way for individuals to communicate their social status and identity. Affective aspects are related to feelings of pleasure, excitement, dominance, freedom, and independence. These elements can become intertwined with personal, social, and even national identities (Edensor, 2004; Mattioli et al., 2020; Schwanen & Lucas, 2011). In short, someone for whom these perceived benefits are important and whose lifestyle and identity are highly linked to the car and the access it provides is less likely to adopt other modes even if those modes exist and are improved (e.g., Piras et al., 2022; Soza-Parra & Cats, 2024).

It is thus critical to understand and act on the psychosocial factors and obstacles hindering change at both societal and individual levels. As previously described, modern life has been mostly reorganized around the ownership and use of cars, and so the shift to sustainable mobility requires substantial changes in daily practices and lifestyles. In short, changing behaviour is a highly complex process and while providing alternatives to the car is essential for change to be possible, it is often insufficient on its own to prompt rapid behaviour change across diverse segments of the population.

Travel Demand Management (or Mobility Management) is a category of strategies that apply social psychology to help with travel behaviour change (Bamberg et al., 2011). While a large variety of approaches fit within the broader TDM and supply side categories, this chapter focuses on *personalized travel behaviour change* strategies. We present how an important model of change, the Stage model of behaviour change (SSBC; Bamberg, 2013b) can be applied by practitioners to enhance the effectiveness of such interventions.

2 A Theory of Change

2.1 The Stage Model of Self-Regulated Behaviour Change

Many models of behaviour change exist. One of the most famous and extensively used is the Transtheoretical Model of Change (TTM; Prochaska & DiClemente, 1982; Prochaska et al., 1997; Prochaska & Velicer, 1997) originally developed to modify health-related behaviour such as smoking cessation or increasing physical activity. It has since been used repeatedly in a variety of behavioural change contexts. More recently, Bamberg (2013b) proposed the Stage Model of Self-Regulated Behaviour change (SSBC) specifically designed to encourage pro-environmental behaviour, including a shift towards low-carbon transport modes.

Both models conceptualize the process of behavioural change as a non-linear progression through a sequence of distinct stages, with the possibility of stagnation or relapse to earlier stages. While the TTM postulates five different stages of behaviour change (i.e., Precontemplation, Contemplation, Preparation, Action, and Maintenance), the SSBC, proposes four. These reflect the process of becoming aware of an issue (Predecision), deciding to act (Preaction), implementing behavioural changes (Action), and maintaining the new behaviour over time (Postaction).

In the TTM, Stage 1 (Precontemplation) is characterized by amotivation, which can manifest itself either as a lack of desire to change or as seeing no reason to change at all. Contrastingly, Stage 2 (Contemplation) denotes the emergence of a certain degree of motivation for change, albeit accompanied by the belief that change is unattainable. In the SSBC, this resistance to change and low motivation are amalgamated into the initial predecision stage (1). Reflecting on this difference, Olsson et al. (2018) introduced the terms *predecisional denial* (1.a) and *predecisional inhibition* (1.b) to differentiate between types of resistance. *Predecisional denial* stems from perceiving no reason to change, while *predecisional inhibition* is associated with the belief that change would be beneficial but impossible to achieve. We shall use this distinction for the remainder of the chapter as we believe these substages are conceptually and psychologically different, thus requiring different strategies to facilitate progression towards higher stages.

Another distinction between the TTM and the SSBC is the time-specific component of TTM stages that is absent from the SSBC. For example, in Stage 2, the TTM specifies that individuals intend to take action within the next 6 months (Prochaska & Velicer, 1997). However, some have criticized these timeframes, notably describing them as arbitrary (e.g., Sutton, 2000). Another distinction is that, unlike the TTM, the SSBC proposes three *transition points*, representing the formation of three types of intention: goal, behavioural, and implementation intentions. A final yet crucial distinction that fortifies the theoretical underpinnings of the SSBC is its integration of the constitutive variables of the Norm activation model (NAM; Schwartz, 1977) and the Theory of planned behaviour (TPB; Ajzen, 1985, 1991). Specifically, the SSBC has the advantage of explicitly outlining the social, cognitive, and affective factors that differentially influence the three types of intention and thus promote stage transition. These psychological mechanisms underlying the progression through the stages are described below.

Because the SSBC is specifically designed for sustainable behaviour, including the transition to low-carbon transportation modes, and because it specifies the psychological mechanisms underlying self-regulated change, this chapter focuses on this model.

2.2 *Psychological Mechanisms That Support Moving from One Stage to the Next*

Change is a multifaceted process, characterized by its gradual, dynamic, and non-linear nature, where setbacks and relapses are not uncommon. Importantly, not everyone initiates change simultaneously or progresses through its stages at the same pace. Furthermore, within each individual, a complex interplay of socio-cognitive factors influences their willingness to change in interaction with external factors such as structural, economic, social, and political conditions. The very nature of the behaviour, along with the perceived or objective degree of difficulty, also plays an important role. Complementary behaviours aimed at the same problem often exist (e.g., reducing one's GHG emissions). For instance, while someone may be contemplating a transition to a vegetarian diet, they may simultaneously be at a more advanced stage when it comes to reducing their reliance on cars. Each stage of change is qualitatively distinct, comprising unique motivations and cognitive processes that shape how individuals perceive information and make decisions. As individuals navigate through these stages, they may find themselves shifting their focus, modifying their priorities, and reevaluating their strategies, all in pursuit of meaningful and sustainable change. The proposed sets of stage-specific socio-psychological variables in the SSBC determine the formation of intentions or transition points (Bamberg, 2013b). The model is presented in Fig. 10.1, including Olsson et al.'s (2018) proposed distinction between predecisional *denial* and *inhibition*.

2.2.1 Stage 1: The Why

While the *predecisional denial* stage (1.a) of the SSBC is characterized by the lack of recognition of the need to change, the *predecisional inhibition* stage (1.b) begins when the individuals become aware of the problem's significance, its impact on themselves (or society), and their personal role in it. For instance, they might think, "the number of cars on the roads is unsustainable. This causes many problems. I use my car too much." When individuals see a discrepancy between their actions (contributing to the problem) and their broader life goals (protecting the environment, staying healthy, spending less, contributing to the common good, etc.), they experience discomfort and unpleasant emotions (guilt, sadness, surprise, questioning). They can start comparing themselves to others (social norms) and feel a growing moral obligation to act (personal norms). In some cases, comparing oneself to others can hinder change if the norm is problematic such as driving a car in our car-dependent societies. At this point, subjective social norms, meaning the judgments of significant people close to us, become important. A positive influence of subjective social norms would be having a network of friends and family that value other modes of transportation over cars. The idea of changing their habits appears increasingly feasible and appealing. They know they would feel better about themselves if

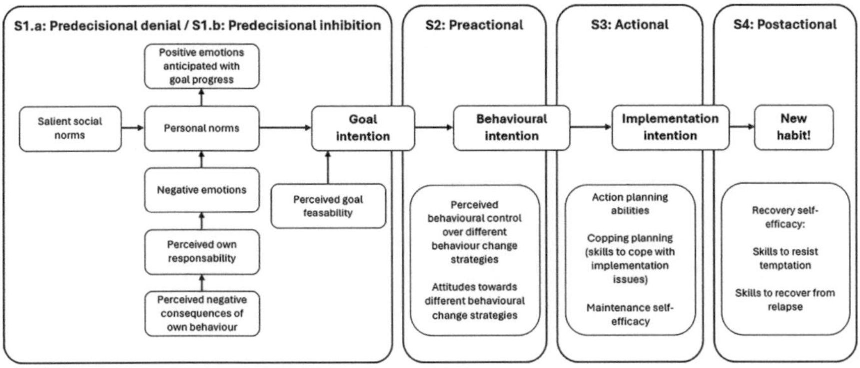

Fig. 10.1 The stage model of self-regulated behavioural change (adapted with permission from Bamberg, 2013b; Bamberg & Schulte, 2018) with the Stage 1 modification suggested by Olsson et al. (2018)

they made changes in their life (anticipated positive emotions). By the end of this phase, they know they want to improve their behaviour (goal intention). They might say at this point, "I would like to reduce my car dependence." However, they are unsure at this stage what behaviour(s) to adopt to achieve this new goal. It still seems very difficult to them.

2.2.2 Stage 2: The What

The individual knows that their behaviour is problematic, but they do not yet know what avenues of change are possible. They will therefore pay attention to information about different solutions and actions they could take. For example, taking the bus, cycling more, and moving to a neighbourhood that favours utilitarian trips on foot. At this stage, direct changes in travel behaviour might not be evident, but actions such as supporting progressive sustainable mobility policies could reduce the internal conflict. They will evaluate the pros and cons of behavioural changes for themselves, as well as their feasibility: "Is it difficult for me to make this change? Do I have the means, control, time, knowledge, and resources? Are the necessary services and infrastructure available in my area? What are the benefits for me? How will they impact the problem I'm trying to address?" By the end of this stage, a behavioural intention forms: the individuals know which action(s) to try in order to reduce their car dependency.

2.2.3 Stage 3: The How

Here, individuals understand why changing their habits is important and have chosen a way to do it. They are now figuring out how to implement their plan. They think about the specifics: how, when, where, and with whom? Planning skills are crucial in this stage. The more they can elaborate and anticipate obstacles, the easier the action will seem to them when they implement it. For example, one could think, "If I want to take public transport to work, I need to wake up at X time, walk to the bus stop located here, and get off at the right stop/station. If the bus is late, I have a backup plan." At this stage, they are attentive to information that supports their self-efficacy (the belief in oneself that they can undertake an action) and helps them plan their action, like guides, apps for route planning, or friends who can accompany them. By the end of this phase, they start trying out the new behaviour. Practising this behaviour at a non-crucial time (e.g., on a day when it does not matter) can be helpful as they can gain experience without the stress of arriving on time in a situation of uncertainty.

2.2.4 Stage 4: Keep Going

Maintaining the new behaviour over time is challenging. There are many reasons to revert to old travel habits: lack of motivation, rushed mornings, bad weather, family obligations, errands, or route changes. At this stage, individuals must resist the temptation to revert to their old ways, which is not easy. Maintaining the new behaviour is a *process* that involves setbacks, mistakes, hesitations, and failures. This is normal and expected. The ability to recover from these relapses is thus important. Just because the new behaviour is not applied every day does not mean they should stop trying to improve. The quality of motivation to maintain this lifestyle change is also essential. Thus, a person who is internally (intrinsically) motivated will be more likely to succeed and will recover from relapses more easily (Pope et al., 2018). External support can also be important at this stage such as friends or colleagues with similar habits, feedback and encouragement systems, and adequate services and infrastructure that support the new behaviour.

In summary, a deep understanding of these mechanisms is crucial to fostering and supporting the changes we wish to see in our society. As explained below, proposed programs, interventions, and solutions must be tailored to the specific stage of change.

2.3 *The Importance of Message Framing*

Framing in communication refers to the way information is presented or "framed" to influence how it is perceived by the audience. It involves selecting and emphasizing certain aspects of an issue or topic while downplaying or omitting others. It can significantly impact how people interpret information, make decisions, and form opinions. By framing information in a particular way, communicators can shape the audience's understanding, attitudes, and behaviours related to the topic at hand. There are various types of framing, including risk framing, economic framing, health framing, gain vs. loss framing.

An individual who is doing a desirable behaviour and an individual who sees no reason to do such a behaviour do not need the same information. In fact, giving reasons to do a behaviour and expecting behaviour change is asking too much of that information. As described in the previous section, a stage-based approach can help practitioners know what type of information is required to help individuals move towards the desired behaviour (Bamberg, 2013b; Pope et al., 2018; Waygood et al., 2012). One common problem that people fall into, is believing that "if only others knew what I know they would do as I do". People have different motivations and ways of viewing problems. It is critical to create messaging that captures people's attention (i.e., that individuals "attend to" that information) and that speaks to them. Furthermore, messaging that relates to an individual's self-determined motives will be more effective in the long term (Pope et al., 2018).

Consideration to what intrinsically motivates people should be applied. For example, some individuals might be highly motivated by the environment, while for others, this might be reducing cost or other social factors such as improving health (Waygood et al., 2012). How this information is itself framed can impact how effective it is in influencing choices (Daziano et al., 2021; Wang et al., 2023). If this occurs, they will search for possible actions that they could take.

This final point is crucial: how information is framed will have a considerable influence on how much influence it has on different populations (Wang et al., 2023). It is not possible to go into great detail here, but examples of information framing for climate change emissions could help inform framing decisions. As an example, loss framing is generally found to be more influential than neutral or positive framing (Waygood & Avineri, 2018). Framing information with respect to societal goals can be more effective overall, though economic impacts can be highly influential for certain parts of the population (Daziano et al., 2021).

2.4 *Efficacy of Personalized Travel Planning and Model-Based Behaviour Change Interventions*

In our view, and based on the scientific literature, *personalized behaviour change programs* based on the models of behaviour change represent a promising avenue for achieving the desired changes in travel behaviour. But before describing how to design this kind of program, it is important to explain what they are and why they are effective at changing travel behaviour. To achieve this, recent and robust reviews on the matter are examined. Additionally, studies that specifically investigate the limited applications of Stage models of behaviour change (SSBC or TTM) to improve travel behaviour change interventions are also explored.

2.4.1 What Is a "Personalized Behaviour Change Intervention"?

Various types exist, but in general, a personalized behaviour change or individualized marketing program directly targets individuals either at work, at school or in their neighbourhood to offer them information, assistance, motivation and incentive to support them in *voluntarily* altering their travel choices (Bamberg & Rees, 2017). When targeted at work or school, the objective is usually to reduce solo driving for commuting. When neighbourhoods are targeted, people or households are encouraged to adopt active travel, public transit, and shared mobility options for their daily trips more broadly.

Taking the neighbourhood example, the idea is that an organization or a municipality will target a specific sector of a city and distribute by mail to all homes within the area a travel information package or bundle. The package provides area-specific information about surrounding walking, cycling, public transit, and shared mobility options available. This can include practical information such as the location of transit stops, bikesharing stations, or bike paths, as well as information about *how* to subscribe and pay for those services. It can also contain motivational messages, information on the benefits of using sustainable mobility, social norms in the neighbourhoods regarding each mode, and information about local neighbourhood groups and events such as exploratory walks. It can also include direct incentives such as free public transport tickets or an invitation to go online to order further incentives and goodies for specific modes (free public transit or bikesharing tickets, bike lights, etc.). Registering to receive those incentives and/or goodies can include commitments to change and to set goals. The ordered incentives can be delivered to one's door by a travel ambassador or agent that can further assist in trying out a new mode of transport. This simplified example based loosely on the Portland (Oregon) SmartTrips program[2] can be adapted to any context, any mode available and can make use of a wide range of persuasion technique, message framing and more. In the next section, further explanation will be provided on the Portland program as an illustration of a long-running large-scale North American case study and describe how the program could be upgraded by incorporating elements of the stage models of behaviour change.

2.4.2 Does It Work? What Are the Evidence?

For a municipality or non-profit organization, building such a program from "scratch" can seem like a daunting task. Thus, a critical question is: is it worth the trouble? These types of programs have been tested in various contexts since the late 1990s with numerous studies on their efficacy. Those studies include narrative or systematic reviews (Chatterjee, 2009; Graham-Rowe et al., 2011; Petrunoff et al.,

---

[2] Seehttps://www.portland.gov/transportation/news/2018/12/26/news-blog-moving-portland-new--year-pbots-here-help-you-find-your-way for a short overview of the current program.

2016; Scheepers et al., 2014; Zarabi et al., 2024) as well as meta-analyses (Arnott et al., 2014; Bamberg & Rees, 2017; Fujii et al., 2009; Möser & Bamberg, 2008; Semenescu et al., 2020). The consensus is that "yes, they work", though improvements can be made to increase their impact (Laviolette, 2020, p. 38). Four meta-analyses found that the overall effect was a 5–7% reduction in driving mode share (Bamberg & Rees, 2017; Fujii et al., 2009; Möser & Bamberg, 2008; Semenescu et al., 2020).

Is a 5–7% modal share reduction of car use a significant achievement? It's all about perspective. If travel behaviour change programs are extended to entire neighbourhoods or cities, reaching tens of thousands of participants annually, a 5–7% decrease in modal share could noticeably impact the city-wide modal share. Because developing a program from scratch entails an important start-up cost, a report on individualized marketing to practitioners in the United States recommends going big and planning for a minimum of 10,000 households per year (Alta Planning + Design & Transit Center, 2017). Results from larger scale projects in the UK, Australia, Europe and Portland (OR), although harder to evaluate with the highest quality design due to the high cost of such evaluation, indicate that similar results of 5–10% reduction in car use can be achieved among the targeted populations (AGO, 2006; Chatterjee, 2009; Dill & Mohr, 2010; Sloman et al., 2010). The cost of such programs is typically orders of magnitude cheaper than the hard (infrastructure) interventions and should be a popular complement to such changes.

2.4.3 Behaviour Change Programs Based on Models of Behaviour Change

In the field of transport, two reviews looked at this question. Friman et al. (2017) conducted an integrative review of the application of the Transtheoretical Model of Change (TTM). Analysing 13 intervention-based studies, they concluded that interventions based on the TTM processes were successful in triggering a behaviour change (reduction in car use or an increased in sustainable mode use). However, due to large variability in measured outcomes, it was not possible to provide clear numbers on those changes. To the authors' surprise, only three of the 13 studies measured the progression to higher stages, which is baffling considering that one strength of stage-based model is to provide more information on the steps between not even considering change (stage 1.a) and fully integrating the change in one's life (stage 5). Overall, the review concluded that more research is needed to determine how the specific processes of the TTM lead to travel behaviour change.

A second review by Keller et al. (2019) focused on ten studies applying Bamberg's Stage Model of Self-Regulated Behavioural Change (SSBC). From these, only three were intervention studies and two were related to travel behaviour: one by Bamberg (2013a) applying a stage-tailored telephone campaign and one by Sunio et al. (2018) where the stage-based intervention was delivered using a mobile app. Both studies as well as the non-travel one (related to beef consumption) showed that the tailored interventions were effective in moving people towards higher stages and performed better than the mismatch-information interventions or the no-information interventions.

In summary, both reviews conclude on promising results, although with many caveats, and suggests more robust, and optimally, longitudinal studies to improve our understanding of how each aspect of a tailored intervention is effective in moving people up the stages towards the desired behaviour.

3 Translating the Theory into Practice

3.1 Conditions for Maximizing Success

The extensive literature evaluating the effectiveness of soft interventions, including personalized behaviour change programs, can provide guidance on how to design interventions to maximize their effectiveness at changing behaviour away from driving. We outline four conditions of success that, in our opinion, practitioners should consider when developing their own interventions.

3.1.1 Don't Copy Paste: Adapt Each Intervention to the Mobility Context of the Targeted Area

This might sound trivial, but promoting and encouraging people to use a product that does not exist, does not meet people's needs, or does not satisfy the basic conditions for safety and convenience is unlikely to work. This does not mean that practitioners have to "wait" for their city to have world-class cycling and/or transit infrastructure to start implementing travel behaviour change programs. A lot comes down to managing expectation: do not oversell something that you cannot deliver. As we will argue later, travel behaviour change interventions can contribute to a positive loop of promoting a sustainable mobility culture that will make it easier to implement and improve the structural conditions which in turn will further strengthen the sustainable mobility culture. Practitioners should start with the low-hanging fruit—individuals who are at Stage 2 and up. They should plan their interventions, first for neighbourhoods which are best served by transit, active travel infrastructure and shared mobility options so that infrastructure barriers are minimized. This should be done in parallel with improvements to infrastructure and services in other neighbourhoods to reduce inequalities and eventually make alternative modes of transport accessible to the entire population. However, evidence from Australia suggests that both neighbourhood behaviour change programs (Ma et al., 2017) and workplace programs (Petrunoff et al., 2016) can work in suburban areas, although, according to Ma et al. (2017) such programs will have higher, longer-lasting effects, in more walkable neighbourhoods. In addition to area targeting, this also means focusing on the promotion of modes that already offer good conditions. If a neighbourhood is well served by public transit but has no protected cycling infrastructure, then the focus should be on the former not the latter. If a small town has a vibrant "core" with local shops and walkable streets, then the focus should be made on promoting walking for people's regular errands. What might motivate individuals to do so? Examine how social, environmental, health, and other messages resonate with your target population. These internal motivators are essential to start. This could be combined with designated parking lots to encourage residents living further away to park and walk the last hundred or so meters to their destination, thus reducing congestion and car nuisance on the commercial street and enhancing its appeal to pedestrians. This latter example brings up the next condition.

3.1.2 Combine Structural and Soft Intervention for Maximum Effectiveness

A company launching a new product is unlikely to be successful if no one knows about it, if its benefits are unclear, or if people do not know how it works. The same principle can be applied to new mobility options. This is critical when: there is a lack of social norms associated with the use of the mode (e.g., bikes, e-bikes); the mode is a new "system" that people are unfamiliar with (e.g., bikesharing or carsharing); the new option improves an aspect of an existing system but is not very visible (e.g., a new express bus line, a new bikesharing or carsharing station). The "build it and they will come" adage in transportation can be true, but to a certain extent only, and can vary drastically by context. A new mass transit rail line, like the Réseau express metropolitan (REM) in Montréal, is highly visible (both physically and in the media environment), but its construction requires considerable capital and cost billions of dollars. Due to its high cost, it can be highly beneficial for the operator, but also for society more broadly, to "catalyse" the adoption curve of the new transit line and boost ridership early on to maximize the "return on investment" by attracting new users and not just existing transit users. In a very similar example, Piras et al. (2022) evaluated the impacts on motorists of a light rail extension in Cagliari, Italy. They tested whether adding personalized travel planning (PTP) interventions to the hard measure was more effective at changing the behaviour of drivers to adopt the rail line than just having the hard intervention. Using a longitudinal survey design, they found that 46% of drivers who received the PTP opted for the light rail compared to 34% in the control group. As a critique to the existing literature, very few studies have specifically investigated the combination of hard and soft interventions. Furthermore, it has to be demonstrated that it can work in a North American context.

3.1.3 Leverage the Habit Discontinuity Hypothesis

Travel behaviour, including mode choice, is often conducted in a stable context, and thus can be initiated by environmental cues without much deliberation or thinking (Gärling & Axhausen, 2003). Over time, this becomes a habitual response, a behaviour performed repeatedly without much conscious effort. Someone commuting to the same workplace by car does not reassess every day if the car is still the "optimal" (or even simply the "good enough") option, even if some small elements of the environment change such as a new protected bike lanes, a new, faster, bus service or if traffic conditions deteriorate slowly over the years. Habits have been acknowledged as a major barrier to behaviour change because it weakens the link between intention and the behaviour (Javaid et al., 2020). It also limits the extent to which people acquire and process new information about alternative transport choices (Verplanken & Roy, 2016).

The *habit discontinuity hypothesis* states that when a major change to individual context happens, habits are broken, at least temporarily. This provides a "window of opportunity" for change where the individual is more open and ready to acquire and process new information about a new behaviour (Verplanken & Roy, 2016). It also makes the choice process more deliberate; that is the pros and cons are reevaluated to determine if the past option is still the best option. Various changes could be leveraged such as starting work/school at a new location (e.g., Ralph & Brown, 2019; Thøgersen, 2012; Walker et al., 2015) or temporary closures of major car infrastructure (e.g., Fujii et al., 2001). However, the most studied context change in transport is residential relocation (e.g., Ampt et al., 2006; Bamberg, 2006; Ralph & Brown, 2019; Verplanken & Roy, 2016). Verplanken and Roy (2016) estimated empirically that the opportunity window lasts about 3 months after a relocation. Targeting recent movers has also been done in large-scale travel behaviour change programs such as the SEGMENT project in Europe (Intelligent Energy Europe, 2015) and the Portland SmartTrips program since 2014 (Dill & Mohr, 2010). Targeting individuals who recently moved might not be easy. SmartTrips Portland for example bought an inexpensive mailing list based on the U.S. Postal Service's National change of address database. Bamberg and Rees (2017) were provided a list of addresses by the Munich municipal office where new residents must register.

In the authors' local context of Québec, several options could be explored. At the provincial level, Hydro-Québec, the state electricity provider, have a database of people who have changed their home address. The *Société de l'assurance automobile du Québec* (SAAQ) is also notified when someone with a driver's license moves to a new address, despite some delay. At the municipal level, any new homeowner must register their new address for real estate and "welcome" tax purposes. However, this does not cover renters. Another possibility is to target those who request a residential parking permit. To get the permit for the correct zone, they must supply their home address. It would be a great opportunity to reach out to motorists who are about to pay to park their vehicle in the public domain.[3]

[3] We would like to thank Pascal Priori from the non-profit Solon for this idea.

3.1.4 Personalize Interventions to Target the Right People with the Right Psychological Mechanisms

A one-size-fits-all approach rarely works best. In addition to geographical segmentation described in section 3.1.1, it is also recommended to plan the behaviour change programs with interventions for various segments of the population (Davies, 2012). A basic approach is to segment the targeted audience based on sociodemographic and household-level characteristics such as age, gender, household composition or occupation status. Students in their 20s, a young family with a child, or a retired couple all have very different mobility needs but also different barriers to change. Preparing for this in the development of a program is a step forward to a more personalized approach that could deliver better results. Over the last decade or so, there has been an increased interest in developing interventions based on psychographic segmentation built on a set of psychological variables (e.g., Gousse-Lessard & Laviolette, 2022; Haustein & Hunecke, 2013). A large-scale travel behaviour change program called SEGMENT was tested for 3 years in several European cities from 2010 to 2013 (Intelligent Energy Europe, 2015). A classification algorithm based on 12 "golden questions" asked in a pre-intervention questionnaire was developed (Anable & Wright, 2013). This allowed the identification of eight psychographic segments which could then be targeted with segment-specific information, recommendation, incentives and more.

In this chapter, we argue for a simpler segmentation approach that simply classifies the targeted population based on their current stage of change, either away from car use or towards specific transport modes. This approach builds on the theoretical constructs supporting changes at each stage as described in section 0. Furthermore, within each stage of change, the strategies and information delivered can be further targeted towards specific demographics. For example, a mother of young children in Stage 1.a could be more receptive than a young man in his 20s to information designed to raise awareness about the need to collectively reduce road violence and traffic-related risks for children by reducing car traffic in their neighbourhoods. We provide examples of psychological mechanisms to target at each stage of change through specific message framing and incentives in 0 below.

3.2 A Level-Based Approach in Integrating the Stage-Based Model in Travel Behaviour Interventions

To facilitate the development of behaviour change interventions, we propose a framework containing four levels of integration of stage-based models. The higher the level, the more complete the integration, but also the more complex and resource-intensive to both develop and deliver. Each level of integration needs to incorporate the lower-level elements to be effective. For each level of integration, and for illustration purpose, we suggest how the Portland SmartTrips program could be improved by using a stage-based model approach.

Portland's current SmartTrips program started in 2003, first as a pilot project before moving to full scale in 2004. The program was inspired by similar successful travel behaviour change programs in Europe and Australia based on the TravelSmart concept developed by Socialdata (see Dill & Mohr, 2010). It is one of the largest such programs in North America. In its first 10 years of full-scale existence (2004–2014), the program targeted roughly 20,000 households in one specific Portland neighbourhood each year. In 2014, the program was transformed to target specifically 10–15% of households having recently moved to or within Portland (Boddy & Kassirer, 2013).

3.2.1 Level 1: An Effective Diagnostic Tool

In this first level of integration, the stage of change models is only used to classify or segment the population based on their current behaviour and readiness to change. This can be done by including one or more stage assessment question into a pre-intervention survey administered to the targeted population. Due to its strong impact on the potential to change, it is essential to also assess the quality of alternatives available. This means mapping out active travel infrastructure and assessing the transit level-of-service (not simply if a bus stop exists). Optimally, this should be done by identifying and calculating accessibility indicators using various modes for the targeted neighbourhood or city or to compute accessibility to a specific destination like a workplace or school/university where the intervention is planned. For further guidance on measuring accessibility see Boisjoly (2023).

As described by Keller et al. (2019) in their review of SSBC applications, there are two approaches in designing stage-based evaluation and interventions. The first is to focus on moving people *away* from their current problematic behaviour. For sustainable transport, this generally means reducing single-occupancy car use. The second approach is to focus on modelling change *towards* a specific desirable behaviour like walking, taking public transport, or cycling. The first approach might work better when several alternatives to the car are available as it leaves participants the freedom to choose. The second approach likely works best when the entire intervention promotes a specific mode. Furthermore, because each mode has its own benefits and barriers, this can help develop even more tailored interventions addressing mode-specific barriers at each stage of change. Finally, both approaches can be combined. This means assessing stages of change in reducing car use combined with either stage-assessment questions for specific modes or, more broadly, willingness to use specific modes for daily trips.

Although standardized methods do not exist (e.g., Adams & White, 2005), one method to capture the stage of an individual is through self-assessment. Reviewing the literature applying the SSBC or the TTM to travel behaviour change, we suggest the following set of statements based on Bamberg (2013a, 2013b) and further adapted by Olsson et al. (2018). For a French-speaking example applicable in Québec, see Chouinard et al. (2021). The first set is to identify the stage of change *away* from car use. We provide an example specific for commuting to work in

Table 10.1 based on Olsson et al. (2018). Follow-up questions can capture whether those not driving intend to start (are they captive non-driver aspiring to drive or not driving by choice?) The second set is to identify the stages *towards* alternative modes for daily trips in general (Table 10.2). In both cases, the stage name or number should not be included in the questionnaire.

There are two ways this could be integrated in a program like Portland SmartTrips. The first is to simply include the relevant behaviour change question within a pre-intervention questionnaire distributed in the targeted neighbourhoods. The second is to ask this self-assessment question to people who visit the online portal to order incentive and material. The question could be made mandatory to order the free material.

3.2.2 Level 2: A More Complete Evaluation of Change

This second level of integration goes one step further by incorporating the stage-assessment question in a post-evaluation survey, in addition to the pre-intervention survey. Whether or not the intervention itself is based on the stage-based model of change, measuring pre- and post-intervention stage membership offers the benefits of capturing more information than simply assessing behaviour before and after. This can be very useful to understand if the targeted population has been successfully moved to higher stages of change, even if that does not mean an observable change in behaviour. An example would be to evaluate the effect of an awareness campaign designed to motivate people who *want* to reduce their car use, thus moving them from the *predecisional inhibition* stage (S1.a, seeing no reason to change) to stage 1.b or 2 (Tables 10.1 and 10.2). The next efforts for those individuals would be to facilitate a change. In short, a behaviour change program could have some success even if it doesn't lead yet to observable behaviour change as people in

Table 10.1 Examples of a car-use reduction stage of change question for commuting to work

#	SSBC Stage name	Question: Which of those statements best describe your intention and use of the car to go to work? Statements:
S1.a	Predecisional denial	I mostly drive alone to go to work. I am pleased with this and see no reason to reduce my car use
S1.b	Predecisional inhibition	I mostly drive alone to go to work. I would like to reduce my car use but feel it is impossible
S2	Preactional	I mostly drive alone to go to work. I am thinking about reducing my car use but I'm not sure how or when to do this
S3	Actional	I mostly drive to go to work. My aim is to reduce my current car use. I know which journey to replace, and which mode to use, but I have not started to do so on a regular basis
S4	Postactional	I will maintain or reduce my already low car use to go to work in the coming months. I try to use other modes than the car as much as possible
	Not applicable	As I do not drive or have access to a car, this question does not apply

Table 10.2 Examples of sustainable mode stage of change question for daily trips

#	SSBC Stage name	Question: Which of those statements best describe your intention and use of [walking/cycling/taking public transit] for your daily trips? Statements:
S1.a	Predecisional denial	I don't do any trips **[on foot[a]/by bike/by public transit]**. I am satisfied with this and see no reason to start
S1.b	Predecisional inhibition	I don't do any trips **[on foot[a]/by bike/by public transit]**. I would like to start but feel it is impossible
S2	Preactional	I don't do any trips **[on foot[a]/by bike/by public transit]**. I would like to start but I'm not sure how or when to do this
S3	Actional	I rarely do trips **[on foot/by bike/by public transit]**. My aim is to start doing more soon. I know which trips to replace, which itinerary to take, but have not started to do so on a regular basis
S4	Postactional	I regularly travel **[on foot/by bike/by public transit]**, I will maintain or increase the trips I do **[on foot/by bike/by public transit]** in the coming months

[a] Because not walking at all for transportation can be rare, the statements for S1.a, S1.b, and S2 can be modified to "I do few or no trips on foot […]"

stages 1.b, 2 and 3 might have slowly developed a higher intrinsic motivation to change which would make them more ready to change when future interventions (infrastructural or soft) are implemented. Furthermore, those at higher stages of change are more likely to be more supportive of transport policies that would allow them to operationalize those changes (Gousse-Lessard & Laviolette, 2022; Kirschner & Lanzendorf, 2020), although more research is needed to validate this hypothesis.

There are two ways of conducting a pre- and post-evaluation of a target population. The first is by using repeated cross-sectional surveys. A first survey is distributed within the targeted population before the intervention and then another at one or more point in time after the intervention (e.g., 1 and 6 months). In that first approach, a different sample of respondents is recruited for the pre- and post-surveys. A critique formulated by Bamberg and Rees (2017) is that this approach of evaluation is too often used in transportation while it does not allow analysis of intra-personal changes resulting from the interventions. For this reason, researchers should consider a longitudinal design where the same individuals are surveyed two or more times to test the effectiveness of interventions. A second critique formulated is that the effect of behaviour change interventions is often not evaluated in the long run (Bamberg & Rees, 2017), leading to the question: are the changes "permanent" or long-lasting? However, there are evidences that behaviour change following well-designed interventions is maintained in time (Semenescu et al., 2020), including in the case of Portland's SmartTrips (Dill & Mohr, 2010).

In summary, from a practitioner's perspective, it is likely sufficient to know whether the population is moving in the desired direction of change (by using repeated cross-sectional surveys). If the budget allows it, practitioners should consider a third evaluation 1–2 years later to see if the changes are maintained in time. See Alta Planning + Design and Transit Center (2017)'s report for practitioners for further recommendations of evaluations. Linking this level to the Portland SmartTrips program, this would mean integrating one or more stage of change questions in the post-intervention surveys and measure the progress from the pre-surveys.

3.2.3 Level 3: Stage-Specific Intervention Design

In this third level of integration, interventions are designed to target people at specific stages of change. Due to the difficulties of tailoring and delivering the information to each individual based on their own stage of change, the goal is to design interventions more broadly to appeal to people in specific stages. Considering that programs are often designed around limited budgets, a first approach would be to design the messages and strategies, such as incentives, to appeal to those individual "ripe" for change: people in Stage 2, early Stage 3 (people have only started to try out the new behaviour) and to a lower extent, Stage 1.b. A second approach can also be to vary the targeted stages based on geographical location. For example, if a sector has a strong majority of people in Stage 1.a and another a larger share of people in Stage 1.b or 2, than the first sector could receive an intervention designed for people in Stage 1.a (to bring them to consider change) and the second sector should apply an intervention with Stage 1.b and 2 in mind even if in each sector there will be people in the other stages. A third possibility when it is impossible to know in which stages targeted individuals find themselves is to use a menu-based approach on a website where people can choose the information and interventions that best fit their current readiness to change (Abraham, 2008). Such approach can be enhanced with a few quick self-assessment questions, such as the stage-of-change questions (Tables 10.1 and 10.2) to then direct the individuals to a reduced set of messages and interventions to choose from. Drawing from various works (Alta Planning + Design & Transit Center, 2017; Bamberg, 2013a, 2013b; Pope et al., 2018; Sunio et al., 2018; Waygood et al., 2012), Table 10.3 presents strategies and message framing for each stage.

3.2.4 Level 4: The Right Intervention to the Right Person

The "ultimate" level of integration of a stage-based approach consists of a program that delivers stage-tailored information and (dis)incentives to individuals based on their current stage. As discussed in section 0, this approach ticks all the benefits of using stage-based individual-tailored approach of a behaviour change intervention. It requires greater effort, though the benefits should be greater. The complexity of delivering such individual-tailored interventions might explain why there are only a few transport-related studies in the literature reaching this level of integration (Bamberg, 2013a). Furthermore, to the best of our knowledge, no large-scale program in transportation has ever been implemented at this level of integration.

Table 10.3 Intervention strategies recommended for each stage of change

Stage	Intervention strategy
1.a Predecisional denial	**Raise awareness about the consequences of car use:** a diversity of information is required as people are motivated in different ways; individual negative impacts such as cost, health, stress; negative societal impacts (environment, climate, road violence, etc.); attention should be given to framing (see section 0)
	Make injunctive social norms more salient: Make it clear that a majority of people think that overuse of cars in our cities is problematic. Make the link with societal objectives (e.g., increasing road safety for all) and the current problematic behaviour particularly those that are irrefutable
	Ascription of responsibility: Consequences (for self and others) should be linked to the individual's own action and that they can alleviate those consequences through a new behaviour
	State the inconveniences related to car use: Make people think and say what bothers them when using their private car, and for which daily trip(s) do these disadvantages apply particularly. Those that relate to negative emotions are likely strong motivations to change
1.b Predecisional inhibition	**Make descriptive social norms more salient:** Ideally, this should be done based on real, credible, statistics and emphasize the behaviour to be performed (new mode) and not the behaviour to change (car use). For example, a message crafted based on results from a pre-intervention survey could state that a majority of people in the neighbourhood are either already or planning to walk/cycle/take public transit for some of their trips
	Promote goal formation to change while preventing reactance: This could be done by asking the person to think of a regular trip that they could do on foot, by bike, or by using transit. Asking for a small step, stating that even small steps matter, while recognizing that this can be difficult and asking what barriers they see in using their preferred alternative mode can reduce reactance (Bamberg, 2013a)
	Highlight advantages of modal shift: The objective is to increase positive attitudes and emotions towards change so that the idea to carry out the new behaviour will outweigh the perceived difficulty of trying something new. People might be motivated by different benefits for the same mode. For some, the physical and mental health argument of walking, cycling and even using transit might work well, others might be more convinced by practical, economic or environmental arguments

(continued)

Table 10.3 (continued)

Stage	Intervention strategy
2. Preactional	**Highlight advantages of alternatives:** At this stage, the advantages of the specific changes the person is considering should be highlighted. Again, the advantages should (as far as possible) be linked to the individual's motivations. If not possible, a diversity of information is recommended
	Increase perceived behavioural control: The idea is to lower the perceived barriers to using alternative modes of travel. Provide quick tips on how to use other modes (e.g., where are the bus/bikesharing/carsharing access points, illustrate how to easily pay for mobility services, suggest downloading the *app* to unlock a shared bike, etc.), provide tools to plan itinerary by bicycle or transit, etc. incentives could also work well to increase PBC at this stage, see below for tips
	Highlight social norms or trends that are supportive: If available, provide relevant examples of the desired behaviour among peer social groups (neighbours, coworkers, classmates). Use inspiring images of people from various demographics doing the desired behaviour
	Offer incentives to try possible alternatives: This is the stage where the incentives will be the most efficient. The idea is to increase perceived behavioural control and allow them to test an alternative at a low cost and with minimal risks. Someone is interested in transit? Offer them a free daily, weekly or monthly pass (see Zarabi et al., 2024 for a review). Interested in cycling? Do they have a bike? Does it need a tune-up? Offer a rebate on tune-ups or on a new bike at a local bike shop. Have they ever tried bikesharing? Offer them a free pass or discount code on the app. The same can work for carsharing. Programs such as Équiterre's Vélovolt that allows workers to try out an e-bike for 2–4 weeks for free also produced great results (Chevalier et al., 2024)
	Establish social relationships that will be a critical source of support and information: Providing social support can seem time-consuming but might be critical in providing the confidence, the support, and the motivation to start the new behaviour. Let the individual know that others are also trying, offer direct support (accompany the person on their first transit or bike trip), link them to neighbourhood groups, invite them to discovery walks or rides in their neighbourhood, ask if they have someone they know that is already performing the behaviour and could help them

(continued)

Table 10.3 (continued)

Stage	Intervention strategy
3. Actional	**Provide practical information on alternative mode use:** This is the "what, when and how". Bring the person to think about which mode should be used, when and which trip should be replaced and how is the behaviour going to be operationalized. As described before, this should be specific to the context (geographic *and* individual). Offer information on how to plan their itinerary, how to dress for uncertain weather, how to pay, where are located the nearest bus stops, bikesharing/carsharing stations, protected bike lanes, etc.
	Offer incentives: Incentives could also be very useful here to bring people to try out their selected alternative
	Acknowledge the person's goals (why they want to change): At this stage, they have positively committed to changing. They might even have already tried the new behaviour. Thank them for their commitment, congratulate them for trying out
	Action planning and commitment: Do they have a plan to enact the behaviour? Bring them to commit: when and where are they going to try the new mode, for which trip? Do they have everything needed?
	Contingency management for when problems will occur: Provide help on how to plan if something goes wrong. Ask them what their plan is if the bus is late, if they were planning to cycle and it rains, what to do if they need to hurry back home for an urgent matter. Guidance can be provided with a quick "how to" guide to the most common problems with each mode
	Restate the benefits of the chosen alternative: Get them to increase positive anticipation of the new mode of transport by reminding them of their benefits and why it matters to them. Overall, information should be as temporally close as possible (i.e., the feedback should be given as close to the behaviour as possible)

(continued)

Table 10.3 (continued)

Stage	Intervention strategy
4. Postactional (maintenance)	**Provide positive feedback after the new behaviour has been accomplished:** Positive, non-criticizing and non-controlling feedback help people feel more competent. If it's possible to follow people's progress throughout the intervention (through an app-based intervention for example, see next section), providing information on how far they have come by recognizing the challenge that this represented could strengthen their determination to keep going. For example, one might say: "You've made great progress, how do you feel about the next step? Here's some information that might help you with your goals. Feel free to use it as you see fit. I noticed you have been working hard on this. Is there anything you think would help you improve more?"
	Help them prevent and recover from relapses: Share information on how to persist in the face of adversity: State obstacles experienced and help people set a coping plan that includes detailed solutions for each barrier; Provide resources into which people can keep track of their progression, goals and difficulties (e.g., log book, calendar, apps); Encourage people to visualize their trips, possible challenges, and successes, as well as to practise self-compassion (positive self-talk) in their efforts (Pope et al., 2018)
	Acknowledge their behaviour that contributes positively to society: Thank them for contributing to reducing excessive car use and its consequences on society. Remind them that they are making an important contribution to society's goals (e.g., reduce GHG emission, to transition to a healthier society, or to increase safety and quality of life for all) by reducing car traffic in their neighbourhood
	Help them plan to maintain sustainable mobility behaviours through life course changes: A next step is to provide them with tips and guidance on how to maintain their low car use through important life transitions such as having a new job in a location not as well served by sustainable mobility options, expecting a child, or relocating to a new home. In short, help them think of what they can do to preserve their high-quality and low-car lifestyle, by providing examples for life transitions that might apply to them
	Helping others: Finally, the ultimate step is to help them contribute more broadly to society shifting away from car dependency. This can be done by suggesting that they help people in their social networks to use active travel and public transit. Providing them with "group" incentives such as public transit/bikesharing tickets that would allow them to bring their friends and family to free, thus helping their close ones try out new behaviour. Going one step further is to suggest ways to get involved at the local level to ask for better active travel and transit infrastructure and safer streets. Suggests a list of local organizations that they can volunteer with to further help their community embrace a fair transition away from car dependency

In his first application of the SSBC, Bamberg (2013a) developed a phone-based module with scripts designed for each of the four stages. The intervention had three steps: (1) People were recruited for the behaviour change campaign via invitation letters sent by mail. To participate, they had to mail back a prestamped postcard which contained a stage-diagnosis question and a phone number where they could be contacted. (2) The main intervention happened when they were contacted by phone by the research team who used the stage-tailored messages. (3) At the end of the phone call, participants were asked if they were interested in receiving material and incentive to help them try out the new behaviour. If a phone-based only approach today (in 2024) might not go as far as when Bamberg conducted his study, the idea of a direct contact, recommended in community-based social-marketing approach (McKenzie-Mohr, 2011), should be considered within a larger program to have a significant impact on a small number of people's behaviour change process (Davies, 2012). Such one-on-one conversations could be conducted through a door-to-door campaign or through local "mobility ambassadors".

More recently, the idea of using information and communication technology (ICT) has been suggested and tested as a means to deliver highly tailored behaviour change interventions (e.g., Ahmed et al., 2020; Briant et al., 2023; Bucher et al., 2019; Sunio et al., 2018). All four levels of integration described so far can be implemented within a web-based or smartphone app. For example, both Ahmed et al. (2020) and Sunio et al. (2018) developed smartphone apps and web tools with the potential to deliver such tailored intervention. Their apps were used to conduct detailed pre and post interventions mobility diagnosis (behaviour, stage-of-change, etc.). While their interventions were personalized to each individual by recommending, for example, realistic alternatives for specific trips taken, neither study appeared to take full advantage of their app to deliver stage-tailored messaging to participants. While they report positive results in shifting people to higher stages, this classifies their interventions as a level 3 integration that also includes the first two levels. However, the potential exists to add this next level of tailoring. Sunio et al. (2018) demonstrated that a technology-based intervention, even when non-tailored to specific stages, seems particularly effective in helping people translate intention into action by supporting self-efficacy, confirming previous research in the health domain (Dallery et al., 2015). However, it seems less effective in changing entrenched beliefs and attitudes (about cars and alternatives for example), thus less likely to bring "unconvinced" individuals to form a goal and behavioural intention to reduce their car use (moving from S1.a to higher stages). It thus remains to be tested if a true level 4 integration with stage-tailored intervention delivered through a phone app would be more effective in moving people at all stages to higher stages. But as described by Briant et al. (2023) in a comprehensive report on using Mobility as a Service (Maas) to support behaviour change, an important challenge with such app-based approach is to bring people to download the app and then to come back to it and keep using it. This might be a particular challenge for motorists in lower stages which might not see any benefits in using a multimodal trip planning app. In such cases, more subtle techniques may be required that nudge people gradually towards more desirable behaviours and tie rewards to such shifts in both behaviour and beliefs.

Finally, a simpler and cost-effective way for a large behaviour change program such as Portland's SmartTrips to deliver stage-tailored intervention would be to integrate a quick self-assessment question on the website where people are invited to order incentives. Based on their stage of change, participants would then be redirected to different pages containing information and incentives tailored to their respective stage (Table 10.3) before being able to order their incentive. While relatively low cost to implement, this approach would have the downside of only reaching people who take the step of ordering incentives (about 10% in the Portland program). These are likely to be people in higher stages. An approach worth considering to attract more people from lower stages (S1 and S2) to the program's website would be to include in the mail-in mobility package an invitation to complete a short online survey about their general travel behaviour (that would include stages of change questions) to be eligible to a lottery to win cash prizes. After survey completion, respondent would then be redirected to stage-tailored information and incentive pages.

4 Conclusion

Changing behaviour is a highly complex process and while providing alternatives to the car is essential, it is often insufficient on its own to prompt rapid behaviour change across diverse segments of the population. In this chapter, we argued that personalized behaviour change strategies should be more systematically included in policy packages to further leverage the impact of increasing the quantity and quality of sustainable transportation option while facilitating public acceptance and demand for structural changes away from cars. In a North American context, we see this as an essential component of a holistic approach to a socio-ecological transition in the transport sector away from car dependency. Unfortunately, too few large-scale applications voluntary travel behaviour change programs have been implemented in North America. Their application typically brings about a 5–7% reduction in car use among participants, which if applied yearly will help cities make great strides towards sustainable transport goals. This chapter was structured as a guide to practitioners who wish to develop such behaviour change programs. To enhance the effectiveness of transport behaviour change interventions, we recommend integrating components of the Stage model of self-regulated behaviour change (SSBC). We propose a four-level integration framework: (1) using the self-assessment question of the SSBC to enhance diagnosis of the targeted audience and measure its readiness to reduce car use or switch to alternative modes; (2) reuse the self-assessment question in post-intervention surveys to measure its effect on behaviour, intention and motivation; (3) developed stage-specific strategies and messages to be delivered, relying on either a "menu-based" approach or the most likely stages of the targeted population; (4) fully integrate the model by delivering individual stage-tailored strategies and messages using one-on-one methods (e.g., door-to-door) or a technology-based approach. The latter remains to be tested in large-scale experiments in a North American context to evaluate its efficacy over simpler interventions.

To help achieve the greatest success, practitioners should: (1) carefully adapt the intervention to the structural context by first prioritizing the individuals most likely to change (stage 2 and up) within the neighbourhoods with the best alternative to the car, and thus the least structural barriers; (2) "advertise" structural changes by combining them with simple behaviour change programs to enhance the benefits of new infrastructure and transport options; (3) use the potential of the habit discontinuity hypothesis to target individuals who have recently undergone important life changes; (4) tailor the programs to individuals as much as possible based on geographic context, socio-demographics and, of course, their current stage of change.

Beyond contributing to reducing car use, such program can also contribute to enhancing a sustainable mobility culture that relies on higher social norms and attitudes towards active travel, public transit and shared mobility and bringing people to fully embrace more multimodal and less car-dependent lifestyles. We hypothesize that this should increase support among the population for more pro-transit and pro-active travel including the politically difficult structural changes that often require taking space and funding away from car infrastructure. In doing so, this contributes to a positive feedback loop of increasing demand for sustainable mobility policies that would further reduce car use and so on. In areas offering the best non-automobile accessibility, such program could even go one step further by helping household voluntarily maintain low car ownership or reduce the number of cars they own.

Finally, from a research perspective, better collaboration is needed between practitioners and researchers from various disciplines to both develop theoretically driven interventions and evaluate them through robust methods to enhance our collective understanding of the pathways leading to sustained changes in travel behaviour.

References

Abraham, C. (2008). Beyond stages of change: Multi-determinant continuum models of action readiness and menu-based interventions. *Applied Psychology, 57*(1), 30–41. https://doi.org/10.1111/j.1464-0597.2007.00320.x

Adams, J., & White, M. (2005). Why don't stage-based activity promotion interventions work? *Health Education Research, 20*(2), 237–243. https://doi.org/10.1093/her/cyg105

AGO. (2006). *Evaluation of TravelSmart projects in the ACT, South Australia, Queensland, Victoria and Western Australia: 2001–2005.* https://www.ttsitalia.it/file/Libreria/Worldwide/evaluation-2005%5B1%5D.pdf

Ahmed, S., Adnan, M., Janssens, D., & Wets, G. (2020). A personalized mobility based intervention to promote pro-environmental travel behavior. *Sustainable Cities and Society, 62*, 102397. https://doi.org/10.1016/j.scs.2020.102397

Ajzen, I. (1985). From intentions to actions: A theory of planned behavior. In J. Kuhl & J. Beckmann (Eds.), *Action control: From cognition to behavior* (pp. 11–39). Springer. https://doi.org/10.1007/978-3-642-69746-3_2

Ajzen, I. (1991). The theory of planned behavior. *Organizational Behavior and Human Decision Processes, 50*(2), 179–211. https://doi.org/10.1016/0749-5978(91)90020-T

Alta Planning + Design, & Transit Center. (2017). *New tools for shaping transportation behavior.* https://altago.com/wp-content/uploads/New-Tools-for-Shaping-Transportation-Behavior.pdf

Ampt, E., Wundke, J., & Stopher, P. (2006). Households on the move: New approach to voluntary travel behavior change. *Transportation Research Record, 1985*(1), 98–105. https://doi.org/10.1177/0361198106198500111

Anable, J., & Wright, S. D. (2013). *Golden questions and social marketing guidance report.* [Report]. SEGMENT. https://aura.abdn.ac.uk/handle/2164/3226

Arnott, B., Rehackova, L., Errington, L., Sniehotta, F. F., Roberts, J., & Araujo-Soares, V. (2014). Efficacy of behavioural interventions for transport behaviour change: Systematic review, meta-analysis and intervention coding. *International Journal of Behavioral Nutrition and Physical Activity, 11*(1), 133. https://doi.org/10.1186/s12966-014-0133-9

Bamberg, S. (2006). Is a residential relocation a good opportunity to change people's travel behavior? Results from a theory-driven intervention study. *Environment and Behavior, 38*(6), 820–840. https://doi.org/10.1177/0013916505285091

Bamberg, S. (2013a). Applying the stage model of self-regulated behavioral change in a car use reduction intervention. *Journal of Environmental Psychology, 33*, 68–75. https://doi.org/10.1016/j.jenvp.2012.10.001

Bamberg, S. (2013b). Changing environmentally harmful behaviors: A stage model of self-regulated behavioral change. *Journal of Environmental Psychology, 34*, 151–159. https://doi.org/10.1016/j.jenvp.2013.01.002

Bamberg, S., & Rees, J. (2017). The impact of voluntary travel behavior change measures—A meta-analytical comparison of quasi-experimental and experimental evidence. *Transportation Research Part A: Policy and Practice, 100*, 16–26. https://doi.org/10.1016/j.tra.2017.04.004

Bamberg, S., & Schulte, M. (2018). Processes of change. In *Environmental psychology* (pp. 307–318). Wiley. https://doi.org/10.1002/9781119241072.ch30

Bamberg, S., Fujii, S., Friman, M., & Gärling, T. (2011). Behaviour theory and soft transport policy measures. *Transport Policy, 18*(1), 228–235. https://doi.org/10.1016/j.tranpol.2010.08.006

Banister, D. (2008). The sustainable mobility paradigm. *Transport Policy, 15*(2), 73–80. https://doi.org/10.1016/j.tranpol.2007.10.005

Boddy, S., & Kassirer, J. (2013). *Portland's Smart Trips Welcome Program.* Tools of change. https://toolsofchange.com/en/case-studies/detail/658

Boisjoly, G. (2023). Planning for people through the lens of accessibility. In *Handbook on transport and land use* (pp. 206–230). Edward Elgar Publishing. https://www.elgaronline.com/edcollchap/book/9781800370258/book-part-9781800370258-19.xml

Briant, L., Marrel, J., Chevereau, L., & Cerema. Centre d'études et d'expertise sur les risques, l'environnement. (2023). *MaaS et Changement de comportement. Le numérique peut-il influer sur les pratiques de mobilité?* Cerema. Bron. https://doc.cerema.fr/Default/doc/SYRACUSE/596816/maas-et-changement-de-comportement-le-numerique-peut-il-influer-sur-les-pratiques-de-mobilite

Bucher, D., Mangili, F., Cellina, F., Bonesana, C., Jonietz, D., & Raubal, M. (2019). From location tracking to personalized eco-feedback: A framework for geographic information collection, processing and visualization to promote sustainable mobility behaviors. *Travel Behaviour and Society, 14*, 43–56. https://doi.org/10.1016/j.tbs.2018.09.005

Chatterjee, K. (2009). A comparative evaluation of large-scale personal travel planning projects in England. *Transport Policy, 16*(6), 293–305. https://doi.org/10.1016/j.tranpol.2009.10.004

Chevalier, H., Bergeron, M., & Tremblay, V. (2024). *Portrait et potentiel du vélo à assistance électrique au Québec: Résultats et recommandations issus de l'expérience Vélovolt* (p. 126). Équiterre. https://www.equiterre.org/fr/ressources/311-velovolt-rapport-final-portrait-et-potentiel-du-velo-a-assistance-electrique-au-quebec?x-craft-preview=kNnhWmQ6w8&token=lXlmT1vnfE79TJOylvJAIcIYkgGQNa8B&utm_campaign=311_V%C3%A9lovolt&utm_medium=email&_hsmi=307055280&utm_content=307055280&utm_source=hs_email

Chouinard, M.-P., Dupont, B., Dupont-Rachiele, C., & Laurin, V. (2021). *Chantier auto-solo—Synthèse de connaissance*. Jalon Montréal. https://praxis.encommun.io/media/notes/note_9568/guide_auto-solo_septembre2021-compressed6689.pdf

Dallery, J., Jarvis, B., Marsch, L., & Xie, H. (2015). Mechanisms of change associated with technology-based interventions for substance use. *Drug and Alcohol Dependence, 150*, 14–23. https://doi.org/10.1016/j.drugalcdep.2015.02.036

Davies, N. (2012). What are the ingredients of successful travel behavioural change campaigns? *Transport Policy, 24*, 19–29. https://doi.org/10.1016/j.tranpol.2012.06.017

Daziano, R., Waygood, E. O. D., Patterson, Z., Feinberg, M., & Wang, B. (2021). Reframing greenhouse gas emissions information presentation on the Environmental Protection Agency's new-vehicle labels to increase willingness to pay. *Journal of Cleaner Production, 279*, 123669. https://doi.org/10.1016/j.jclepro.2020.123669

Dill, J., & Mohr, C. (2010). Long-term evaluation of individualized marketing programs for travel demand management. In *Urban studies and planning faculty publications and presentations*. https://doi.org/10.15760/trec.132

Edensor, T. (2004). Automobility and national identity: representation, geography and driving practice. *Theory, Culture & Society, 21*(4–5), 101–120. https://doi.org/10.1177/0263276404046063

Ewing, R., & Cervero, R. (2010). Travel and the built environment: A meta-analysis. *Journal of the American Planning Association, 76*(3), 265–294. https://doi.org/10.1080/01944361003766766

Friman, M., Huck, J., & Olsson, L. E. (2017). Transtheoretical model of change during travel behavior interventions: An integrative review. *International Journal of Environmental Research and Public Health, 14*(6), 6. https://doi.org/10.3390/ijerph14060581

Fujii, S., Gärling, T., & Kitamura, R. (2001). Changes in drivers' perceptions and use of public transport during a freeway closure: Effects of temporary structural change on cooperation in a real-life social dilemma. *Environment and Behavior, 33*(6), 796–808. https://doi.org/10.1177/00139160121973241

Fujii, S., Bamberg, S., Friman, M., & Gärling, T. (2009). Are effects of travel feedback programs correctly assessed? *Transportmetrica, 5*(1), 43–57. https://doi.org/10.1080/18128600802591277

Gallez, C., & Dupuy, G. (2018). La dépendance automobile. Retour sur la genèse du concept et ses enjeux politiques. *Flux, 111–112*(1–2), 104–110. https://doi.org/10.3917/flux1.111.0104

Gärling, T., & Axhausen, K. W. (2003). Introduction: Habitual travel choice. *Transportation, 30*(1), 1–11. https://doi.org/10.1023/A:1021230223001

Gärling, T., & Steg, L. (2007). *Threats from car traffic to the quality of urban life: Problems, causes and solutions*. Emerald Group Publishing Limited.

Glazener, A., Sanchez, K., Ramani, T., Zietsman, J., Nieuwenhuijsen, M. J., Mindell, J. S., Fox, M., & Khreis, H. (2021). Fourteen pathways between urban transportation and health: A conceptual model and literature review. *Journal of Transport & Health, 21*, 101070. https://doi.org/10.1016/j.jth.2021.101070

Gössling, S. (2020). Why cities need to take road space from cars—And how this could be done. *Journal of Urban Design, 25*(4), 443–448. https://doi.org/10.1080/13574809.2020.1727318

Gousse-Lessard, A.-S., & Laviolette, J. (2022). Transformation des villes et mobilité durable: Regard sur les déterminants psychosociaux de l'attachement à l'auto solo. *VertigO—la revue électronique en sciences de l'environnement, Hors-série 36*, Article Hors-série 36. https://doi.org/10.4000/vertigo.37073

Gouvernement du Québec. (2022). *Inventaire québécois des émissions de gaz à effet de serre*. Direction des inventaires et de la gestion des halocarbures du ministère de l'Environnement, de la Lutte contre les changements climatiques, de la Faune et des Parcs (MELCCFP). https://www.environnement.gouv.qc.ca/changements/ges/index.htm

Gouvernement du Québec. (2023, November 8). *Plan de mise en œuvre*. Gouvernement du Québec. https://www.quebec.ca/gouvernement/politiques-orientations/plan-economie-verte/plan-mise-en-oeuvre

Graham-Rowe, E., Skippon, S., Gardner, B., & Abraham, C. (2011). Can we reduce car use and, if so, how? A review of available evidence. *Transportation Research Part A: Policy and Practice, 45*(5), 401–418. https://doi.org/10.1016/j.tra.2011.02.001

Handy, S. (2017). Thoughts on the meaning of Mark Stevens's meta-analysis. *Journal of the American Planning Association, 83*(1), 26–28. https://doi.org/10.1080/01944363.2016.1246379

Haustein, S., & Hunecke, M. (2013). Identifying target groups for environmentally sustainable transport: Assessment of different segmentation approaches. *Current Opinion in Environmental Sustainability, 5*(2), 197–204. https://doi.org/10.1016/j.cosust.2013.04.009

Intelligent Energy Europe. (2015). *The SEGMENT toolkit: Resources for creating segmented marketing campaigns for sustainable transport*. https://civitas.eu/sites/default/files/segment_deliverable_7-8.3_social_marketing_toolkit.pdf

Javaid, A., Creutzig, F., & Bamberg, S. (2020). Determinants of low-carbon transport mode adoption: Systematic review of reviews. *Environmental Research Letters, 15*(10), 103002. https://doi.org/10.1088/1748-9326/aba032

Keller, E., Eisen, C., & Hanss, D. (2019). Lessons learned from applications of the stage model of self-regulated behavioral change: A review. *Frontiers in Psychology, 10*. https://www.frontiersin.org/journals/psychology/articles/10.3389/fpsyg.2019.01091

Kirschner, F., & Lanzendorf, M. (2020). Support for innovative on-street parking policies: Empirical evidence from an urban neighborhood. *Journal of Transport Geography, 85*, 102726. https://doi.org/10.1016/j.jtrangeo.2020.102726

Laviolette, J. (2020). *Mobilité et psychologie: Comprendre et agir pour soutenir les changements de comportement*. David Suzuki Foundation. https://fr.davidsuzuki.org/publication-scientifique/mobilite-et-psychologie-comprendre-et-agir-pour-soutenir-les-changements-de-comportement/

Laviolette, J., Morency, C., & Waygood, E. O. D. (2020). Persistance de l'automobilité? Analyse en trois perspectives. *Flux, 119–120*(1–2), 142–172. https://doi.org/10.3917/flux1.119.0142

Ma, L., Mulley, C., & Liu, W. (2017). Social marketing and the built environment: What matters for travel behaviour change? *Transportation, 44*(5), 1147–1167. https://doi.org/10.1007/s11116-016-9698-2

Mattioli, G. (2016). Transport needs in a climate-constrained world. A novel framework to reconcile social and environmental sustainability in transport. *Energy Research & Social Science, 18*, 118–128. https://doi.org/10.1016/j.erss.2016.03.025

Mattioli, G., Roberts, C., Steinberger, J. K., & Brown, A. (2020). The political economy of car dependence: A systems of provision approach. *Energy Research & Social Science, 66*, 101486. https://doi.org/10.1016/j.erss.2020.101486

McKenzie-Mohr, D. (2011). *Fostering sustainable behavior: An introduction to community-based social marketing* (3rd ed.). New Society Publishers.

Miner, P., Smith, B. M., Jani, A., McNeill, G., & Gathorne-Hardy, A. (2024). Car harm: A global review of automobility's harm to people and the environment. *Journal of Transport Geography, 115*, 103817. https://doi.org/10.1016/j.jtrangeo.2024.103817

Mishra, N. B., Pani, A., Mohapatra, S. S., & Sahu, P. K. (2023). Decoding private or commercial vehicle ownership decisions for low-carbon mobility transitions: A systematic review of the literature. *Transportation Research Record, 2678*, 87. https://doi.org/10.1177/03611981231194346

Möser, G., & Bamberg, S. (2008). The effectiveness of soft transport policy measures: A critical assessment and meta-analysis of empirical evidence. *Journal of Environmental Psychology, 28*(1), 10–26. https://doi.org/10.1016/j.jenvp.2007.09.001

Nolan, J. M., Schultz, P. W., Cialdini, R. B., Goldstein, N. J., & Griskevicius, V. (2008). Normative social influence is underdetected. *Personality and Social Psychology Bulletin, 34*(7), 913–923. https://doi.org/10.1177/0146167208316691

Olsson, L. E., Huck, J., & Friman, M. (2018). Intention for car use reduction: Applying a stage-based model. *International Journal of Environmental Research and Public Health, 15*(2), 2. https://doi.org/10.3390/ijerph15020216

Petrunoff, N., Rissel, C., & Wen, L. M. (2016). The effect of active travel interventions conducted in work settings on driving to work: A systematic review. *Journal of Transport & Health, 3*(1), 61–76. https://doi.org/10.1016/j.jth.2015.12.001

Piras, F., Sottile, E., Tuveri, G., & Meloni, I. (2022). Does the joint implementation of hard and soft transportation policies lead to travel behavior change? An experimental analysis. *Research in Transportation Economics, 95*, 101233. https://doi.org/10.1016/j.retrec.2022.101233

Pope, J. P., Pelletier, L., & Guertin, C. (2018). Starting off on the best foot: A review of message framing and message tailoring, and recommendations for the comprehensive messaging strategy for sustained behavior change. *Health Communication, 33*(9), 1068–1077. https://doi.org/10.1080/10410236.2017.1331305

Prochaska, J. O., & DiClemente, C. C. (1982). Transtheoretical therapy: Toward a more integrative model of change. *Psychotherapy: Theory, Research & Practice, 19*(3), 276–288. https://doi.org/10.1037/h0088437

Prochaska, J. O., & Velicer, W. F. (1997). The transtheoretical model of health behavior change. *American Journal of Health Promotion: AJHP, 12*(1), 38–48. https://doi.org/10.4278/0890-1171-12.1.38

Prochaska, J. O., Diclemente, C. C., & Norcross, J. C. (1997). *In search of how people change: Applications to addictive behaviors* (p. 696). American Psychological Association. https://doi.org/10.1037/10248-026

Ralph, K. M., & Brown, A. E. (2019). The role of habit and residential location in travel behavior change programs, a field experiment. *Transportation, 46*(3), 719–734. https://doi.org/10.1007/s11116-017-9842-7

Scheepers, C. E., Wendel-Vos, G. C. W., den Broeder, J. M., van Kempen, E. E. M. M., van Wesemael, P. J. V., & Schuit, A. J. (2014). Shifting from car to active transport: A systematic review of the effectiveness of interventions. *Transportation Research Part A: Policy and Practice, 70*, 264–280. https://doi.org/10.1016/j.tra.2014.10.015

Schwanen, T., & Lucas, K. (2011). Chapter 1—Understanding auto motives. In *Auto motives: Understanding car use behaviours*. Emerald Publishing Limited. http://ebookcentral.proquest.com/lib/polymtl-ebooks/detail.action?docID=713463

Schwartz, S. H. (1977). Normative influences on altruism. In L. Berkowitz (Ed.), *Advances in experimental social psychology* (Vol. 10, pp. 221–279). Academic Press. https://doi.org/10.1016/S0065-2601(08)60358-5

Semenescu, A., Gavreliuc, A., & Sârbescu, P. (2020). 30 Years of soft interventions to reduce car use—A systematic review and meta-analysis. *Transportation Research Part D: Transport and Environment, 85*, 102397. https://doi.org/10.1016/j.trd.2020.102397

Sheller, M. (2004). Automotive emotions: Feeling the car. *Theory, Culture & Society, 21*(4–5), 221–242. https://doi.org/10.1177/0263276404046068

Sheller, M., & Urry, J. (2000). The city and the car. *International Journal of Urban and Regional Research, 24*(4), 737–757. https://doi.org/10.1111/1468-2427.00276

Sloman, L., Cairns, S., Newson, C., Anable, J., Pridmore, A., & Goodwin, P. (2010). *The effects of smarter choice programmes in the sustainable travel towns: Summary report* (GRE) [Report]. Department for Transport. https://webarchive.nationalarchives.gov.uk/ukgwa/20111005180138oe_/http://www.dft.gov.uk/publications/blog/publication/the-effects-of-smarter-choice-programmes-in-the-sustainable-travel-towns-summary-report/

Soza-Parra, J., & Cats, O. (2024). The role of personal motives in determining car ownership and use: A literature review. *Transport Reviews, 44*(3), 591–611. https://doi.org/10.1080/01441647.2023.2278445

Steg, L. (2005). Car use: Lust and must. Instrumental, symbolic and affective motives for car use. *Transportation Research Part A: Policy and Practice, 39*(2), 147–162. https://doi.org/10.1016/j.tra.2004.07.001

Sunio, V., Schmöcker, J.-D., & Kim, J. (2018). Understanding the stages and pathways of travel behavior change induced by technology-based intervention among university students. *Transportation Research Part F: Traffic Psychology and Behaviour, 59*, 98–114. https://doi.org/10.1016/j.trf.2018.08.017

Sutton, S. (2000). A critical review of the transtheoretical model applied to smoking cessation. In *Understanding and changing health behaviour*. Psychology Press.

Thøgersen, J. (2012). The importance of timing for breaking commuters' car driving habits. In A. Warde & D. Southerton (Eds.), *The habits of consumption* (Vol. 12, pp. 130–140). Helsinki Collegium for Advanced Studies.

Verplanken, B., & Roy, D. (2016). Empowering interventions to promote sustainable lifestyles: Testing the habit discontinuity hypothesis in a field experiment. *Journal of Environmental Psychology, 45*, 127–134. https://doi.org/10.1016/j.jenvp.2015.11.008

Walker, I., Thomas, G. O., & Verplanken, B. (2015). Old habits die hard: Travel habit formation and decay during an office relocation. *Environment and Behavior, 47*(10), 1089–1106. https://doi.org/10.1177/0013916514549619

Wang, B., Waygood, E. O. D., Ji, X., Naseri, H., Loiselle, A. L., Daziano, R. A., Patterson, Z., & Feinberg, M. (2023). How to effectively communicate about greenhouse gas emissions with different populations. *Environmental Science & Policy, 147*, 29–43. https://doi.org/10.1016/j.envsci.2023.05.015

Waygood, E. O. D., & Avineri, E. (2018). CO_2 valence framing: Is it really any different from just giving the amounts? *Transportation Research Part D: Transport and Environment, 63*, 718–732. https://doi.org/10.1016/j.trd.2018.07.011

Waygood, E. O. D., Avineri, E., & Lyons, G. (2012). Chapter 12—The role of information in reducing the impacts of climate change for transport applications. In T. Ryley & L. Chapman (Eds.), *Transport and climate change* (Vol. 2, pp. 313–340). Emerald Group Publishing. https://doi.org/10.1108/S2044-9941(2012)0000002015

Zarabi, Z., Waygood, E. O. D., Olsson, L., Friman, M., & Gousse-Lessard, A.-S. (2024). Enhancing public transport use: The influence of soft pull interventions. *Transport Policy*. https://doi.org/10.1016/j.tranpol.2024.05.005

Jérôme Laviolette is a postdoctoral researcher at McGill University (Montréal, Canada) in the geography department. He oversees a research project on behaviour change towards cycling. With his collaborators, Pr. Kevin Manaugh (McGill) and Owen Waygood (Polytechnique Montréal), this project in collaboration with the City of Montréal, Vélo Québec and Équiterre aims to identify the impact of infrastructure and soft behaviour change interventions on cycling stage-of-change among various population groups. The project has also the objective of understanding how to frame cycling infrastructure projects as to minimize opposition and increase acceptability. Jérôme is also a prominent speaker in the field of sustainable mobility in Québec, giving talks at numerous civil society events on the topic of car dependency and behaviour change in transport. He has completed a Ph.D. in transportation engineering at Polytechnique Montréal in 2023 focusing on understanding the various factors (structural, psychological, societal) influencing car ownership and car dependency.

Owen Waygood is a professor of sustainable transport at Polytechnique Montréal. He is interested in how to help shift society towards sustainable transport behaviours. As such, he studies how the built environment affects not just how we travel, but its impact on our lives through social, environmental, and economic impacts. He has also researched the role of information, in particular carbon dioxide, in transportation applications. Owen has published research on children's travel (Canadian, British, Japanese and Swedish), life-cycle stages, cohort effects, information use for more environmentally friendly travel, and psychological impacts on the interpretation of CO_2

information. He was a professor at Laval University from 2012 to 2018. Before that he held a research position at the Centre for Transport & Society in the United Kingdom from 2009 to 2012. He completed a PhD in Transportation Behaviour and Urban Management from Kyoto University in 2009, a Masters in Biomimicry from the University of Toronto in 2005, and undergraduate degrees in Mechanical Engineering and Computer Science from the University of Saskatchewan in 2001.

Anne-Sophie Gousse-Lessard is a professor in the Department of Social and Public Communication at the Université du Québec à Montréal (UQAM) as well as at the Institut des sciences de l'environnement (ISE). Her field of specialization lies at the intersection of psychology and environmental communication. She is the co-director of the Groupe interdisciplinaire de recherche sur les écoémotions et l'engagement citoyen (GIREEC) since its founding in 2020. She is also associated with several groups and research centres, including the InterSectoral Flood Network of Québec (RIISQ), the Centre de recherche en éducation et formation relatives à l'environnement et à l'écocitoyenneté (Centr'ERE) and the Institut santé et société (ISS). Her research focuses on environmental motivation and adaptation, on the psychosocial impacts of climate change (including eco-anxiety), and on sustainable mobility issues. She is particularly interested in collective engagement and the transformation of lifestyles in a socio-ecological transition perspective. In addition, she is involved in educational and participatory initiatives aimed at strengthening the resilience and capacity for action of the population, especially young people, in the face of environmental challenges.

Chapter 11
Fostering Sustainability Through Digital Evolution: Evaluating Industry 5.0 Preparedness in Quebec's Regional SMEs

Stéfanie Vallée, Myriam Ertz, Chourouk Ouerghemmi, and Antoine Périn

Abstract The transformative journey toward Industry 4.0 (I4.0) has revolutionized business operations, requiring firms to adapt to digitalization and environmentally-conscious practices (often termed "green digital," "sustainable digital," "smart green," or "responsible digital"). This adaptation may be challenging, especially for small and medium-sized enterprises (SMEs) that typically lack larger firms' adaptive resources and capacities. This chapter delves into the dynamic intersection of sustainability and digital transformation by focusing on the specific case of SMEs to propose a preliminary framework for evaluating SMEs' Industry 5.0 (I5.0) maturity levels (digital), which has a broader scope by including green and social practices (sustainable) with digitalization. The chapter first provides some background by highlighting the push for organizations to integrate sustainable practices into their digital transformation endeavors seamlessly. In a world grappling with environmental and social challenges, aligning technological advancements with responsibility is increasingly relevant for SMEs. The *schwerpunkt* of the chapter is an extensive (but non-systematic) literature review of the digital maturity level assessment domain, coupled with insights from extant research in the sustainable digital (5.0) assessment area. The objective of the review is to construct a preliminary framework for evaluating SMEs' digital preparedness while concurrently measuring their commitment to sustainable practices. This framework incorporates critical parameters such as resource efficiency, circular economy principles, and the incorporation of renewable energy sources in digital control operations. The chapter further sheds light on SMEs' challenges in achieving a coherent balance between digitalization and sustainability. These challenges span technological, organizational, and regulatory dimensions, emphasizing the multifaceted nature of the transition. Notably, the chapter identifies these challenges and proposes practical strategies and best practices to overcome them.

S. Vallée · M. Ertz (✉) · C. Ouerghemmi · A. Périn
University of Quebec at Chicoutimi, Chicoutimi, QC, Canada
e-mail: stefanie.vallee1@uqac.ca; Myriam_Ertz@uqac.ca; couerghemm@etu.uqac.ca; aperin2@etu.uqac.ca

M. Cheriet et al. (eds.), *Accelerating the Socio-Ecological Transition*,
https://doi.org/10.1007/978-3-031-82896-6_11

Keywords SMEs · Digitalization · Sustainability · Digital maturity level · Sustainability framework

1 Introduction

Industry 4.0 (I4.0) has initiated a profound transformation in how small- and medium-sized enterprises (SMEs) create value within the socio-economic landscape (Schwab, 2016; Blanchet, 2016; Berger-Douce, 2014). The technological, organizational, and social innovations at the core of I4.0 target enhanced efficiency and improved innovation for SMEs while posing mounting challenges concerning environmental and social impacts. Meanwhile, in addition to I4.0, the circular economy—meant to replace the linear one (Mondejar et al., 2021)—represents another crucial industrial paradigm shift that has taken place over recent years (Antikainen et al., 2018; Suárez-Eiroa et al., 2021; Rosa et al., 2020). While different, they share common ground since sustainability and digitalization are both transformative in nature and are crucial to harmonizing economic advancement (Nosratabadi et al., 2023). Consequently, political, technological, and economic shifts compel manufacturing firms to realign their strategic focus toward digitized and green (or sustainable) processes (Münnich et al., 2022).

Despite their transformative resemblance, the capacity to handle both the sustainable and digital transitions requires firms to become "ambidextrous" (Nyagadza, 2022). More specifically, The sustainable digital transformation is initially triggered by digital disruption, which entails fundamental structural changes, novel value creation opportunities, and double-edged implications for sustainability: (1) on the one hand, digitalization enables closer monitoring and acting upon sustainability-related factors (e.g., Artificial intelligence improves predictive and prescriptive capacities regarding firm's equipment lifetime (Ertz et al., 2022)); (2) on the other, digitalization poses mounting challenges regarding the responsible and ethical management of data (e.g., privacy issues) and equipment (e.g., electronic and electric waste arising from increased usage of electronic and electric devices) (Ertz et al., 2024).

Integrating maturity-based assessment merged with sustainability assessment seems paramount to navigate these pressing challenges and capitalize on the opportunities of I4.0 (Guillaume et al., 2022). This integration was dubbed "green computing" by the U.S. Environmental Protection Agency (upon launching of the Energy Star labeling program) (Barnett et al., 2018), "ecological information and communication technologies (ICT)" or "green ICT" (Klimova et al., 2016), also fits squarely with the next level of industrial transformation dubbed Industry 5.0 (I5.0). Such an integration provides SMEs with a comprehensive view of their readiness for I4.0 and sustainability, enabling them to align their digital transformation efforts with sustainability goals (Ghobakhloo et al., 2021). The pivotal role of SMEs in driving economic growth and innovation suggests that they are a crucial component in shifting toward I5.0, motivating the need for a framework evaluating their maturity levels in coupling digital transformation (e.g., technology adoption, organizational readiness, skills development) and sustainability practices (e.g., environmental impact, social responsibility).

In light of these challenges, the overall objective of the chapter is to develop a preliminary integrated framework that intersects the I4.0 technological readiness self-assessment with sustainability practices self-assessment to gauge SMEs' I5.0 preparedness. The specific objectives of the chapter are threefold:

1. Eliciting the digital implications of the shift from I4.0 to Industry 5.0 (I5.0).
2. Aligning technological advancement with sustainability.
3. Introducing an integrated digital maturity and sustainability assessment framework.

The chapter analyzes the sustainable digital transformation of Quebecer SMEs in the Saguenay-Lac-Saint-Jean region, also known as "Region 02" from the Quebec government region's classification system. The study relies on that territorial focus as a case study for the model, which has possible implications for broader usage beyond regional and possibly provincial boundaries.

The chapter includes a literature review to conceptualize the key concepts under study and examines the shift from I4.0 to I5.0, specifically emphasizing sustainability. As such, the chapter proposes an integrated framework for assessing SMEs' readiness for I5.0, outlines the research methodology, presents the results through the digital sustainable maturity model, discusses theoretical and managerial implications, and suggests future research directions.

2 Literature Review

2.1 Conceptualization

Drawing on extant research, this sub-section constructs a comprehensive framework for evaluating SMEs' digital preparedness while concurrently measuring their commitment to sustainable practices. This framework incorporates critical parameters such as smart manufacturing, which stands for resource efficiency, circular economy principles, and the incorporation of sustainable ICT in digital control operations.

Although the concepts of green ICT, ecological ICT, green digital, or smart green tend to reduce the "sustainability" concept to its "environmental" dimension, the discussion on the role of digital technologies in advancing sustainability also includes the social dimension above and beyond the economic one. Scholars have examined this aspect through sociological lenses, leading to discussions on how digital technologies can potentially reduce inequality (Dale & Kyle, 2016). Others have also proposed a critical evaluation of the social impact of ICT (e.g., O'Donnell & Henriksen, 2002). The literature generally indicates a significant link between digitalization and human well-being (Torres & Augusto, 2020), with studies suggesting that these technologies can improve quality of life and reduce poverty, although sustainable entrepreneurship remains a scattered topic across the literature (Gino & Staats, 2012; Mora et al., 2021).

Evidence and examples support the fact that introducing I4.0 is a significant shift in the industrial world (Ansari et al., 2018). Notably, lean production was a steppingstone to smart manufacturing (Ansari et al., 2018), highlighting its impact on sustainable competitiveness and advanced technology adoption (Al-Swidi et al., 2023). For example, Rauch and Cochran (2021) show how additive manufacturing and artificial intelligence illustrate the accessible strategies for SMEs to engage with I4.0, thus demonstrating practical approaches to digital transformation with moderate effort. The development of maturity level-based assessment tools for I4.0 in SMEs grounded in a thorough literature review and the creation of models to evaluate I4.0 implementation, providing a structured methodology for sustainability analysis is in high demand, and some of such models have already been developed (e.g., Nagano, 2019; Nosratabadi et al., 2023; Rosa et al., 2020; Brozzi et al. 2020).

Table 11.1 provides an overview of the main themes and sub-topics addressed in the literature in the context of digital transformation and sustainability (of SMEs) and the main findings identified. The "digital transformation" theme is composed of two important sub-themes: (1) one pertaining to the understanding of SMEs' digital transformation by integrating advanced technologies (e.g., cyber-physical systems, IoT, Cloud Computing); and (2) the tools for assessing I4.0 readiness 4.0 among SMEs. The second theme, "opportunities and challenges of digitalization for sustainability," relates to SMEs' use of digitalization to track their efforts with sustainability, focusing on green relationships, green operations, and sustainable management systems. The third "sustainable impact of digitalization" theme relates to strategic planning and refers to the environmental and, more broadly, the sustainable impacts of digitalization. The last theme concerns "digital maturity and sustainability assessment," which merges SMEs' digitalization with the sustainable transition to fast-track and smooth the transition from I4.0 to I5.0.

In addition, some sub-themes tend to explore the same aspects of digital transformation and sustainability. For example, the sub-themes "green operations" and "sustainability assessment" agree on the sustainability goals with a focus on the environmental impact aspect of the digital transition. This sub-theme encompasses, among others, resource optimization, waste reduction, and carbon emission assessment. In summary, these two sub-themes analyze the contribution of digital transformation to the achievement of sustainability goals. Last but not least, we believe there are additional sub-topics to be required, such as "management" and "aligning digitalization with the sustainable transition." Indeed, the managerial challenges identified in the first sub-theme contribute to deepening the discussion on digitalization opportunities for sustainability through human adoption of both transitions within SMEs.

Table 11.1 Summary of the themes on digital transformation and sustainability of SMEs

Theme	Sub-themes	Key findings	Relevant studies
Digital transformation	I4.0 for SMEs	I4.0 represents integrating cyber-physical systems, IoT, and Cloud Computing into manufacturing processes for SMEs	Schwab (2016), Piccarozzi (2018)
	SME I4.0 Maturity Assessment	Concrete tool enabling SMEs to position themselves in terms of their technological maturity 4.0	Schumacher et al. (2016), Mittal et al. (2018), Basl (2018), Kolla et al. (2019), Kotler et al. (2019), Priyono et al. (2020), Hizam-Hanafiah et al. (2020), Hoyer (2020), Ma (2023), Rauch et al. (2020)
Opportunities and challenges of digitalization for sustainability	Green Relationships	Strong collaborative relationships among supply chain partners facilitated by digital platforms can promote sustainability initiatives, such as joint product development, shared transportation, and waste reduction programs	de Sousa Jabbour et al. (2018), Lerman (2022)
	Green Operations	Digital transformation can enhance green operations within supply chains by optimizing resource utilization, reducing waste, and improving energy efficiency through real-time monitoring and data-driven insights	Rauch and Cochran. (2021), Lai et al. (2023), Nureen et al. (2023)
	Sustainable Management Systems	Due to lower resources, technological expertise, and organizational capabilities, SMEs face unique challenges in adopting digital transformation initiatives to improve human life	Brozzi et al. (2018), Blanchet (2016), Berger-Douce (2014), Gino and Staats (2012), Mora et al. (2021)

(continued)

Table 11.1 (continued)

Theme	Sub-themes	Key findings	Relevant studies
Sustainable impact of digitalization	Environmental Impact	Assessing the environmental impact of digital transformation initiatives, including reductions in carbon emissions, energy consumption, and waste generation, contributes to overall sustainability goals within supply chain operations	Antikainen et al. (2018), Korhnen (2018), Ertz et al. (2022), Mondejar (2021), Guillaume et al. (2022)
	Sustainability Assessment	Digital transformation initiatives can be assessed for their environmental impact, including reductions in carbon emissions, energy consumption, and waste generation, contributing to overall sustainability goals within supply chain operations	Khan and Porras (2018), de Sousa Jabbour et al. (2018), Lerman (2022)
Digital maturity assessment and sustainability assessment	Aligning digitalization with the sustainable transition	Aligning both digital maturity and sustainability assessments will support SMEs' technological readiness for I4.0 while identifying opportunities for promoting circular economy principles and sustainable management practices regardless of their activity sector	Müller and Pfleger (2014), Nagano (2019), Nosratabadi et al. (2023), Rosa et al. (2020), Brozzi et al. (2020)

Finally, this research is referring to expressions that seem similar but have some important differentiations in the scientific literature. We are referring to digitalization and environmentally-conscious practices often termed "green digital," "sustainable digital," "smart green," or "responsible digital" and must clarify the roots that apply to our research. We are classifying them according to our research perspective, which must serve regional SMEs to pursue their digital transformation (DT) while embedding sustainable management and eco-friendly practices in their

DT strategy. Therefore, green digital cannot be seen as a way to promote eco-friendly practices across industries (Hu et al., 2022; Xi & Wang, 2024; Bai et al., 2020) in our opinion. Considering our research has an impact on local businesses, it is more relevant to their physical and socio-ecological ecosystem to address "green digital" as the integration of digital technologies to support environmental sustainability and reduce ecological footprints (Yin, 2019; Gholami et al., 2023; Suh & Cho, 2022) and as a way to create sustainable business models and operations (Radu, 2020; Brennen & Kreiss, 2016; Zhu et al., 2022). Green digital as such can be synonymous of "sustainable Digital" respectively from Seele and Lock (2017), Kuntsman and Rattle (2019), and Ullah et al. (2023) in a sense that it implies strategic planning. Indeed, these authors mention that sustainable digital requires the strategic implementation of digital tools to support long-term ecological, social, and economic sustainability goals. The long-term component is also an added value to that definition, relating to the concept of "transformation" which has long-lasting and often non-reversable effects. Sustainable digital can also be considered in our research as the use of digital solutions to address sustainability challenges while minimizing the negative environmental impacts of technology itself (George et al., 2020; Lopes et al., 2022; Beier et al., 2018). Smart Green is an expression very similar to Sustainable digital and can be considered as synonymous, as referred to by authors such as Petersen et al. (2023), Lytras et al., (2019), and Vinuesa et al. (2020): it refers to the combination of intelligent technologies and environmentally friendly practices to optimize resource use and reduce environmental impact. The definitions respectively from Taghizadeh-Hesary and Hyun (2022), Kshetri (2021), and Pappas et al. (2023) are rather pertaining to producers of technologies which are less relevant to our research and therefore won't be retained. Lastly, the term "Responsible Digital" refers to the balance achieved when introducing technological advancements without compromising human rights, privacy, or the ecological environment (Buhmann et al., 2020; Linkov et al., 2021; Stahl et al., 2017). Alternative definitions of this expression are not relevant to our research, as they primarily address societal impacts beyond the scope of local SMEs. However, it could be pertinent to larger corporations if included in our study. See our selection of choices highlighted in bold inside Table 11.2.

In conclusion, we are using all synonymous expressions of green digital, attributed to clusters of authors whose concept definitions align with the realities of our research corpus.

2.2 Transition from Industry 4.0 to Industry 5.0

As industry shifts from mass production to mass customization, the production of goods is becoming more responsive to demand in a flexible manner. I4.0 promotes innovation in organizing and controlling the entire value chain life cycle. The advancement of technology provides insights into the future of digital transformation, aiming for high flexibility, high productivity, and resource efficiency, leading to a new level of human-machine relationships (Hamidi et al., 2018).

The next step, I5.0, involves manufacturers creating products on demand, reducing inventories, and quickly adapting to market changes (Jefroy et al., 2022). This

Table 11.2 Classification of definitions for green digital and synonyms

Expression	Definition 1	Definition 2	Definition 3
Green Digital	The integration of digital technologies to support environmental sustainability and reduce ecological footprints	The use of digital tools and platforms to promote eco-friendly practices and drive green innovation across industries	The convergence of digital transformation and environmental consciousness to create sustainable business models and operations
	Yin (2019), Gholami et al. (2023), Suh and Cho (2022)	Hu et al. (2022), Xi and Wang (2024), Bai et al. (2020)	Radu (2020); Brennen and Kreiss (2016), Zhu et al. (2022)
Sustainable Digital	The development and application of digital technologies that balance economic growth, social inclusion, and environmental protection	The use of digital solutions to address sustainability challenges while minimizing the negative environmental impacts of technology itself	The strategic implementation of digital tools to support long-term ecological, social, and economic sustainability goals
	Taghizadeh-Hesary and Hyun (2022), Kshetri (2021), Pappas et al. (2023)	George et al. (2020), Lopes et al. (2022), Beier et al. (2018)	Seele and Lock (2017), Kuntsman and Rattle (2019), Ullah et al. (2023)
Smart Green	The combination of intelligent technologies and environmentally friendly practices to optimize resource use and reduce environmental impact	The application of smart digital solutions to enhance energy efficiency, waste reduction, and sustainable urban development	The integration of IoT, AI, and other advanced technologies to create eco-friendly products, services, and infrastructure
	Petersen et al. (2023), Lytras et al., (2019), Vinuesa et al. (2020)	Bibri and Krogstie (2020), Yigitcanlar et al. (2020), Allam et al. (2022)	Nižetić et al. (2020), Schiavone et al. (2022), Esmaeilian et al. (2020)
Responsible Digital	The ethical and sustainable development, deployment, and use of digital technologies with consideration for social and environmental impacts	The practice of leveraging digital innovations to address societal challenges while adhering to principles of transparency, accountability, and inclusivity	The approach to digital transformation that prioritizes human rights, privacy, and environmental stewardship alongside technological advancement
	Lobschat et al. (2021), Dwivedi et al. (2021), Vial (2019)	Taddeo and Floridi (2018), Gupta et al. (2021), van den Buuse and Kolk (2019)	Buhmann et al. (2020), Linkov et al. (2021), Stahl et al. (2017)

Source: Vallée (2025)

approach emphasizes predictive and self-correcting production processes focusing on product usage (Prajapati et al., 2019). The approach also promotes flexible work structures, making work environments more appealing and adaptable (Destouet et al., 2023). I5.0 focuses markedly on human-technology interaction (e.g., human-machine systems, human-robot collaboration, collaborative robots/technologies) (Jefroy et al., 2022). This shift represents a significant change in economic strategy, requiring companies to rethink production and customer engagement (Blanchet, 2016). Philbin et al. (2022) explore how digital transformation is pivotal in helping SMEs achieve sustainable development. A meticulous examination of the existing literature reveals that digitalization intersects with sustainability concerns, encompassing challenges related to information systems and change management (Varenne, 2020). The integration of technological tools with change management processes is essential in both the I4.0 and I5.0, except for the fact that, in the latter, the human factor is a key component.

Looking beyond I4.0, SMEs should now consider integrating the new paradigmatic shifts of I5.0 into their strategic visions. I5.0 seeks to transform manufacturing toward sustainability through technology and collaboration (Jefroy et al., 2022). It represents a significant shift from the traditional "manufacturing to consumer" model, with a revised focus on "customer to manufacturing," which is a complete U-turn approach. This shift emphasizes the importance of customer-centricity and sustainability in manufacturing operations. According to Ghobakhloo et al. (2023), I5.0 seeks to decrease environmental impact while enhancing economic, environmental, and social outcomes. The study identifies 12 key functions critical to I5.0's ability to enhance sustainable manufacturing, encapsulated in four dimensions. First is the *integration of value networks*, which is the connection between industrial systems and their micro-components, including machines, humans, and processes, to communicate more reliably and effectively manage the flow of information during the supply chain. Secondly, *sustainable technology governance* facilitates decision-making through industrial automation solutions that promote real-time communication. These solutions aim to improve efficiency and productivity by reducing human error and expenses. The third element, the *innovative business models*, refers to the opportunity to design innovative modular models adapted to components of production processes, which, in turn, instills innovation in business models while applying them in sectors other than manufacturing. Finally, *skill development* is crucial to prepare staff for the requirements of I5.0.

2.3 Sustainability in Industry 5.0

Integrating digitization and sustainability to prepare for I5.0 provides valuable insights into current research trends. Sustainability, as defined by UNESCO (2015), entails the development that meets current needs without compromising the ability of future generations to meet their own. In an industrial context, firms must be accountable for resource usage (Kleindorfer et al., 2009) and the resulting environmental impacts. This evaluation encompasses production, transportation, employee health, safety, and community quality of life (Gimenez et al., 2012).

Philbin et al. (2022) emphasize that I5.0's environmental focus centers on preserving the earth's biosphere and natural resources by moving away from the traditional linear take-make-waste economic model. Instead, I5.0 promotes a circular economy, integrates sustainable innovation into products and processes, incorporates renewable resources, and enhances human-centricity. This comprehensive approach to environmental sustainability includes waste and emission reduction, resource efficiency, and renewable energy integration. Industry 5.0 also addresses broader global sustainability challenges such as shorter product life cycles, limited product recyclability, and rebound effects (Ghobakhloo et al., 2023). This paradigm shift is crucial in fostering long-term environmental stewardship and resource conservation.

The concept of "green digitalization" is a critical asset for transforming SMEs in the contemporary economy, which requires resilience and adaptability. According to the OECD (2023), leveraging digital technologies can help SMEs enhance their operational efficiencies, reduce environmental impacts, and adapt to changing market conditions. By embracing digital tools and platforms, SMEs can better monitor and manage their resource usage, streamline production processes, and implement sustainable practices.

The transition from I4.0 to I5.0 and adopting a circular economy model represents significant steps toward sustainability. This holistic approach ensures that technological advancements are aligned with sustainable development goals, fostering a more resilient and sustainability-conscious industrial sector.

3 The Seamless Integration of Sustainable Practices into Digital Transformation: A Conceptual Framework

3.1 Alignment of Technological Advancements with Sustainability

The concepts of "digital transformation," or "digital shift," encapsulate the proactive adoption of digital technologies by organizations or individuals to bolster efficiency and adaptability in response to the evolving digital era. This paradigm shift, a direct outcome of I4.0 (Hoyer et al., 2020), heralds the Fourth Industrial Revolution (Schwab, 2016), characterized by high-technology industries using modern assets such as artificial intelligence (AI), Big Data, Internet of Things (IoT), Cloud Computing, System Integration, and Digital Twin, nonexclusively.

Green digitization refers to using digital technologies to enhance a company's environmental innovation, aiming to achieve environmental and economic sustainability by improving practices like R&D investment (Guo et al., 2022). For instance, a company could use big data and data modeling to create a digital twin of their organization and see its evolution with sustainable goals and parameters that lead to predictive outcomes.

The United Nations' Sustainable Development Goals (SDGs) influence SMEs differently, affecting operational strategies, market dynamics, product/service alignment with SDGs, access to financial resources, regulatory compliance, and collaborative partnerships (Islam et al., 2022). Aligning with SDGs not only improves SMEs' competitiveness, resilience, and long-term sustainability but also positions them at the forefront of evolving market dynamics.

Under I4.0, industrial strategy undergoes a profound transformation, altering economic rationales and mechanisms governing value creation (Blanchet, 2016; Schwab, 2016; Priyono et al., 2020). The capacity of SMEs to transition toward a sustainable development-driven economy within the context of digital transformation depends on overcoming pivotal challenges, such as mastering data usage in a Cloud environment.

This alignment of sustainable development and digital transformation, as defined by Schumacher (2016), is threefold:

1. Affordable Technology for Low-Income Areas: SMEs can innovate and promote sustainable development using cost-effective and suitable technologies, particularly in low-income regions.
2. Overcoming Challenges for Sustainable Transformation: SMEs must address challenges such as accessing the right technology and innovating within existing infrastructural and environmental limits.
3. Emerging Economies Leading Innovation: The shift of innovative capabilities to emerging economies like China and India suggests that SMEs in these regions could take the lead in developing technologies that meet the needs of the poor, contributing to sustainable development.

The research underlines the vital role of integrating sustainability and digitalization at the micro-level to bolster SME resilience against disruptions from sustainability crises and digital transformation up to the highest maturity level. It encompasses aligning global community interests, national agendas, and enterprise objectives across economic sectors (Martynov, 2023). SMEs' digital transformation reengineers business models by industrializing low-value-added processes (Blanchet, 2016; Schwab, 2017; Pachenko & Dovhenko, 2023), delivering digitized products or services through digital artifacts, which improves performance and transforms user behaviors (Pachenko & Dovhenko, 2023; Varenne, 2020; Brozzi et al., 2018; Blanchet, 2016). This transformation requires timely progress and ROI measures.

Infrastructure plays a crucial role in both concepts, as seen in the manufacturing sector with smart green supply chain management (SCM) (Lerman et al., 2022). Various streams address digital transformation in supply chain management using labels such as I4.0, Supply Chain 4.0, IoT-enabled supply chains, and digitized supply chains (Lerman et al., 2022). These studies focus on developing green operations and relationships to enhance green performance within SCM, using constructs like digital strategy and base technologies to measure digital transformation in surveys.

Islam et al. (2022) view digitalization as a means to advance the Sustainable Development Goals (SDGs) by 2030, offering new data sources and analytical capacities while raising ethical, social, and environmental concerns. Incorporating I5.0 principles and focusing on digitalization can significantly enhance SME sustainability efforts. SMEs adopting a circular economy model and integrating sustainable innovations can meet current environmental standards and contribute to a more sustainable future, aligning technological advancements with ethical and sustainability goals.

Irimiás and Mitev (2020) consider digitalization an efficient means to support sustainable environmental, social, and economic development. ICT can help measure organizational change to reach sustainability goals, which is especially challenging for many SMEs to implement. Stakeholder engagement and employee involvement are crucial in fostering a sustainable performance culture within SMEs.

In summary, research highlights the pivotal role of digitalization in advancing SDGs and fostering sustainable SMEs. It emphasizes the challenges SMEs face in integrating digitalization and sustainability, requiring further research and a holistic approach to enhance SME resilience and achieve sustainable development. The intersection of digitalization and sustainability presents a multifaceted landscape for SMEs, necessitating an intertwined model between traditional maturity assessment and sustainable assessment. Collectively, these studies align with our stated objectives to illuminate SMEs' sustainable digitization evolution through a self-assessment framework on digital maturity and sustainable practices.

3.2 Integrating Digital and Sustainable Maturity Models: Toward a Unified Framework

Assessments are key practices for addressing significant issues. Digital maturity model assessments have advanced in the 2010s, with models now focusing on 4–8 dimensions and 8–10 items per dimension. Maturity levels range from "not initiated" to "fully mature" over 5–6 levels, but the models vary in practicality and transparency (Brozzi et al., 2018; Hizam-Hanafiah et al., 2020).

There is no consensus on the number and content of dimensions for a conventional maturity model. However, a new Green Government IT index has been proposed for assessing environmentally responsible IT use (Lokshin & Widmar, 2023). This aligns with the need for sustainable practices, given forecasts of significant increases in ICT electricity usage by 2030 and a trend toward green governance and reporting.

Comparative analyses of Green ICT maturity models suggest a roadmap for organizations to become eco-friendlier. However, current models lack clarity regarding the intersection of digital transformation and sustainability, highlighting the need for hybrid models that integrate I4.0 principles with sustainability dimensions. Some studies hinted at that endeavor, though.

Islam et al. (2022) see digitalization as promising for advancing the SDGs and propose a Smart Quadruple Bottom Line (SQBL) model to help SMEs achieve long-term survival. They emphasize integrating digital paradigms like Big Data and AI to address SDG research gaps. Focusing on (semi-)durable goods, Ertz et al. (2022) suggest that organizations (SMEs or not) may foster product lifetime extension (PLE) strategies using technologies like IoT, Additive Manufacturing, Big Data, and AI. Piloting and continuous monitoring will help identify areas for improvement and drive continuous innovation in SMEs.

3.3 Assessing SMEs' Industry 4.0 Maturity Levels and Sustainability Practices

To seamlessly integrate I4.0 maturity levels and sustainable practices in SMEs, it is essential to assess both technological and sustainability dimensions (Müller & Pfleger 2014). This dual approach offers SMEs a structured way to analyze their maturity in sustainability efforts, ensuring a holistic understanding of how they integrate advanced technologies and sustainable practices to become more innovative and resilient.

Elia et al. (2020) emphasize the crucial role of technological innovation in shaping entrepreneurship, particularly through Digital Entrepreneurship Ecosystems (DEEs). DEEs represent a new model for understanding digital entrepreneurship dynamics, where digital technologies play a significant role as both enablers and outcomes of entrepreneurial activities. The concept of Collective Intelligence provides a valuable framework for analyzing DEEs, showcasing how success in digital entrepreneurship can stem from diverse approaches, whether process-, resource-, or product-driven. Isensee et al. (2023) build on this foundation by introducing the concept of sustainable digital entrepreneurs, categorizing them into four main types. Their research emphasizes integrating sustainability and digitalization at the micro-level to enhance SME resilience against disruptions from sustainability crises and digital transformation up to the highest maturity level.

Eisner (2022) offers a self-assessment framework for corporate environmental sustainability. Their model includes questions to measure the extent to which a company is green, with scores ranging from 0 to 5. The framework provides a detailed check with 64 questions per area, helping companies identify areas for improvement. Experts suggested making the model easier to understand and use, enabling companies to learn where they need to improve and how they compare to others in being green.

Khan and Porras (2018) introduced a framework and web application designed for SMEs to self-assess their sustainable green ICT practices. The model includes five maturity levels, starting with level 0 for no knowledge or capability and ending

Table 11.3 The maturity levels of sustainable green ICT practices self-assessment for SMEs

Level	State	Description
0	Initial	No awareness, no implementation of Green ICT actions
1	Repeatable	Minimal awareness of Green ICT, immature initiatives that limit non-sustainable actions
2	Defined	Informal initiatives of Green ICT strategies and eliminating the cause of non-sustainable effects
3	Quantitatively Managed	Formal, measured, and controlled Green ICT strategies for favorable sustainability effects
4	Optimized	Optimized environmental impacts for sustainable development

Source: Khan and Porras (2018)

with level 4 for full comprehension and implementation of sustainable ICT. This model is highly relevant to our study and provides a basis for comparison (Table 11.3).

The maturity model comprises eight parameters: (1) power management of computers; (2) reduction of computers for power management; (3) power management of imaging equipment; (4) paper and energy saving from ICT equipment; (5) power management of video conferencing suites; (6) IT load reduction management of servers; (7) cooling management of ICT equipment; and (8) e-waste disposal management. Each level delineates specific characteristics showcasing technology usage, with examples such as manual computer shutdowns (level 0) to automated shutdowns and display sleep modes (level 4).

The review emphasizes integrating information systems and technology with human resources in SMEs and considers sustainability in terms of human well-being. It suggests that management control in SMEs, supported by technology, should encompass all these factors. Additionally, Islam et al.'s (2022) study introduces the Smart Quadruple Bottom Line (SQBL) model, highlighting four dimensions for analysis: (1) Organization, (2) Technology, (3) Energy, and (4) Personnel.

In sum, this review provides a comprehensive analysis of integrating digital tools to track and measure sustainability practices in SMEs. It offers valuable insights for researchers and practitioners, showcasing the positive impact of these practices on both environmental and financial performance. By combining I4.0 principles with sustainability dimensions, SMEs can enhance their operational efficiency and align with sustainable practices, ensuring long-term resilience and ethical compliance. Yet, the optimal hybrid framework has to be conceptualized and demonstrated.

3.4 *Proposed Integrated Framework*

The integrated framework by our team combines digital maturity levels and sustainability practices following the same assessments scheme to evaluate SMEs' readiness for I4.0. It draws on traditional maturity models but is designed to be adaptable across all SME sectors, as Brozzi et al. (2018) recommended. This framework named PME 4.0 (in French or SME 4.0, Vallée, 2023, 2025) assesses the adoption of digital technologies and its embedding into the organization's internal processes. It also evaluates the environmental, social, and economic impacts of SMEs' operations by tracking a "sustainable management index." Initially, we wanted to gage the relevance to eventually merge the two strategic goals, DT and Sustainable practices among SMEs, within just one assessment. The research have shown the relevance while addressing four dimensions for analysis implicitly suggested by Ertz (2021) and as suggested by Islam et al.'s (2022): (1) Organization, (2) Technology, (3) Energy, and (4) Personnel.

Following the recommendations of this chapter analysis, we are taking this initiative to the next level, offering a framework that could be validated within a qualitative research as a first step. We have tentatively merged levels and practices, from Khan and Porras' (2018) model to ours, PME 4.0, also according to our respondent's answers. We have changed the "State" column's nomenclature to fit our model; then we have updated the column of sustainable practices and, finally, we have added our digital model by description levels. It then turns our initial digital assessment framework into a potential hybrid roadmap suiting SMEs in Region 02, pursuing two strategic goals for SMEs: increasing digital maturity while increasing management sustainable practices. Table 11.4 is a demonstration of this new merged framework project, called Sustainable SME 4.0.

The labeling of each level follows an indicatively and psychologically positive approach (Csikszentmihalyi, 1997) as it is used by SMEs that need to be encouraged to move toward the next maturity level. A creative approach also has influenced the conceptualization of the model, in order to make it more attractive to team management in charge of both transformation, using color gradation following scheme related from discomfort (yellow = moving fast with a sense of emergency) to full comfort (blue purple = fully trustworthy), (Alberts & Van Der Geest, 2011) (Fig. 11.1).

By integrating these maturity level assessment models seamlessly, we are merging best practices in both expertise domains while offering an optimal vision of how digital tools can support sustainable goals for SMEs. Thus, the framework provides a comprehensive view of readiness, highlighting areas for improvement in both digital and sustainability domains. This approach aligns with the circular economy concept, which aims at optimizing resource use and contributing to the well-being of individuals and communities (Winans et al., 2017).

The framework evaluates technology presence and adoption, organizational readiness, workforce skills, and environmental and social practices. It aims to enhance digitization among SMEs and offers a practical methodology for operational support, such as the DSIFAT (Discovery, Awareness, Integration, Training, Support, Transformation) proposed by Varenne (2020). This methodology can be readily implemented in practical settings to promote sustainability in managerial practices.

Integrating digital maturity and circular economy concepts introduces an important intersection between I4.0 and sustainable development, aligning with the

Table 11.4 Sustainable SME 4.0 Assessment Model: merged application of digital maturity levels showing increasingly sustainable green ICT practices for SMEs

Level	State	Description of sustainable maturity level practices	Description of digital maturity level practices
0	Traditional	No metrics tracking traditional implementation of Green ICT actions: recycling, creating employment, buying locally	No awareness of digital transformation and what 4.0 technologies can bring to improve competitive level
1	Initiated	Emerging awareness of Green ICT, traditional initiatives that limit sustainable actions impact	Emerging awareness of what digital tool can offer as competitive advantage and beginning of exploring them
2	Engaged	Formal initiatives of Green ICT strategies that are showing measurable results although not fully eliminating the cause of non-sustainable effects	Technologies and digital tools are fully used although not fully suited, exploration of more suitable next-level technology and practices are priorities
3	Interconnected	Formal, measured, and controlled Green ICT strategies for favorable sustainability effects with an increasing used	Technologies and digital tools are supporting most of the operations, structuring data, and providing a beginning of change in organizational overall performance culture
4	Integrated	Fully measured, controlled Green ICT strategies that have optimized environmental impacts for sustainable SME development and competitiveness	Technologies and digital tools are fully supporting the operations, structuring data for more performing I4.0 technologies, business culture has transformed into fully accountable for overall performance
5	Mature	Sustainable practices are embedded into predictive processes and a part of the organization's performance culture and business model	Optimized integrated process management and management controls with digital tools that support both productivity, environmental impacts for sustainable development and business model

Source: Vallée (2025)

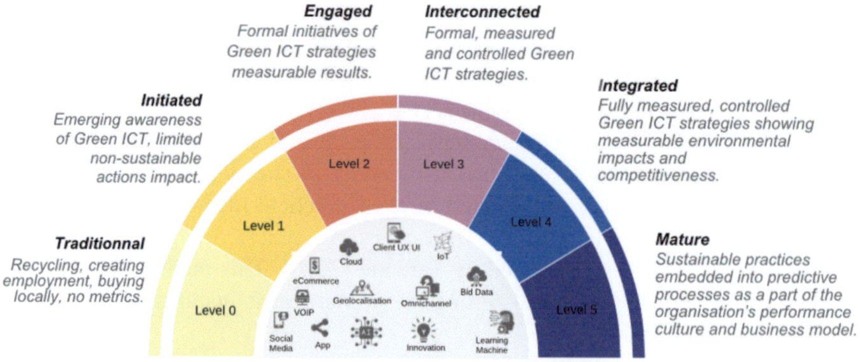

Fig. 11.1 Barometer of sustainable SME 4.0. (Source: Vallée (2025))

principle of "overall performance" in management controls as defined by St-Pierre and Cadieux (2011). The framework is designed to be flexible and resource-friendly, aiming to improve SMEs' market sustainability and profitability (Hamidi et al., 2018).

The framework can be applied using qualitative and quantitative methods, including surveys, interviews, and sustainability assessments. SMEs can use it to assess their current maturity level and sustainability performance, identify improvement areas, and develop roadmaps for digital transformation and sustainability improvement. Policymakers and industry associations can also use the framework to develop targeted support programs promoting digital and sustainability goals.

4 Methodology

4.1 Participative Action-Research Project

The current research is part of a broader research program entitled "PME 4.0" (i.e., SME 4.0) that took place between January 2021 and September 2023 with SMEs from the Saguenay-Lac-Saint-Jean region in the primary, secondary, and tertiary sectors (Charmillot & Dayer, 2007; Paillé & Mucchielli, 2003). The research was initially aimed at constructing a comprehensive framework to support SMEs' digital transition while gradually evolving toward enabling SMEs to implement their digital transition responsibly. Hence, the study matured into examining the state of SMEs' current or anticipated digital transformation along with their sustainable practices (MELCCFP, 2023). Deployed in the framework of a participative research-action (Ozanne & Saatcioglu, 2008), the investigative efforts concur to answer the following research questions:

RQ1: What is the degree of technological maturity of SMEs in Saguenay–Lac-Saint-Jean (SLSJ) in 2023?
RQ2: What is the degree of sustainable technological practices of SMEs in Saguenay–Lac-Saint-Jean (SLSJ) in 2023?

A pilot study was conducted in five stages: (1) pilot unstructured interviews; (2) online intention survey and pre-interview questionnaire; (3) qualitative semi-structured interviews; (4) thematic qualitative analysis; and (5) elaboration of a preliminary version of the digital maturity model coupled with sustainability framework. The stage is described in more detail below.

First, a pilot of unstructured interviews was carried out with three socio-economic organizations, two certified 4.0 auditors from the Ministry of Economics, Innovation, and Energy (MEIE), and two SMEs to validate the need and research questions.

Second, 55 SMEs were initially approached by the six socio-economic partners of the research, i.e., economic development organizations and agencies. Of these 55 SMEs, 36 completed the preselection and ethical consent form electronically via the LimeSurvey platform. This form included six questions, one of which asked about the company's intention to undergo digital transformation within the next 12 months, offering options such as focusing on technological tools, human resources, or the way of working, and reasons for not pursuing transformation, such as lack of time, human resources, or funding. The 30 participants were selected based on their responses to the preselection form cited above, and then they completed a pre-interview questionnaire on Microsoft Forms. This questionnaire contained 50 questions, divided into four segments:

1. Company introduction and governance.
2. Digital connectivity and technological tools dashboard.
3. Knowledge of 4.0 technologies and technology-assisted management practices.
4. Sustainable practices and future vision.

Third, each business owner or managing employee participated in a private semi-structured interview with the researcher, either in person or via videoconference, recorded on Zoom. Lasting 90–120 min, these interviews delved deeper into the preselection form and pre-interview questionnaire. The researcher aimed to understand each company's specific issues and realities concerning internal and external management practice changes related to I4.0 from 2019 to 2023. The year 2019 was chosen as an anchor due to the Covid-19 pandemic, which appeared in the pilot as a cornerstone prompting SMEs to fasten their digital transformation pace or giving it an impetus. The questions addressed strategic and operational business practices, major challenges, achievements in digital transformation, whether aligned with sustainable development or not, and digital literacy. The recorded interviews were transcribed verbatim using artificial intelligence with Amberscript, then verified and anonymized by assigning codes to each SME participant according to their SCIAN and interview number.

Fourth, a thematic analysis involved carefully reviewing the data to identify key themes and patterns. This involved several steps, beginning with familiarization, where the researcher immersed themselves in the data by reading and re-reading the transcripts. Initial codes were then generated systematically across the entire dataset. These codes were then collated into potential themes, which were reviewed and refined to ensure they accurately captured the essence of the data. The themes were defined and named, providing clear definitions and labels.

Fifth, the themes were woven together to construct a coherent narrative addressing the research questions through a digital maturity model and sustainability framework. The researcher's observation notes were integrated with the interview responses to enrich the thematic analysis. If necessary, dated memos were added to provide additional context and depth, such as respondent's emotions within context.

4.2 *Digital Maturity Model Coupled with a Sustainability Framework*

The pilot research has successfully developed a comprehensive maturity model assessment meant to support SMEs of all activity sectors and all sizes, which includes a sustainable digital framework. This avenue has been chosen over integrating a full algorithm, crossing all questions with the sustainability dimension at this stage. However, this framework has not yet demonstrated a correlation between digital transformation and sustainable practices due to the low literacy awareness of many respondents (Vallée, 2023). While it might be early for Quebec SMEs to integrate these two concepts fully, it is nonetheless time to introduce the possibility of measuring sustainability with digitized control management all across the four assessment sections: (1) Company introduction and governance; (2) Digital connectivity and technological tools dashboard; (3) Knowledge of 4.0 technologies and technology-assisted management practices; (4) Sustainable practices and future vision. The timing is crucial to embrace this opportunity, as many respondents already indicate engagement in sustainable practices. Yet, few respondents were aware of sustainable practices pertaining to human-centric management. It appeared more obvious to most of them that sustainable practices are rather environmental-centric practices.

The proposed conceptual framework can thus be used to assess SMEs' I4.0 maturity levels and sustainable practices across all sectors, sizes, and organizational life cycles. Pertaining to maturity levels, we draw on Khan and Porras (2018) to assess maturity levels by state and description (see Table 11.5), according to a rigorous examination of themes emerging from the respondents' interviews.

We have chosen to score digital maturity based on a maturity scale that suits SMEs of all sectors. Therefore, the dimensions that appear as the most prevalent in the respondent verbatims are:

1. Technology.
2. Interconnection.
3. Data.

Each of these three dimensions has six levels of maturity, scoring from 0 to 5, while some items considered "Traditional" score at 0 or negatively if paired with two or more low-level dimensions items. The assessment questionnaire spans six sections over fifty (50) questions, with the first (governance) and last (sustainability) sections not contributing to the digital maturity level score. This is because respondents exhibit an over- or under-assessment bias in the first section addressing governance based on the interviews. Data is scored according to its level of structure fit to be used by AI supported within the Cloud and an operation system. As for sustainability, we aimed to pilot this dimension and items to validate which are the most relevant. Therefore, we aimed to build an assessment framework that will emerge from our qualitative research. Here are the six sections of the assessment model: (1) Governance, (2) Connectivity, (3) Literacy, (4) Technology, (5)

Table 11.5 Comparison between Khan and Porras (2018) model and the proposed model

Level	State (Khan & Porras, 2018)	State (Vallée, 2023)	Description (Khan & Porras, 2018)	Description (Vallée, 2023)
0	Initial	Traditional	No awareness, no implementation of Green ICT actions	No awareness of I4.0, Fourth Industrial Revolution, no digital literacy, no measurement of sustainable parameters
1	Repeatable	Initiated	Minimal awareness of Green ICT, immature initiatives that limit non-sustainable actions	Minimal awareness of I4.0, Fourth Industrial Revolution, low digital literacy, limited technology adoption, low measurement of sustainable digital
2	Defined	Engaged	Informal initiatives of Green ICT strategies and eliminating the cause of non-sustainable effects	Formal practices are modified by technologies that change processes and reduce causes of productivity loss. Formal practices on sustainable digital
3	Quantitatively managed	Interconnected	Formal, measured, and controlled Green ICT strategies for favorable sustainability effects	Formal, measured management control strategies encompassing sustainable digital. Technologies transform jobs
4	Optimized	Integrated	Optimized environmental impacts for sustainable	The business model presents new opportunities through innovation induced by digital transformation
5	–	Mature	–	Innovation is a common practice, and sustainable digital is tracked and embedded in productivity as an asset

Interconnection, and (6) Sustainability. The questionnaire framework currently adds up all answer choices divided by the number of answers possible, multiplied by hundreds. It is represented as a percentage of efforts put into all the most relevant answers for SMEs in the region.

The multiple-choice question asked about sustainability reads as follows:

1. Has your company/organization put in place sustainable development practices? The expression "Sustainable practices" is explained below the question, as defined by the Government of Quebec as a practice that (1) Maintains the integrity of the environment, (2) Ensures social equity, and (3) Encourages economic viability."
2. The answer options are derived from Ertz's (2021) model and identified as the most relevant to SMEs in the SLSJ region. These options read as follows: (1)

Table 11.6 Sector analysis: detailed sustainable management practices

Sector's Analysis Detailed sustainable management practices	Secondary	Tertiary	Total	Proportion
Creates/Maintains jobs in the region	9	17	26	16%
Performs equipment maintenance	10	16	26	16%
Creates partnerships	7	18	25	16%
Offers advice on the use of its products	6	14	20	13%
Increases productivity without compromising respect for the environment	4	15	19	12%
Saves energy	9	7	16	10%
Offers training on how to use the product or how to use the service	5	11	16	10%
Conducts consultations on product or service development/improvement	4	6	10	6%
Total	54	104	158	100%

Source: Vallée (2023)

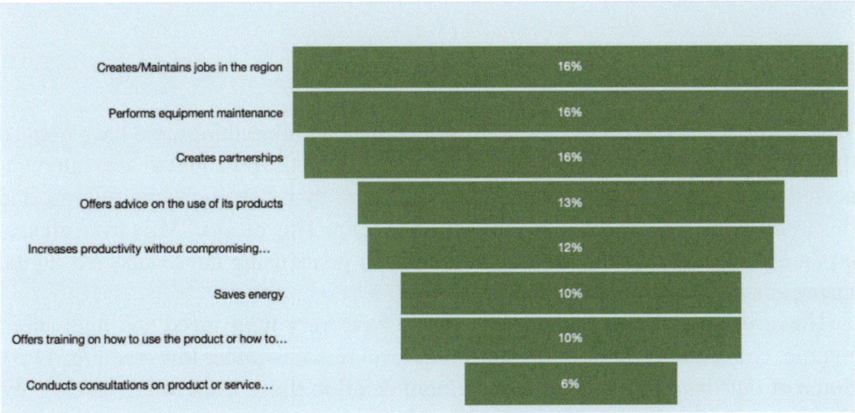

Fig. 11.2 Sustainable management practices in SLSJ SMEs. (Source: Vallée (2023))

Creates/maintains jobs in the region; (2) Performs equipment maintenance; (3) Creates partnerships; (4) Offers advice on the use of its products; (5) Increases productivity without compromising respect for the environment; (6) Saves energy; (7) Offers training on how to use the product or how to use the service and; (8) Conducts consultations on product or service development or improvement. The choice answer did not include any recycling initiatives intentionally in order to tap into the least expected side of sustainability known by SMEs in the region, which is the so-called "sustainable management practices" based upon our pilot conducted in 2022. It is interesting to mention that all respondents had at least three of the eight sustainable initiatives in place and that the tertiary sector is performing ahead of the secondary sector in this matter (see Table 11.6 and Fig. 11.2).

The most prevalent initiatives scoring on average at 16% in all answer choices and across sectors are: (1) Creates/maintains jobs in the region; (2) Performs equipment maintenance; and (3) Creates partnerships. While the answers that came in first and second place seem obvious from an economic standpoint, the third one is quite interesting. It indeed indicates an economic growth potential factor in the region, whereas partnerships can be viewed as a strategy to enhance the sustainability of SMEs.

Also, interestingly, during the interviews, none of the respondents disclaimed measuring their sustainable initiatives in any cross-questioning addressing their data management habits.

Nonetheless, the above results show the potential to measure sustainable practices with digital tools embracing modern management control to serve SMEs' profitability. The following results will demonstrate that potential from the digital capability standpoint and by industry sectors. The results below were collected through a digital questionnaire and validated during the individual semi-directed interviews.

5 Results

The results yield several interesting insights. Among other things, we have noticed that Internet connectivity among SMEs in SLSJ offers the potential to host interconnected technological tools, sometimes from more than one connectivity type, and this trend is not limited to the manufacturing sector. This means SMEs from all sectors are potentially enabled to track sustainable goals using interconnected digital management control tools (Table 11.7).

However, while most respondents (69%) have very high-speed and high-speed Internet connections, their interconnection level remains rather low (see Fig. 11.3). Some of our data can be compared in more detail to those from MEIE (2020). We found that the interconnection of information management systems within manufacturing companies is highly variable, and surprisingly, the tertiary sector scores higher in this regard. Regarding digital systems and applications, they are currently used by a minority of companies (28–58%). Those using these tools report an average interconnection level of only 35–42% compared to their other systems. This indicates room for improvement in both human interaction with technology and the interconnection of technology itself with API, IoT, or else to serve sustainable objectives.

Notably, the interconnection of information systems externally and internally for the SLSJ region is 79.9% and 47.3%, respectively (MEIE, 2020). The high external interconnection rate is primarily influenced by the local business practice adopted by SMEs that use government online services for tax purposes. Meanwhile, internal interconnection is just below the average for all SMEs (MEIE, 2020).

Table 11.7 Type of connectivity across all sectors

Types of connectivity across all sectors	Total (*n*)
High-speed Internet (5 Mbps and above)	12
Very high-speed Internet (100 Mbps and above)	15
Regular Internet	4
Mobile Internet connection	8
Total	39

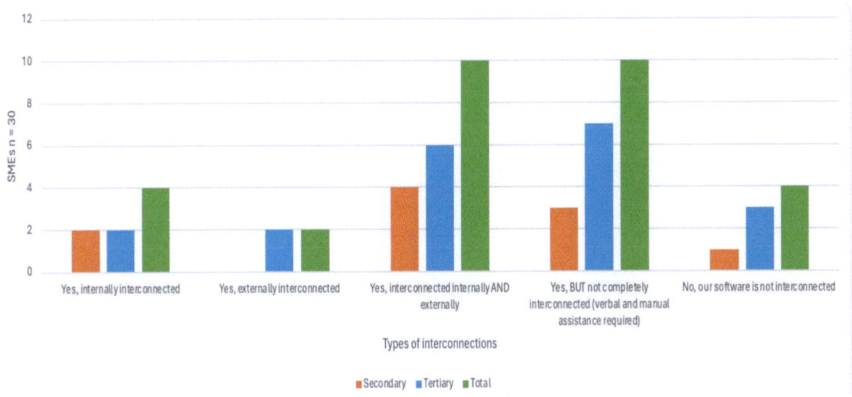

Fig. 11.3 Interconnectivity in SLSJ for SMEs. (Source: Vallée (2023))

In any digital maturity and sustainability assessment cases, the respondents' awareness of management control (Sloan, 1990) supported by technology should include critical indicators related to sustainable management practices (Berger-Douce, 2014). These would also become a part of practical strategies and best practices for overcoming challenges pertaining to both DT and sustainable practices. While SMEs have implemented sustainable practices, they tend to lack the tracking of sustainability-related key performance indicators (KPIs) on their dashboards. This indicates that although respondents have high connectivity and are well interconnected internally and externally (see Fig. 11.3), sustainability is not prioritized for inclusion on management dashboards. Some additional choice answers going in that direction might help educate the respondents on that possibility.

As our results show, the respondents are moving toward the Cloud to store their data but are not using this data to build new tools or document tasks and job descriptions, which would concur with sustainable practices related to the human factor item in our model. Figure 11.4 represents respondents' answer combinations from a multiple-choice answer question. The question is: "What are you using the Cloud mostly for?". As it appears, the most frequent answers aggregated are "A) Consulting documents, data." On that specific answer item, three other aggregated answers are to be chosen such as "D) Sharing documents, consulting documents, data and viewing training"; "E) Sharing documents, archiving documents, data, working collaboratively on documents, consulting documents, data, developing new management tools, documenting tasks and jobs"; and "F)" which is the same aggregated answers group, adding "viewing training" answer (Table 11.8). In all these answer combinations most frequently chosen by the respondents, we argue that data tracking on sustainability practices among the SMEs should be the norm, supported by the Cloud in real-time. Sustainable management practices were addressed in the interviews as part of a follow-up question on consulting data and documenting tasks. This topic sparked interest among some respondents who were curious about achieving these tasks at a low cost. More interestingly, the challenge of manually

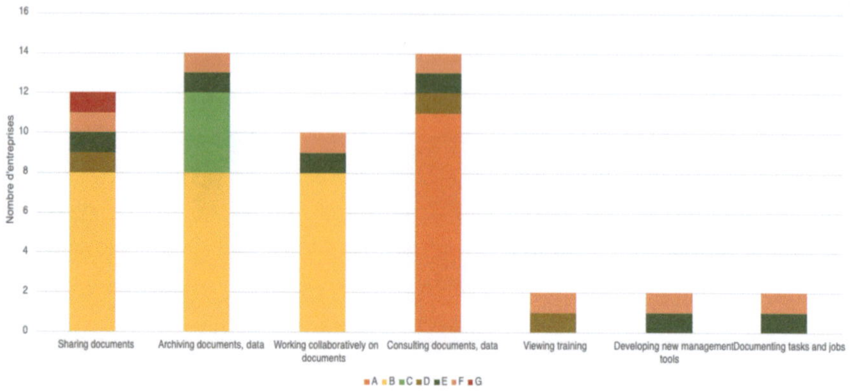

Fig. 11.4 Aggregated responses on Cloud usage. (Source: Vallée (2023))

Table 11.8 Combined answer occurrences

A	Consulting documents, data
B	Sharing documents/Archiving documents, data/Working collaboratively on documents
C	Archiving documents, data
D	Sharing documents/Consulting documents, data/Viewing Training
E	Sharing documents/Archiving documents, data/Working collaboratively on documents/ Consulting documents, data/Developing new management tools/Documenting tasks and jobs
F	Sharing documents/Archiving documents, data/Working collaboratively on documents/ Consulting documents, data/Developing new management tools/Viewing training/ Documenting tasks and jobs
G	Sharing documents

Source: Vallée (2023)

tracking measurements emerged as an additional task that SMEs could not afford. Therefore, implementing an assessment tool that can measure these sustainable management practices and convert all related efforts into key performance indicators (KPIs) would enhance the relevance and effectiveness of these initiatives.

The aggregated responses on Cloud usage show varying levels of adoption and integration within the surveyed companies. The data reveals that Cloud services are utilized to differing extents, highlighting a range of digital maturity levels across response combinations (the "F" answer being the highest). The detailed results, as presented in the graph, provide insights into the current state of Cloud technology implementation and its impact on operational efficiency and technological advancement. Without sufficient structured data on the Cloud, organizational diagnostics cannot easily be supported by AI internally, even less predictive or prescriptive management measures applied to sustainable practices within the business. Sustainable KPIs are part of the predictability measurement supported by AI. Therefore, without sufficient usage of the Cloud, measuring and tracking sustainable initiatives becomes more complex, adding an additional task to the human workforce.

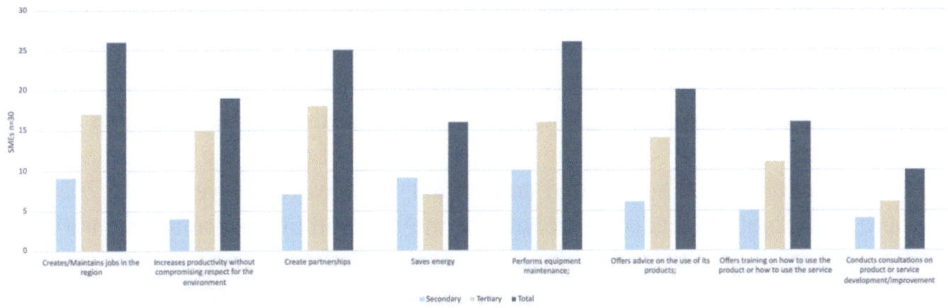

Fig. 11.5 Sustainable management practices by sectors. (Source: Vallée (2023), Ertz (2021))

In terms of sustainable practices, when looking more closely at each of the following sustainable management practices offered in Fig. 11.5 (Ertz, 2021), we have analyzed them by activity sector and noticed that the tertiary sector adopts more frequently sustainable practices against the secondary sector. Besides, the secondary sector is the least prompted to increase productivity without compromising environmental respect. Finally, the item shows the largest gap with the tertiary sector. It is otherwise coming first in saving energy and close to the tertiary sector in conducting consultations on product or service development and improvement. This shows some relevance pertaining to I4.0 in terms of producing a more relevant fit to the market needs, which has to be linked to some sales and inventory key performance indicators.

Also, 7 out of 8 practices show total engagement from more than 30% of the respondents. Therefore, addressing these practices as part of a global management strategy is relevant, stressing the importance of embracing digital transformation while adopting management control KPIs on these.

The last relevant view on our results is the score percentage of each respondent ($n = 30$) on sustainable practices (Ertz, 2021). Again, Fig. 11.5 shows that SMEs from SLSJ are highly engaged in sustainable practices of two forms: environmental and social. Governance was not considered in the scoring at this stage of our study because of the lack of KPI and interconnected tools to measure the engagement level without bias.

6 Theoretical Implications

Integrating sustainability into management controls is crucial for SMEs operating in modern sustainability landscapes globally, emphasizing the need for sustainable practices alongside digital transformation efforts. This research introduces a new analytical reference by incorporating characteristics of sustainable management practices into an assessment tool, highlighting the importance of aligning business practices with sustainability principles for SMEs. The study intersects global and sustainable management with Industry 4.0 (i.e., I5.0), emphasizing themes such as business digitization, use of big data for sustainable production, alternative production methods, and collaboration principles, thereby expanding the theoretical understanding of sustainable digital transformations in SMEs.

Revisiting traditional management controls allowed for a reassessment of how modern technologies are disrupting management practices in the era of the Fourth Industrial Revolution. The original conceptual framework (Hoyer et al., 2020) of digital transformation needed to include the intersection of two complementary and well-documented concepts: the circular economy and Industry 4.0. This research contributes two significant theoretical insights. First, sustainability as a factor in management controls has become crucial for any SME operating within the modern economic landscape, regionally and globally. Second, by integrating characteristics of sustainable goods and services into the interview questionnaire and the self-assessment tool, the research provides a new analytical reference. Additional themes intersecting global and sustainable management with I4.0 were also incorporated, such as business digitization, the use of big data to enable sustainable production and consumption, the principle of added value through alternative production methods, and the collaborative principle.

Firms need to become "ambidextrous" to effectively navigate the dual transitions of sustainable and digital transformations, as Nyagadza (2022) highlighted in the paper. Being "ambidextrous" allows firms to handle the fundamental structural changes and value creation opportunities brought about by sustainable digital transformation while also addressing the challenges related to responsible and ethical management of increased digitalization. This approach enables firms to leverage the benefits of digitalization, such as closer monitoring of sustainability-related factors through technologies like Artificial Intelligence, while mitigating the negative implications, such as the rise in electronic waste due to increased usage of electronic devices.

7 Managerial Implications

Small- and medium-sized enterprises (SMEs) must undergo digital transformation to remain competitive with larger enterprises. Often, SMEs lack information and awareness about new digital possibilities (Kergroach, 2020), and the lengthy process of government assistance in digital auditing can sometimes require resources that SMEs do not have (Vallée, 2023). One of SMEs' key challenges is competitiveness regarding large corporations in a global world market. Many countries have established support systems to assist SMEs in this process, such as subsidy programs, support from specific-area central organizations, loans with moratoriums, and so on. Assessing the digital maturity of an enterprise is crucial for providing the appropriate support. While there are numerous models and tools for assessing digital maturity, many of them are either theoretical, incomplete, designed for specific vendors, or tailored for larger enterprises (Kljajić Borštnar & Pucihar, 2021; Brozzi et al., 2018; Vallée, 2023). The tool developed in this chapter aims thus to better equip SMEs to assess their level of sustainable digital maturity by offering a framework application and fostering a common language for collaboration.

8 Limitations and Future Research Opportunities

The participative research-action project faced challenges related to participants' perceptions of digital transformation and their unfamiliarity with I4.0, which could introduce biases. Besides, the subject of the digital maturity self-audit combined with sustainable development for regional SMEs across all sectors in the SLSJ is unique, with no direct equivalents for comparison. However, regarding future research avenues, this model should be extended through a longitudinal study supported by developing a full-fledged sustainable digital self-assessment tool. This chapter offers preliminary insights into such endeavors.

Finally, contrary to other models developed in the literature, this preliminary framework is proprietary since it is developed in partnership with field partners who retain conjoint ownership over the model along with the researchers. Limitations thus exist regarding the dissemination and communication surrounding the developed tool, especially in relation to its items, questions, and labeling dimensions.

9 Conclusion

The research underscores the imperative for SMEs to embrace digital transformation to remain competitive, particularly within the realms of Industry 4.0 and sustainability initiatives. Despite the evident benefits, many SMEs lack awareness of the digital possibilities available to them, placing them at a disadvantage compared to larger corporations. Governments and support organizations are crucial in this process, providing subsidies, loans, and support systems that help SMEs navigate their digital transformations. The self-assessment tool developed in this research is designed to empower SMEs to evaluate their sustainable digital readiness, addressing gaps in existing models. Future research should focus on extending this model through longitudinal studies, supported by sustainable digital self-assessment tools, to ensure continuous improvement and adaptation to the ever-evolving digital landscape. Overcoming challenges related to participants' perceptions and unfamiliarity with Industry 4.0 will require a multidisciplinary and innovative approach, emphasizing the importance of courage and innovation in research efforts. Future research opportunities in digital transformation for SMEs are rich with potential. Extending the developed model through longitudinal studies, supported by sustainable digital self-assessment tools, can provide valuable insights into the long-term impact of digital transformation on SMEs' sustainability efforts. Exploring Industry 5.0 and its implications for sustainability practices offers a forward-looking perspective on the evolution of digital technologies in SMEs, focusing on human-centric and advanced technological integration. Investigating the role of innovation systems in facilitating socio-ecological transitions can help understand how SMEs can leverage technological advancements for sustainable development, examining the supportive role of policies, networks, and institutional frameworks. Examining the advantages of Industry 4.0 applications for sustainability across different sectors and regions can highlight best practices and success factors for SMEs. Finally, conducting comparative studies across regions or countries to assess the effectiveness of various support systems and policies in promoting sustainable digital transformations among SMEs can offer valuable benchmarks and recommendations for policymakers and stakeholders. These research avenues collectively can significantly contribute to the knowledge base and practical strategies for enhancing SMEs' digital and sustainable development.

References

Alberts, W. A., & Van Der Geest, T. M. (2011). Color matters: Color as a trustworthiness cue in websites. *Technical Communication, 58*(2), 149–160.

Al-Swidi, A. K., Hair, J. F., & Al-Hakimi, M. A. (2023). Sustainable development-oriented regulatory and competitive pressures to shift toward a circular economy: The role of environmental orientation and Industry 4.0 technologies *Business Strategy and the Environment, 32*(7), 4782–4797.

Allam, Z., Sharifi, A. Bibri, S.E. Chabaud, D. (2022). Emerging trends and knowledge structures of smart urban governance. *Sustainability, 14*(9), 5275. https://doi.org/10.3390/su14095275

Ansari, F., Erol, S., & Sihn, W. (2018). Rethinking human-machine learning in Industry 4.0: How does the paradigm shift treat the role of human learning? *Procedia Manufacturing, 23*, 117–122. https://doi.org/10.1016/j.promfg.2018.04.003

Bai C, Feng C, Du K, Wang Y, & Gong Y (2020). Understanding spatial-temporal evolution of renewable energy technology innovation in China: Evidence from convergence analysis. Energy Policy, 143, 111570.

Antikainen, M., Uusitalo, T., & Kivikytö-Reponen, P. (2018). Digitalisation as an enabler of circular economy. *Procedia CIRP, 73*, 45–49. https://doi.org/10.1016/j.procir.2018.04.027

Barnett, T., Jain, S., Andra, U., & Khurana, T. (2018). *Cisco Visual Networking Index (VNI) complete forecast update, 2017–2022* (pp. 1–30). Americas/EMEAR Cisco Knowledge Network (CKN) Presentation.

Basl, J. (2018). Analysis of Industry 4.0 readiness indexes and maturity models and proposal of the dimension for enterprise information systems. In *Research and practical issues of entreprise information systems: 12th IFIP WG 8.9 Working Conference, CONFENIS 2018, Held at the 24th IFIP World Computer Congress, WCC 2018, Porznan, Poland, Proceedings* (Vol. 12, No. 327, pp. 57–68). Springer. https://doi.org/10.1007/978-3-319-99040-8_5

Beier, G., Niehoff, S., & Xue, B. (2018). More sustainability in industry through industrial internet of things? *Applied Sciences, 8*(2), 219. https://doi.org/10.3390/app8020219

Berger-Douce, S. (2014). Capacité dynamique d'innovation responsible et performance globale: Etude longitudinale dans une PME industrielle. *Revue Interdisciplinaire sur le Management et l'Humanisme, 12*(3), 10–28. https://doi.org/10.3917/rimhe.012.0010

Bibri, S. E., & Krogstie, J. (2020). The emerging data–driven smart city and its innovative applied solutions for sustainability: The cases of London and Barcelona. *Energy Informatics, 3*(1), 5. https://doi.org/10.1186/s42162-020-00108-6

Blanchet, M. (2016). Industrie 4.0 Nouvelle donne industrielle, nouveau modèle économique. *Outre-Terre, 46*, 62–85. https://doi.org/10.3917/oute1.046.0062

Brennen, J. S., & Kreiss, D. (2016). Digitalization. In K. B. Jensen, R. T. Craig, J. D. Pooley, & E. W. Rothenbuhler (Eds.), The International Encyclopedia of Communication Theory and Philosophy, *Wiley-Blackwell.* 556–566. https://doi.org/10.1002/9781118766804.wbiect111

Brozzi, R., D'Amico, R. D., Pasetti Monizza, G., Marcher, C., Riedl, M., & Matt, D. (2018). Design of self-assessment tools to measure Industry 4.0 readiness. A methodological approach for craftsmanship SMEs. In *Product Lifecycle Management to Support Industry 4.0. 15th IFIP WG 5.1 International Conference*, PLM, Turin (pp. 566–578). https://doi.org/10.1007/978-3-030-01614-2_52

Brozzi, R., Forti, D., Rauch, E., & Matt, D. T. (2020). The advantages of Industry 4.0 applications for sustainability: Results from a sample of manufacturing companies. *Sustainability, 12*(9), 3647. https://doi.org/10.3390/su12093647

Buhmann, A., Paßmann, J., & Fieseler, C. (2020). Managing algorithmic accountability: Balancing reputational concerns, engagement strategies, and the potential of rational discourse. *Journal of Business Ethics, 167*, 87–104. https://doi.org/10.1007/s10551-019-04226-4

Charmillot, M., & Dayer, C. (2007). Démarche compréhensive et méthodes qualitatives : clarifications épistémologiques. *Recherches Qualitatives, 3*(1), 126–139.

Csikszentmihalyi, M. (1997). *Creativity: Flow and the psychology of discovery and invention.* HarperPerennial. https://doi.org/10.1037/e404942005-009

Dale, J., & Kyle, D. (2016). Smart humanitarianism: Re-imagining human rights in the age of entreprise. *Critical Sociology, 42*(6), 783–797. https://doi.org/10.1177/0896920516640041

de Sousa Jabbour, A. B. L., Jabbour, C. J. C., Foropon, C., & Filho, M. G. (2018). When titans meet—Can Industry 4.0 revolutionize the environmentally-sustainable GEM manufacturing wave? The role of critical success factors. *Technological Forecasting and Social Change, 132*, 18–25. https://doi.org/10.1016/j.techfore.2018.01.017

Destouet, C., Tlahig, H., Bettayeb, B., & Mazari, B. (2023). Flexible job shop scheduling problem under Industry 5.0: A survey on human reintegration, environmental consideration and resilience improvement. *Journal of Manufacturing Systems, 67*, 155–173. https://doi.org/10.1016/j.jmsy.2023.01.004

Dwivedi, Y. K., Ismagilova, E., Hughes, D. L., Carlson, J., Filieri, R., Jacobson, J., Jain, V., Karjaluoto, H., Kefi, H., Krishen, A. S., Kumar, V., Rahman, M. M., Raman, R., Rauschnabel, P. A., Rowley, J., Salo, J., Tran, G. A., & Wang, Y. (2021). Setting the future of digital and social media marketing research: Perspectives and research propositions. *International Journal of Information Management, 59*, 102168. https://doi.org/10.1016/j.ijinfomgt.2020.102168

Eisner, E., Hsien, C., Mennenga, M., Khoo, Z. Y., Dönmez, J., Herrmann, C., & Low, J. S. C. (2022). Self-assessment framework for corporate environmental sustainability in the era of digitalization. *Sustainability, 14*(4), 2293. https://doi.org/10.3390/su14042293

Elia, G., Margherita, A., & Passiante, G. (2020). Digital entrepreneurship ecosystem: How digital technologies and collective intelligence are reshaping the entrepreneurial process. *Technological Forecasting and Social Change, 150*, 119791. https://doi.org/10.1016/j.techfore.2019.119791

Ertz, M. (2021). *Marketing responsable*. Éditions JFD.

Ertz, M., Sun, S., Boily, E., Kubiat, P., & Quenum, G. G. Y. (2022). Transitioning to Industry 4.0 promotes circular product lifetimes. *Industrial Marketing Management, 101*, 125–140. https://doi.org/10.1016/j.indmarman.2021.11.014

Ertz, M., Tandon, U., Sun, S., Torrent-Sellens, J., & Sarigöllü, E. (2024). *The Palgrave handbook of digitalization for sustainable development in society*. Palgrave Macmillan.

Esmaeilian, B., Sarkis, J., Lewis, K., & Behdad, S. (2020). Blockchain for the future of sustainable supply chain management in Industry 4.0. *Resources, Conservation & Recycling, 163*, 105064. https://doi.org/10.1016/j.reconrec.2020.105064

George, G., Merrill, R. K., & Schillebeeckx, S. J. D. (2020). Digital sustainability and entrepreneurship: How digital innovations are helping tackle climate change and sustainable development. *Entrepreneurship Theory and Practice, 45*(5), 999–1027. https://doi.org/10.1177/1042258719899425

Gimenez, C., Sierra, V., & Rodon, J. (2012). Sustainable operations: Their impact on the triple bottom line. *International Journal of Production Economics, 140*(1), 149–159. https://doi.org/10.1016/j.ijpe.2012.01.035

Ghobakhloo, M., Iranmanesh, M., Grybauskas, A., Vilkas, M., & Petraitė, M. (2021). Industry 4.0, innovation, and sustainable development: A systematic review and a roadmap to sustainable innovation. *Business Strategy and the Environment, 30*(8), 4237–4257. https://doi.org/10.1002/bse.2867

Ghobakhloo, M., Iranmanesh, M., Tseng, M. L., Grybauskas, A., Stefanini, A., & Amran, A. (2023). Behind the definition of Industry 5.0: A systematic review of technologies, principles, components, and values. *Journal of Industrial and Production Engineering, 40*(6), 432–447. https://doi.org/10.1080/21681015.2023.2216701

Gholami, H., Abdul-Nour, G., Sharif, S., & Streimikiene, D. (Eds.). (2023). Sustainable Manufacturing in Industry 4.0: Pathways and Practices. Springer Nature.

Gino, F., & Staats, B. R. (2012). *Samasource: Give work, not aid.* Harvard Business School NOM Unit Case (912-011), 2013-12. https://ssrn.com/abstract=2014220

Gupta, S., Langhans, S., Domisch, S., Fuso-Nerini, F., Felländer, A., Battaglini, M., Tegmark, M., & Vinuesa, R. (2021). Assessing whether artificial intelligence is an enabler or an inhibitor of sustainability at indicator level. *Transportation Engineering, 4*, 100064. https://doi.org/10.1016/j.treng.2021.100064

Guillaume, B., Benjamin, D., & Vincent, C. (2022). Review of the impact of IT on the environment and solution with a detailed assessment of the associated Gray literature. *Sustainability, 14*(4), 2457. https://doi.org/10.3390/su14042457

Guo, Q., Geng, C., & Yao, N. (2022). Environmental innovation and green digitalization: An analysis of Chinese listed firms. *Journal of Environmental and Economic Sustainability, 1*(1), 1–15. https://doi.org/10.3390/ijerph192316303

Hamidi, S. R., Aziz, A. A., Shuhidan, S. M., Aziz, A. A., & Mokhsin, M. (2018). SMEs maturity model assessment of IR4. 0 digital transformation. In *Proceedings of the 7th International Conference on Kansei Engineering and Emotion Research 2018: KEER 2018*, Kuching, Sarawak, Malaysia (Vol. 739, pp. 721–732). Springer. https://doi.org/10.1007/978-981-10-8612-0_75

Hizam-Hanafiah, M., Soomro, M. A., & Abdullah, N. L. (2020). Industry 4.0 readiness models: A systematic literature review of model dimensions. *Information, 11*(7), 364. https://doi.org/10.3390/info11070364

Hoyer, C., Gunawan, I., & Reaiche, C. H. (2020). The implementation of Industry 4.0: A systematic literature review of the key factors. *Systems Research and Behavioral Science, 37*(4), 557–578. https://doi.org/10.1002/sres.2701

Hu, J., Zhang, Y., Tian, Y., & Jiang, Y. (2022). Green digital transformation: A systematic literature review. *Sustainability, 14*(3), 1323. https://doi.org/10.3390/su14031323

Irimiás, A., & Mitev, A. (2020). Change management, digital maturity, and green development: Are successful firms leveraging sustainability? *Sustainability, 12*(10), 4019. https://doi.org/10.3390/su12104019

Isensee, C., Teuteberg, F., & Griese, K. M. (2023). Success factors of organizational resilience: A qualitative investigation of four types of sustainable digital entrepreneurs. *Management Decision, 61*(5), 1244–1273. https://doi.org/10.1108/MD-03-2022-0326

Islam, A., Wahab, S. A., & Latiff, A. A. (2022). Annexing a smart sustainable business growth model for small and medium enterprises (SMEs). *World Journal of Entrepreneurship, Management and Sustainable Development, 18*(2), 185–209. https://doi.org/10.47556/J.WJEMSD.18.2.2022.2

Jefroy, N., Azarian, M., & Yu, H. (2022). Moving from Industry 4.0 to Industry 5.0: What are the implications for smart logistics? *Logistics, 6*(2), 26. https://doi.org/10.3390/logistics6020026

Kergroach, S. (2020). Giving momentum to SME digitalization. *Journal of the International Council for Small Business, 1*(1), 28–31. https://doi.org/10.1080/26437015.2020.1714358

Khan, F., & Porras, J. (2018, September). A framework and a web application for self-assessment of sustainable green ICT practices in SMEs. *Preprint*. https://doi.org/10.20944/preprints201809.0002.v1

Klimova, A., Rondeau, E., Andersson, K., Porras, J., Rybin, A., & Zaslavsky, A. (2016). An international Master's program in green ICT as a contribution to sustainable development. *Journal of Cleaner Production, 135*, 223–239. https://doi.org/10.1016/j.jclepro.2016.06.032

Kleindorfer, P. R., Singhal, K., & Van Wassenhove, L. N. (2009). Sustainable operations management. *Production and Operations Management, 14*(4), 482–492. https://doi.org/10.1111/j.1937-5956.2005.tb00235.x

Kljajić Borštnar, M., & Pucihar, A. (2021). Multi-attribute assessment of digital maturity of SMEs. *Electronics, 10*(8), 885. https://doi.org/10.3390/electronics10080885

Kolla, S., Minufekr, M., & Plapper, P. (2019). Deriving essential components of LEAN and Industry 4.0 assessment model for manufacturing SMEs. *Procedia CIRP, 81*, 753–758. https://doi.org/10.1016/j.procir.2019.03.189

Korhnen, J. J. (2018). *Entreprise transformation capability for the digital era—Demands for organizations and CIOs*. https://urn.fi/URN:ISBN:978-952-60-8013-0

Kotler, P., Kartajaya, H., & Setiawan, I. (2019). Marketing 3.0: From products to customers to the human spirit. In K. Kompella (Ed.), *Marketing wisdom* (pp. 139–156). Springer. https://doi.org/10.1007/978-981-10-7724-1_10

Kshetri, N. (2021). Blockchain and sustainable supply chain management in developing countries: Insights from case studies. *International Journal of Information Management, 60*, 102350. https://doi.org/10.1016/j.ijinfomgt.2021.102376

Kuntsman, A., & Rattle, I. (2019). Towards a paradigmatic shift in sustainability studies: A systematic review of peer reviewed literature and future agenda setting to consider environmental (Un) sustainability of digital communication. *Environmental Communication, 13*(5), 567–581.

Lai, K., Yunting, F., & Zhu, Q. (2023). Digital transformation for green supply chain innovation in manufacturing operations. *Transportation Research Part E: Logistics and Transportation Review, 174*, 103145. https://doi.org/10.1016/j.tre.2023.103145

Lerman, L. V., Benitez, G. B., Müller, J. M., de Sousa, P. R., & Frank, A. G. (2022). Smart green supply chain management: A configurational approach to enhance green performance through digital transformation. *Supply Chain Management, 27*(6), 1–17. https://doi.org/10.1108/SCM-08-2022-0354

Lobschat, L., Müller, B., Eggers, F., Brandimarte, L., Diefenbach, S., et al. (2021). Corporate digital responsibility. *Journal of Business Research, 122*, 875–888. https://doi.org/10.1016/j.jbusres.2019.10.006

Lokshin, M., & Widmar, E. (2023). *Toward environmentally sustainable public institutions: The green government IT index*. Policy Research Working Papers https://doi.org/10.1596/1813-9450-10361

Linkov, I., Trump, B. D., & Keisler, J. (2021). Resilience and risk analysis for cyber–physical–social systems. *Environment Systems and Decisions, 41*, 3–5. https://infoscience.epfl.ch/server/api/core/bitstreams/87cfe245-c138-43cb-87c9-4062dc1a0519/content#page=6

Lopes, J. M., Farinha, L., & Ferreira, J. J. M. (2022). Digital transformation and smart entrepreneurship: A bibliometric review. *Sustainability, 14*(1), 347. https://doi.org/10.3390/su14010347

Lytras, M. D., Visvizi, A., & Sarirete, A. (2019). Artificial Intelligence (AI) and smart cities: Challenges and opportunities. *Sustainability, 11*(24), 6614. https://doi.org/10.3390/su11236614

Ma, X. (2023). Self-assessment of digital transformation. Dans methodology for digital transformation. In *Management for Professionals* (pp. 63–86). Springer. https://doi.org/10.1007/978-981-19-9111-0_7

MEIE (2020). Tableau de bord du numérique : Intégration des processus d'affaires au Québec [Digital Dashboard: Integration of Business Processes in Quebec]. Government of Quebec. https://www.economie.gouv.qc.ca/fileadmin/contenu/documents_soutien/strategies/economie_numerique/tableau_bord_economie_numerique/tableau_internet_aux_processus_affaires_Quebec.pdf

Martynov, V. V. (2023). Problems of digital transformation management. *Scientific Research and Development Economics, 11*(3), 41–45. https://doi.org/10.12737/2587-9111-2023-11-3-41-45

MELCCFP (2023). Le développement durable au coeur de votre modèle d'affaires [Sustainable development at the heart of your business model]. *Government of Quebec*. https://www.environnement.gouv.qc.ca/developpement/entreprises/index.htm

Ministère de l'Économie, de la Science et de l'Innovation. (2016). *Plan d'action en économie numérique : pour l'excellence numérique des entreprises et des organisations québécoises*. http://collections.banq.qc.ca/ark:/52327/2635942

Mittal, S., Khan, M. A., Romero, D., & Wuest, T. (2018). A critical review of smart manufacturing & Industry 4.0 maturity models: Implications for small and medium-sized enterprises (SMEs). *Journal of Manufacturing Systems, 49*, 194–214. https://doi.org/10.1016/j.jmsy.2018.10.005

Mondejar, M. E., Avtar, R., Diaz, H. L. B., Dubey, R. K., Esteban, J., Gómez-Morales, A., Hallam, B., Mbungu, N. T., Okolo, C. C., Prasad, K. A., She, Q., & Garcia-Segura, S. (2021). Digitalization to achieve sustainable development goals: Steps towards a Smart Green Planet. *Science of the Total Environment, 794*, 148539. https://doi.org/10.1016/j.scitotenv.2021.148539

Mora, H., Mendoza-Tello, J. C., Varela-Guzmán, E. G., & Szymanski, J. (2021). Blockchain technologies to address smart city and society challenges. *Computers in Human Behavior, 122*, 106854. https://doi.org/10.1016/j.chb.2021.106854

Müller, A. L., & Pfleger, R. (2014). Business transformation towards sustainability. *Business Research, 7*(2), 313–350. https://doi.org/10.1007/s40685-014-0011-y

Münnich, M., Stange, M., Sübe, M., & lhlenfeldt, S. (2022). Integration of digitalization and sustainability objectives in a maturity model-based strategy development process. In *Global conference on sustainable manufacturing* (pp. 918–926). Springer. https://doi.org/10.1007/978-3-031-28839-5_102

Nagano, A. (2019). An integrated index towards sustainable digital transformation. In *2019 Third world conference on smart trends in systems security and sustainability (WorldS4)* (pp. 228–234). IEEE. https://doi.org/10.1109/WorldS4.2019.8904031

Nižetić, S. Šolić, P., López-de-Ipiña González-de-Artaza, D., & Patrono, L. (2020). Internet of Things (IoT): Opportunities, issues and challenges towards a smart and sustainable future. *Journal of Cleaner Production, 274*, 122877. https://doi.org/10.1016/j.jclepro.2020.122877

Nosratabadi, S., Atobishi, T., & Hegedus, S. (2023). Social sustainability of digital transformation: Empirical evidence from EU-27 countries. *Administrative Sciences, 13*(5), 126. https://doi.org/10.3390/admsci13050126

Nureen, N., Sun, H., Irfan, M., Nuţă, A. C., & Malik, M. (2023). Digital transformation: Fresh insights to implement green supply chain management, eco-technological innovation, and collaborative capability in manufacturing sector of an emerging economy. *Environmental Science and Pollution Research, 30*(32), 78168–78181. https://doi.org/10.1007/s11356-023-27796-3

Nyagadza, B. (2022). Sustainable digital transformation for ambidextrous digital firms: Systematic literature review, meta-analysis and agenda for future research directions. *Sustainable Technology and Entrepreneurship, 1*(3), 100020. https://doi.org/10.1016/j.stae.2022.100020

OECD. (2023). *SME digitalization for resilience and sustainability*. Research report. https://www.oecd.org/digital/sme/D4SME%20PoW%202023-24.pdf

Ozanne, J. L., & Saatcioglu, B. (2008). Participatory action research. *Journal of Consumer Research, 35*(3), 423–439. https://doi.org/10.1086/586911

O'Donnell D, & Henriksen LB. (2002). Philosophical foundations for a critical evaluation of the social impact of ICT. Journal of information Technology, 17(2);89–99.

Pachenko, V., & Dovhenko Y. (2023). Digitalization as an innovative modern factor business development: prospects and threats. The development of innovations and financial technology in the digital economy: monograph. *OÜ Scientific Center of Innovative Research. 230*, 87–106. https://doi.org/10.36690/DIFTDE2023-87-106

Pappas, I. O., Mikalef, P., Dwivedi, Y. K., Jaccheri, L., & Krogstie, J. (2023). Responsible digital transformation for a sustainable society. *Information Systems Frontiers*. https://doi.org/10.1007/s10796-023-10406-5

Paillé, P., & Mucchielli, A. (2003). *Analyse qualitative en sciences humaines et sociales*. Armand Colin.

Petersen, K., Becker, C., Kern, E., & Penzenstadler, B. (2023). Smart green software engineering: Concepts, practices, and perspectives. *IEEE Software, 40*(2), 15–23. https://doi.org/10.1109/MS.2022.3218999

Philbin, S., Viswanathan, R., & Telukdarie, A. (2022). Understanding how digital transformation can enable SMEs to achieve sustainable development: A systematic literature review. *Small Business International Review, 6*(1), e473. https://doi.org/10.26784/sbir.v6i1.473

Piccarozzi, M., Aquilani, B., & Gatti, C. (2018). Industry 4.0 in management studies: A systematic literature review. *Sustainability, 10*, 3821. https://doi.org/10.3390/su10103821

Prajapati, A., Arno, R., Dowling, N., & Moylan, W. (2019, May). Enhancing reliability of power systems through iiot-survey and proposal. In *2019 IEEE/IAS 55th Industrial and Commercial Power Systems Technical Conference (I&CPS)* (pp. 1–7). IEEE. https://doi.org/10.1109/ICPS.2019.8733363

Priyono, A., Moin, A., & Putri, V. N. A. O. (2020). Identifying digital transformation paths in the business model of SMEs during the COVID-19 pandemic. *Journal of Open Innovation: Technology, Market, and Complexity, 6*, 104. https://doi.org/10.3390/joitmc6040104

Radu, L.-D. (2020). Big Data and smart city: A bibliometric analysis. *Smart Cities, 3*(3), 1142–1159. https://doi.org/10.3390/smartcities3030042

Rauch, E., & Cochran, D. S. (2021). Sustainable introduction of Industry 4.0: a systematic literature review. In *Proceedings of the International Conference on Industrial Engineering and Operations Management: 4th European Rome Conference 2021* (pp. 138–148). IEOM. https://hdl.handle.net/10863/33413

Rauch, E., Unterhofer, M., Rojas, R. A., Gualtieri, L., Woschank, M., & Matt, D. T. (2020). A maturity level-based assessment tool to enhance the implementation of Industry 4.0 in small and medium-sized enterprises. *Sustainability, 12*(9), 3559. https://doi.org/10.3390/su12093559

Rosa, P., Sassanelli, C., Urbinati, A., Chiaroni, D., & Terzi, S. (2020). Assessing relations between Circular Economy and Industry 4.0: A systematic literature review. *International Journal of Production Research, 48*(6), 1662–1687. https://doi.org/10.1080/00207543.2019.1680896

Schumacher, A., Erol, S., & Sihn, W. (2016). A maturity model for assessing Industry 4.0 readiness and maturity of manufacturing enterprises. *Procedia CIRP, 52*, 161–166. https://doi.org/10.1016/j.procir.2016.07.040

Schwab, K. (2016). The fourth industrial revolution. New York, NY: Crown Currency.

Schwab, K. (2017). *The fourth industrial revolution*. Crown Currency.

Seele, P., & Lock, I. (2017). The game-changing potential of digitalization for sustainability: Possibilities, perils, and pathways. *Sustainability Science, 12*(2), 183–185. https://doi.org/10.1007/s11625-017-0426-4

Stahl, B. C., Timmermans, J., & Flick, C. (2017). Ethics of emerging information and communication technologies: On the implementation of responsible research and innovation. *Science and Public Policy, 44*(3), 369–381. https://doi.org/10.1093/scipol/scw069

Schiavone, F., Leone, D., Caporuscio, A., & Lan, S. (2022). Digital servitization and new sustainable configurations of manufacturing systems. *Technological Forecasting and Social Change, 176*, 121441. https://doi.org/10.1016/j.techfore.2021.121441

Sloan, A. (1990). *My years with general motors*. Crown Currency

St-Pierre, J., & Cadieux, L. (2011). La conception de la performance: Quels liens avec le profil entrepreneurial des propriétaires dirigeants de PME ? *Revue de l'Entrepreneuriat, 10*(1), 33–52. https://doi.org/10.3917/entre.101.0033

Suárez-Eiroa, B., Fernández, E., & Méndez, G. (2021). Integration of the circular economy paradigm under the just and safe operating space narrative: Twelve operational principles based on circularity, sustainability, and resilience. *Journal of Cleaner Production, 322*, 129071. https://doi.org/10.1016/j.jclepro.2021.129071

Suh, A., & Cho, S. (2022). Digital transformation and sustainability: A case study on digital technology in environmental innovation. *Sustainability, 14*(8), 4529.

Taddeo, M., & Floridi, L. (2018). How AI can be a force for good. *Science, 361*(6404), 751–752. https://doi.org/10.1126/science.aat5991

Taghizadeh-Hesary, F., & Hyun, S. (2022). Sustainable digital economy and society: Challenges and opportunities. *Sustainability, 14*(9), 5057. https://doi.org/10.3390/su14095057

Torres, P., & Augusto, M. (2020). Digitalisation, social entrepreneurship and national well-being. *Technological Forecasting and Social Change, 161*, 120279. https://doi.org/10.1016/j.techfore.2020.120279

Ullah, S. Luo, R. Nadeem, M. Cifuentes-Faura, J. (2023). Advancing sustainable growth and energy transition in the United States through the lens of green energy innovations, natural resources and environmental 2 policy. *Resources Policy, 85*, 103848. https://doi.org/10.1016/j.resourpol.2023.103848

UNESCO, (2015) Education for sustainable development. https://www.unesco.org/en/sustainable-development/education

Vallée, S. (2023). Offrir aux PME régionales un outil d'autoaudit pour arrimer virage numérique et développement durable: le projet PME 4.0 au Saguenay–Lac-Saint-Jean. *Revue Organisations & territoires, 32*(3), 191–209.

Vallée, S. (2025). La maturité numérique des PME en région au SLSJ (Doctoral dissertation, Université du Québec à Chicoutimi).

van den Buuse, D. Kolk A. (2019). An exploration of smart city approaches by international ICT firms. *Technological Forecasting and Social Change, 142*, 220–234. https://doi.org/10.1016/j.techfore.2018.07.029

Varenne, P. (2020). *La transformation digitale des entreprises: Effectuation et Business Model Digital Dynamique (BMD²)* (Doctoral dissertation, Université de Lyon). Management and administration. https://theses.hal.science/tel-02957670/

Vial, G. (2019). Understanding digital transformation: A review and a research agenda. *The Journal of Strategic Information Systems, 28*(2), 118–144. https://doi.org/10.1016/j.jsis.2019.01.003

Vinuesa, R., Azizpour, H., Leite, I., Balaam, M., Dignum, V., ... & Fuso Nerini, F. (2020). The role of artificial intelligence in achieving the Sustainable Development Goals. *Nature Communications, 11*, 233. https://doi.org/10.1038/s41467-019-14108-y

Winans, K., Kendall, A., & Deng, H. (2017). The history and current applications of the circular economy concept. *Renewable and Sustainable Energy Reviews, 68*, 825–833. https://doi.org/10.1016/j.rser.2016.09.123

Xi, L., & Wang, H. (2024). The influence of green innovation and digital transformation on the high-quality development of enterprises: The mediating role of ESG Management. *Sustainability, 16*, 10923. https://doi.org/10.3390

Yigitcanlar, T., Desouza, K. C., Butler, L., & Roozkhosh, F. (2020). Contributions and risks of artificial intelligence (AI) in building smarter cities: Insights from a systematic review of the literature. *Energies, 13*(6), 1473. https://doi.org/10.3390/en13061473

Yin, J. (2019). Green digital transformation: Concept, key technologies and development trends. *ZTE Technology Journal, 25*(5), 2–7.

Zhu, X. Zhang, B. Yuan, H. (2022). Digital economy, industrial structure upgrading and green total factor productivity – Evidence in textile and apparel industry from China. *PLOS ONE, 17*(11), e0277259. https://doi.org/10.1371/journal.pone.0277259

Stéfanie Vallée is a member of the Laboratory of Research on New Forms of Consumption (LaboNFC), and of the Canada Research Chair on Technology, Sustainability, and Society. She is a candidate for the Ph.D. in Applied Human Sciences, at the Department of Human and Social Sciences, University of Quebec at Chicoutimi (UQAC).

Myriam Ertz is a full professor of marketing, Director of the Laboratory of Research on New Forms of Consumption (LaboNFC), and Holder of the Canada Research Chair in Technology, Sustainability, and Society at the Department of Economics and Administrative Sciences, University of Quebec at Chicoutimi (UQAC). She is also co-responsible for Axis 1 of the Quebec Circular Economy Research Network (RRECQ) and a member of CIRODD.

Chourouk Ouerghemmi is a postdoctoral fellow in marketing and a member of the Laboratory of research on New Forms of Consumption (LaboNFC), and the Canada Research Chair on Technology, Sustainability, and Society, at the Department of Economics and Administrative Sciences University of Quebec at Chicoutimi (UQAC).

Antoine Périn is a member of the Laboratory of Research on New Forms of Consumption (LaboNFC) and of the Canada Research Chair on Technology, Sustainability, and Society. He holds a Master's degree in organization management from the Department of Economics and Administrative Sciences, University of Quebec at Chicoutimi (UQAC).

Index

M. Cheriet et al. (eds.), *Accelerating the Socio-Ecological Transition*,
https://doi.org/10.1007/978-3-031-82896-6

MIX
Papier aus verantwortungsvollen Quellen
Paper from responsible sources
FSC® C105338

FSC
www.fsc.org

If you have any concerns about our products,
you can contact us on
ProductSafety@springernature.com

In case Publisher is established outside the EU,
the EU authorized representative is:
Springer Nature Customer Service Center GmbH
Europaplatz 3, 69115 Heidelberg, Germany

Printed by Libri Plureos GmbH
in Hamburg, Germany